Advances in Mechanics and Mathematics

Volume 32

Series Editors
David Y. Gao
Virginia Polytechnic Institute Department of Mathematics, Blacksburg,
Virgin Islands, USA

Tudor Ratiu
Lausanne, Switzerland

Driven by elaborate modern technological applications, the relationship between mathematics and mechanics is continually developing. The burgeoning number of specialized journals has generated an ever growing duality gap between the partners. *Advances in Mechanics and Mathematics* is a series intending to bridge the gap by providing a platform for the publication of interdisciplinary content with rapid dissemination of monographs, graduate texts, handbooks, and edited volumes, on the state-of-the-art research in the broad area of modern mechanics and applied mathematics. Topics with multi-disciplinary range, such as duality, complementarity and symmetry in mechanics, mathematics, and physics, are of particular interest.

Contributions are aptly reviewed to guarantee high scientific standards. Monographs place an emphasis on creativity, novelty, and innovativeness in the field; handbooks and edited volumes provide comprehensive surveys of the state-of-the-art in particular subjects; graduate texts may feature a combination of exposition and computer/multimedia, downloadable from the web.

New to AMMA is the welcome addition of publications that focus on computational methods. Topics can involve theory, algorithms, programming, coding, numerical simulation, error and uncertainty analysis and/or the novel application of computational techniques to problems throughout many scientific disciplines. Especially encouraged are expositions on mathematical and computational models and methods based on mechanics and their interactions with other fields.

The series is addressed to applied mathematicians, engineers, and scientists, including advanced students at universities and in industry, who are interested in mechanics and applied mathematics.

More information about this series at http://www.springer.com/series/5613

Brian Straughan

Convection with Local Thermal Non-Equilibrium and Microfluidic Effects

 Springer

Brian Straughan
Department of Mathematical Sciences
Durham University
Durham
UK

ISSN 1571-8689   ISSN 1876-9896 (electronic)
Advances in Mechanics and Mathematics
ISBN 978-3-319-37240-2   ISBN 978-3-319-13530-4 (eBook)
DOI 10.1007/978-3-319-13530-4

Mathematic Subject Classification: 76E06 76E30 76S05 76T30 76R50 35Q30 35Q35

Springer Cham Heidelberg NewYork Dordrecht London
© Springer International Publishing Switzerland 2015
Softcover reprint of the hardcover 1st edition 2015

Printed on acid-free paper

Springer is part of Springer Science+Business Media (www.springer.com)

To
Cole, Caleb and Amelie

Preface

This book is devoted to an account of theories of thermal convection which involve local thermal non-equilibrium (LTNE) effects, or are particularly important in a microfluidic situation. The term "local thermal non-equilibrium" refers to thermal convection in a fluid saturated porous material where the fluid temperature and the temperature of the solid skeleton may be different. Microfluidics refers to fluid dynamics on a small scale which may involve thermal convection in a clear fluid, or thermal convection in a fluid saturated porous medium. The areas of microfluidics and nanofluidics are very topical at present.

This is not an attempt to survey the area of convection with local thermal non-equilibrium effects, nor that of convection in a microfluidic scenario. Both topics are extremely popular research areas and such a survey would be a gargantuan task. For example, if one inserts "local thermal non-equilibrium" in the Springer query box, 12,197 entries are found, on 30th August, 2014. Likewise if one enters the same expression in the query box of Science Direct, 123,332 entries are found, on 30th August, 2014. This book is simply an account of what I believe is an appropriate collection of subjects in a very topical area.

Chapters 2–7 deal specifically with LTNE effects whereas chapter 8 contains work with LTNE effects and some microfluidic work employing a single temperature. Chapters 9–15 concentrate mostly on microfluidic situations where a single temperature field is employed although section 15.4 is concerned with LTNE. Sections 6.1, 6.2, 9.4, 12.3, 13.2, 13.3, 14.2, 14.3, 14.5, 15.3 and 15.4 contain new material and/or new numerical results which I believe are not available elsewhere.

I should like to thank three anonymous referees for pointed and very useful comments which led to improvements in this book. It is a pleasure to thank Achi Dosanjh of Springer for her advice with editorial matters. I should also like to thank Jeff Taub and Suresh Kumar of Springer for their help with Latex and production matters.

This research was in part supported by a grant from the Leverhulme Trust, "Tipping points: mathematics, metaphors and meanings", reference number F/00128/BF.

Durham   *Brian Straughan*

Contents

Chapter 1
Introduction

1.1 Microfluidics, Local Thermal Non-equilibrium

1.1.1 Applications, Examples

This book focusses on thermal convection problems which are likely to be of interest in microfluidic situations, i.e. where the dimensions of the spatial configuration of the phenomenon are very small, although not exclusively so since some of the topics will be of interest in their own right at the macro scale.

There are a variety of physical mechanisms which will undoubtedly have a major effect on thermal convection or fluid flows in general when the spatial dimensions of the problem are small. One area involving heat flow is that of second sound, the mechanism whereby temperature travels as a wave, which in itself is also a topic of increasing attention. In particular, as modern technology is creating smaller and smaller devices, the phenomenon of temperature travelling as a wave becomes increasingly important, especially in metallic-like solids. Pilgrim *et al.* [343] develop a mathematical model for finite speed heat transport in semi-conductor devices and they observe that, ... the "hyperbolic description will become increasingly important as device dimensions move even further into the deep sub-micron regime". Since it is believed finite speed heat propagation is important in certain metallic material situations, we believe it is worthwhile considering this aspect in thermal convection flows in porous metallic foams, especially if the device dimensions are small. Various occurrences of finite speed heat transport are reviewed in the book by Straughan [425]. As he points out, most theoretical work involves the model proposed by Cattaneo in [79] to govern the behaviour of the heat flux and the temperature field. The history of the Cattaneo theory is discussed in detail in [425], chapter 1, where he also notes that a similar model, but in dielectric theory, was proposed earlier by Dario Graffi [165]. In this book we do consider second sound effects in thermal convection. Attention is paid to second sound and other physical effects for thermal convection in both a clear fluid and in a fluid saturated porous medium.

B. Straughan, *Convection with Local Thermal Non-Equilibrium and Microfluidic Effects*,
Advances in Mechanics and Mathematics 32, DOI 10.1007/978-3-319-13530-4_1

1

Porous media is a subject well known to everyone. Such materials occur everywhere and influence all of our lives. There are numerous types of porous media and almost limitless applications of and uses for such media. The theory of porous media is driven by the need to understand the nature of the many such materials available and to be able to use them in an optimum way.

A key terminology in the theory of porous media is the concept of *porosity*. The porosity is the ratio of the void fraction in the porous material to the total volume occupied by the porous medium. The void fraction is usually composed of air or some other liquid and since both liquids may be described as fluids we define the *porosity* at position **x** and time t, $\varepsilon(\mathbf{x},t)$ by

$$\varepsilon = \frac{\text{fluid volume}}{\text{total volume of porous medium}}. \tag{1.1}$$

Clearly, $0 \le \varepsilon \le 1$. However, in mundane situations ε may be as small as 0.02 in coal or concrete, see e.g. [312], whereas ε is close to 1 in some animal coverings such as fur or feathers, [117], or in man-made high porosity metallic foams, [56, 189, 240, 497].

We include photographs of some well known porous materials. Figure 1.1 dislays wet sand which is a very well known porous material but one with a relatively low porosity. Figure 1.2 shows natural sandstone found on a beach. This natural sandstone has a higher porosity than sand. Figure 1.3 shows lava from Mount Etna in Sicily, and this lava is another type of porous rock. In figure 1.4 we show a naturally occurring aggregate found on a beach. This aggregate also has a higher porosity than sand. The photograph in figure 1.5 displays concrete which has been weathered by the sea. The concrete has a low porosity. Figure 1.6 shows animal fur which is a good example of a porous medium with high porosity, i.e. porosity close to 1. Figure 1.7 displays another type of rock but one which is highly anisotropic. Figure 1.8 displays a highly grained piece of wood (oak). This is another example which shows that a porous medium may be highly anisotropic. In fact, wood is essentially isotropic in all directions orthogonal to the grain. The anisotropy is clearly evident in the grain direction. Anisotropy such as this where one direction is very different from those directions orthogonal is known as transverse isotropy. Figure 1.9 is a schematic picture of a bidispersive or a double porosity porous medium. The large gaps between the dark objects reveal a macro porosity but the darker objects themselves are composed of small spheres and have between them a micro porosity. Such materials may be man made or can occur naturally, cf. [445] for the latter case.

In addition to these we can cite other examples of porous media, such as biological tissues, e.g. bone, skin; building materials such as sand, cement, plasterboard, brick; man-made high porosity metallic foams such as those based on copper oxide or aluminium, and other materials in everyday use such as ceramics. The types of porous materials we can think of is virtually limitless.

Applications of porous media in real life and their connection to microfluid flows are likewise very many. We could list many, but simply quote some to give an idea of the vastness of porous media theory. Use of copper based foams and other porous

Fig. 1.1 Sand. Photograph taken on Seaham beach, March 2014.

Fig. 1.2 Natural sandstone. Photograph taken on Seaham beach, March 2014.

materials in heat transfer devices such as heat pipes used to transfer heat from such as computer chips is a field influencing everyone, see e.g. [57, 189, 269, 312, 497]. Likewise, porous media are prevalent in combustion heat transfer devices where the porous medium is employed with a liquid fuel in a porous combustion heater, see [200]. Global warming is very topical and porous media are involved there in connection with topics such as ice melting, or carbon dioxide storage, see e.g. [55, 76, 77, 183]. Many foodstuffs are porous materials. Modern technology is involved in such as microwave heating, [112], or drying of foods or other natural materials, see e.g. [500, 501]. Porous media have application in storage of energy or natural convection within the upper region of the Earth, [331, 499]. The latter areas being of particular interest in the field of renewable energy. There are many

Fig. 1.3 Lava from Mount Etna, Sicily. Photograph taken at Capomulini, June 2007.

Fig. 1.4 Natural aggregate. Photograph taken on Seaham beach, March 2014.

other diverse application areas of porous materials, such as heat retention in birds or animals, [117], bone modelling, [125], or the manufacture of composite materials, increasingly in use in aircraft or motor car production, see e.g. [108].

Very much connected with thermal convection on a micro-scale are convective flow problems in a porous medium where the fluid temperature, T_f, may be different from the solid skeleton temperature, T_s. Such problems of thermal convection are being increasingly studied. This situation where the two temperatures may be different is usually referred to as local thermal non-equilibrium, abbreviated to LTNE. One of the driving reasons for the increased attention of LTNE flows in porous media is the numerous amount of applications of this area in real life. For example, there are applications in tube refrigerators in space, [19]; in nanofluid flows,

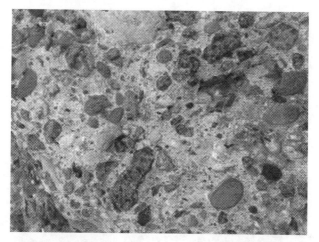

Fig. 1.5 Concrete, weathered by the sea. Photograph taken on Seaham beach, March 2014.

Fig. 1.6 Animal fur is a good example of a high porosity material, as seen in this cat.

[320, 420, 425], chapter 8; in fuel cells [102]; in resin flow, important in processing composite materials, [108]; in nuclear reactor maintenance, [137]; in heat exchangers, [107, 269]; in microwave ablation of the liver, [208]; in biological tissue analysis, [493]; in flows in microchannels, [215]; in flow and heat transfer in porous metallic foams, [189, 240, 241, 269]; in thermovibrational filtration, [388, 389], in textile transport, [486]; and in convection in stellar atmospheres, cf. [425], chapter 8, [426]. An interesting paper analysing various causes of local thermal non-equilibrium situations is that of [470].

Continuum theories for local thermal non-equilibrium effects on flow in porous materials appear to have started in the late 1990's, cf. the work of [291, 307], and [340], and instability in thermal convection taking into account LTNE effects was

Fig. 1.7 Rock strata with an anisotropic structure. Photograph taken at Low Pittington, County Durham. This rock has a permeability which is approximately horizontally isotropic.

Fig. 1.8 Wood is a very good example of a porous medium which exhibits a strong anisotropy. The grain effect in this oak is clearly visible. This is a material which is approximately transversely isotropic.

Fig. 1.9 This is a schematic picture of a bidispersive material. Note the large pore structure in between the darker elements which are themselves composed of small spherical bead like bodies giving a dual porosity raspberry-like structure.

addressed by [27], and by [260, 261]. [416] demonstrated that these instability results were really sharp by showing that they also represented a global nonlinear stability threshold, not just a linear instability one. Since then many studies of thermal convection with LTNE have appeared, employing a variety of geometries and incorporating various other effects, for example rotation, double diffusion, see e.g. [5, 34, 35, 50, 66, 81, 90, 238, 239, 257–259, 264–266, 308, 314, 320, 366, 368, 389, 393, 397, 399, 440, 441, 459–462], and the references therein.

In addition to thermal wave and LTNE effects, in this book we pay special attention to slip at a solid boundary bordering a clear fluid, or a fluid saturated porous medium. There is increasing interest in micro-electro-mechanical-systems (MEMS), and flow in microfluidic channels due to their applications in the electronics and related industries. In particular, at nanoscales there is increasing evidence that boundary conditions of slip type are needed rather than those of no-slip, cf. [14, 24, 83, 118, 119, 234, 295, 350, 358, 382, 383, 402, 408, 489, 494, 496]. An especially important application of microscale flow involving slip boundary conditions is to flow in porous metallic foams. [240] give many industrial examples of this and provide a thorough review of the state of the art.

Other areas of microfluidics we focus attention on are where the properties of the porous medium are anisotropic, on bidispersive materials, on the effect of the inertia term in a porous medium theory, and on nanofluids. Since many real porous materials display strong anisotropy this focus of attention is natural and is studied by e.g. [176, 193, 214, 226, 228, 370, 387, 394, 395, 398, 400, 404, 439] and [485], and the references therein. These articles concentrate mainly on anisotropy of the permeability tensor or thermal conductivity. While it is true that many studies of thermal convection in porous media neglect inertia effects in the momentum equation for the movement of fluid in the porous skeleton, there are increasing instances of where the inertia term can lead to important new physics, and this is likely to be even more so in microfluidic situations. [456] showed that the inertia term leads to oscillatory convection which is not present when the inertia is neglected. Other recent studies also show that inclusion of inertia leads to novel effects, see e.g. [12, 51, 130, 403, 421, 422, 425]. Bidispersive materials are those where basically the porous medium has a double porosity structure, a macro porosity, and a porosity on the micro scale, cf. [153, 429, 431, 442, 443]. Theories for thermal convection in a bidispersive porous medium are developed and analysed by [317–319], with an extension to thermal convection in a fluid saturated porous medium with a tridispersive structure proposed by [227, 322]. Nanofluids are a class of man made materials where a carrier fluid such as ethylene glycol or water contains a suspension of metallic, or metallic-like, particles. Theories to describe the behaviour of a nanofluid, experimental work on nanofluids, and analysis of nanofluids is a hot research topic due to many applications such as heat transfer.

1.1.2 Notation, Definitions

Standard indicial notation is used throughout this book together with the Einstein summation convention for repeated indices. Standard vector or tensor notation is also employed where appropriate. For example, we write

$$u_x \equiv \frac{\partial u}{\partial x} \equiv u_{,x} \qquad u_{i,t} \equiv \frac{\partial u_i}{\partial t} \qquad u_{i,i} \equiv \frac{\partial u_i}{\partial x_i} \equiv \sum_{i=1}^{3} \frac{\partial u_i}{\partial x_i}$$

$$u_j u_{i,j} \equiv u_j \frac{\partial u_i}{\partial x_j} \equiv \sum_{j=1}^{3} u_j \frac{\partial u_i}{\partial x_j}, \quad i = 1, 2 \text{ or } 3.$$

In the case where a repeated index sums over a range different from 1 to 3 this will be pointed out in the text. Note that

$$u_j u_{i,j} \equiv (\mathbf{u} \cdot \nabla)\mathbf{u} \qquad \text{and} \qquad u_{i,i} \equiv \operatorname{div} \mathbf{u}.$$

As indicated above, a subscript t denotes partial differentiation with respect to time. When a superposed dot is used it means the material derivative, e.g.

$$\dot{u}_i \equiv \frac{\partial u_i}{\partial t} + u_j \frac{\partial u_i}{\partial x_j},$$

or

$$\dot{T} = \frac{\partial T}{\partial t} + u_i \frac{\partial T}{\partial x_i},$$

where u_i in the equations above is the velocity field and $T(\mathbf{x}, t)$ is another field such as temperature.

To further clarify the indicial notation used, we note for example, the product of two second order tensors A_{ij} and B_{ij}, say $\mathbf{C} = \mathbf{AB}$, may be written

$$C_{ij} = A_{ik}B_{kj}$$

$$\equiv \sum_{k=1}^{3} A_{ik}B_{kj}.$$

Further, if $\mathbf{x}$ and $\mathbf{b}$ are vectors then $\mathbf{Ax} = \mathbf{b}$ may be written as

$$b_i = A_{ik}x_k$$

$$\equiv \sum_{k=1}^{3} A_{ik}x_k.$$

We now derive some expressions which are useful for later. Recall the definitions of the Kronecker delta, δ_{ij}, and the alternating tensor, ε_{ijk}. Namely,

$$\delta_{ij} = \begin{cases} 1, & \text{if } i = j; \\ 0, & \text{if } i \neq j, \end{cases}$$

i.e. $\delta_{11} = \delta_{22} = \delta_{33} - 1$, the rest are zero. Also,

$$\begin{aligned} \varepsilon_{123} = \varepsilon_{231} = \varepsilon_{312} = +1, \\ \varepsilon_{213} = \varepsilon_{321} = \varepsilon_{132} = -1, \end{aligned} \tag{1.2}$$

the rest are zero.

In terms of the Kronecker delta and the alternating tensor,

$$\varepsilon_{ijk}\varepsilon_{irs} = \delta_{jr}\delta_{ks} - \delta_{js}\delta_{kr}$$

and then for a vector field $\mathbf{\Lambda}$,

$$(\text{curl}\,\mathbf{A})_i = \varepsilon_{ijk}A_{k,j}$$

and

$$\begin{aligned} (\text{curl}\,\text{curl}\,\mathbf{A})_i &= \varepsilon_{ijk}\varepsilon_{krs}A_{s,rj} \\ &= \varepsilon_{kij}\varepsilon_{krs}A_{s,rj} \\ &= (\delta_{ir}\delta_{js} - \delta_{is}\delta_{jr})A_{s,rj} \\ &= A_{j,ij} - \Delta A_i. \end{aligned} \tag{1.3}$$

Thus, if v_i is a divergence free vector field, $v_{i,i} = 0$, and then

$$(\text{curl}\,\text{curl}\,\mathbf{v})_i = -\Delta v_i,$$

where $\Delta = \partial^2/\partial x^2 + \partial^2/\partial y^2 + \partial^2/\partial z^2$ is the Laplace operator.

The vector product (or cross product) of two vectors $\mathbf{A}$ and $\mathbf{B}$ may be written in component form in indicial notation as

$$(\mathbf{A} \times \mathbf{B})_i = \varepsilon_{ijk} A_j B_k.$$

If we let $\mathbf{k} = (0,0,1)$ then a particular vector product which occurs frequently in analysis of thermal convection in a rotating frame is $\mathbf{u} \times \mathbf{k}$ which may be written in component form as

$$(\mathbf{u} \times \mathbf{k})_i = \varepsilon_{ijk} u_j k_k = \varepsilon_{ijk} u_j \delta_{k3} = \varepsilon_{ij3} u_j.$$

Recalling (1.2) we thus find for $\mathbf{u} = (u,v,w)$, $\mathbf{u} \times \mathbf{k} = (v,-u,0)$.
Another frequently used expression is curl $(\mathbf{u} \times \mathbf{k})$ and this is

$$\begin{aligned}
\left[\mathrm{curl}\,(\mathbf{u} \times \mathbf{k})\right]_i &= \varepsilon_{irs}(\mathbf{u} \times \mathbf{k})_{s,r} \\
&= \varepsilon_{irs}\varepsilon_{sab}(u_a k_b)_{,r} \\
&= \varepsilon_{sir}\varepsilon_{sab} k_b u_{a,r} \\
&= (\delta_{ia}\delta_{rb} - \delta_{ib}\delta_{ra})\delta_{b3} u_{a,r} \\
&= (\delta_{ia}\delta_{r3} - \delta_{i3}\delta_{ra}) u_{a,r} \\
&= u_{i,3} - \delta_{i3} u_{a,a}
\end{aligned}$$

and so if div $\mathbf{u} = 0$, i.e. if $u_{i,i} = 0$, then

$$\left[\mathrm{curl}\,(\mathbf{u} \times \mathbf{k})\right]_i = u_{i,3}. \tag{1.4}$$

Let us put $\mathbf{u} = (u,v,w)$ and $\mathbf{x} = (x,y,z)$. We then observe that the third component of expression (1.4) yields,

$$\left[\mathrm{curl}\,(\mathbf{u} \times \mathbf{k})\right]_3 = \frac{\partial w}{\partial z}. \tag{1.5}$$

When dealing with anisotropic media we have recourse to deal with diagonal matrices and so we introduce the notation

$$\mathbf{A} = \mathrm{diag}\,(a_1, a_2, a_3) \tag{1.6}$$

to mean $\mathbf{A}$ is the diagonal matrix

$$\mathbf{A} = \begin{pmatrix} a_1 & 0 & 0 \\ 0 & a_2 & 0 \\ 0 & 0 & a_3 \end{pmatrix}.$$

The letter Ω will denote a fixed, bounded region of 3-space with boundary, Γ, sufficiently smooth to allow applications of the divergence theorem. When we are dealing with convection problems we handle motion in a plane layer, say $\{(x,y) \in$

$\mathbb{R}^2\} \times \{z \in (0,d)\}$. In this case, we usually refer to functions that have an (x,y) behaviour which is repetitive in the (x,y) direction, such as regular hexagons. The periodic cell defined by such a shape and its Cartesian product with $(0,d)$ will be denoted by V. The boundary of the period cell V will be denoted by ∂V.

The symbols $\|\cdot\|$ and $(\cdot,\cdot)$ will denote, respectively, the L^2 norm on Ω or V, and the inner product on $L^2(\Omega)$ or $L^2(V)$, where the context will define whether Ω or V is to be used, e.g.,

$$\int_V f^2 dV = \|f\|^2 \quad \text{and} \quad (f,g) = \int_V fg\, dV,$$

with equivalent definitions for Ω. We sometimes have recourse to use the norm on $L^p(\Omega)$, $1 < p < \infty$, and then we write

$$\|f\|_p = \left(\int_\Omega |f|^p dx \right)^{1/p}.$$

In this book we examine various situations of fluid flows where stability and instability are important. The concepts of stability and of instability are different, see e.g. [414, 417], and for applications it is vital to understand these. We introduce the ideas of stability and instability in the context of the following system of reaction - diffusion equations,

$$\frac{\partial u}{\partial t} = \frac{\partial}{\partial x_j}\left(a_{ij}\frac{\partial u}{\partial x_i}\right) + f(u,v),$$

$$\frac{\partial v}{\partial t} = \frac{\partial}{\partial x_j}\left(b_{ij}\frac{\partial v}{\partial x_i}\right) + g(u,v),$$

(1.7)

where f and g are known nonlinear functions and where $\mathbf{x} \in \Omega \subset \mathbb{R}^3$. (In practice equations (1.7) would be defined with suitable boundary conditions.) The coefficients $a_{ij} = a_{ij}(\mathbf{x})$ and $b_{ij} = b_{ij}(\mathbf{x})$ are symmetric and positive-definite in the sense that $a_{ij} = a_{ji}, b_{ij} = b_{ji}$ and

$$a_{ij}\xi_i\xi_j \geq a_0\xi_i\xi_i, \qquad b_{ij}\xi_i\xi_j \geq b_0\xi_i\xi_i, \qquad \forall \xi_i,$$

and for constants $a_0 > 0$ and $b_0 > 0$.

We introduce notation in the context of a steady solution to (1.7), namely a solution $\bar{u}, \bar{v}$ satisfying

$$0 = \frac{\partial}{\partial x_j}\left(a_{ij}\frac{\partial \bar{u}}{\partial x_i}\right) + f(\bar{u},\bar{v}),$$

$$0 = \frac{\partial}{\partial x_j}\left(b_{ij}\frac{\partial \bar{v}}{\partial x_i}\right) + g(\bar{u},\bar{v}).$$

(1.8)

(We could equally deal with the stability of a time-dependent solution, but many of the problems encountered in this book are for stationary solutions and at this juncture it is as well to keep the ideas as simple as possible.) Let (w,r) be a perturbation to the solution to equation (1.8), i.e. put $u = \bar{u} + w(\mathbf{x},t)$, $v = \bar{v} + r(\mathbf{x},t)$. Then, it is seen from (1.7) and (1.8) that (w,r) satisfies the system

$$\frac{\partial w}{\partial t} - \frac{\partial}{\partial x_j}\left(a_{ij}\frac{\partial(w+\bar{u})}{\partial x_i}\right) - \left[f(\bar{u}+w,\bar{v}+r) - f(\bar{u},\bar{v})\right] = 0,$$

$$\frac{\partial r}{\partial t} - \frac{\partial}{\partial x_j}\left(b_{ij}\frac{\partial(r+\bar{v})}{\partial x_i}\right) - \left[g(\bar{u}+w,\bar{v}+r) - g(\bar{u},\bar{v})\right] = 0. \tag{1.9}$$

To discuss the concept of linearized instability we linearize (1.9) which means we keep only the terms which are linear in w and r. From a Taylor series expansion of f and g we have

$$f(\bar{u}+w,\bar{v}+r) = f(\bar{u},\bar{v}) + w\frac{\partial f}{\partial u}(\bar{u},\bar{v}) + r\frac{\partial f}{\partial v}(\bar{u},\bar{v})$$
$$+ O(w^2, wr, r^2),$$

$$g(\bar{u}+w,\bar{v}+r) = g(\bar{u},\bar{v}) + w\frac{\partial g}{\partial u}(\bar{u},\bar{v}) + r\frac{\partial g}{\partial v}(\bar{u},\bar{v})$$
$$+ O(w^2, wr, r^2). \tag{1.10}$$

Expansions (1.10) are inserted into (1.9) and then with the aid of equations (1.8) we may derive the linearized equations satisfied by w and r, namely

$$\frac{\partial w}{\partial t} - \frac{\partial}{\partial x_j}\left(a_{ij}\frac{\partial w}{\partial x_i}\right) - wf_u(\bar{u},\bar{v}) - rf_v(\bar{u},\bar{v}) = 0,$$

$$\frac{\partial r}{\partial t} - \frac{\partial}{\partial x_j}\left(b_{ij}\frac{\partial r}{\partial x_i}\right) - wg_u(\bar{u},\bar{v}) - rg_v(\bar{u},\bar{v}) = 0, \tag{1.11}$$

where $f_u = \partial f/\partial u$, etc.

Since (1.11) is a system of linear equations we may introduce an exponential time dependence in (w,r) so that $w = e^{\sigma t}s(\mathbf{x})$ and $r = e^{\sigma t}q(\mathbf{x})$. Then (1.11) yields

$$\sigma s - \frac{\partial}{\partial x_j}\left(a_{ij}\frac{\partial s}{\partial x_i}\right) - sf_u(\bar{u},\bar{v}) - qf_v(\bar{u},\bar{v}) = 0,$$

$$\sigma q - \frac{\partial}{\partial x_j}\left(b_{ij}\frac{\partial q}{\partial x_i}\right) - sg_u(\bar{u},\bar{v}) - qg_v(\bar{u},\bar{v}) = 0. \tag{1.12}$$

We say that the steady solution $(\bar{u},\bar{v})$ to (1.8) is *linearly unstable* if

$$Re(\sigma) > 0,$$

where $Re(\sigma)$ denotes the real part of σ. Equations (1.12) (together with appropriate boundary conditions) constitute an eigenvalue problem for σ. For many of the problems discussed in this book the eigenvalues may be ordered so that

$$Re(\sigma_1) > Re(\sigma_2) > \ldots$$

For linear instability we then need only ensure $Re(\sigma_1) > 0$.

We now introduce the concept of nonlinear stability. Let $w_0(\mathbf{x}) = w(\mathbf{x}, 0)$, $r_0(\mathbf{x}) = r(\mathbf{x}, 0)$, be the initial data functions associated to the solution (w, r) of equations (1.9). The steady solution $(\bar{u}, \bar{v})$ to (1.8) is *nonlinearly asymptotically stable* if and only if for each $\varepsilon > 0$ there is a $\delta = \delta(\varepsilon)$ such that

$$\|w_0\|^2 + \|r_0\|^2 < \delta \Rightarrow \|w(t)\|^2 + \|r(t)\|^2 < \varepsilon, \qquad \forall t > 0, \qquad (1.13)$$

and there exists γ with $0 < \gamma \le \infty$ such that

$$\|w_0\|^2 + \|r_0\|^2 < \gamma \Rightarrow \lim_{t \to \infty} \left(\|w(t)\|^2 + \|r(t)\|^2 \right) = 0. \qquad (1.14)$$

If $\gamma = \infty$, we say the solution is *unconditionally* nonlinearly stable (or simply refer to it as being asymptotically stable), otherwise for $\gamma < \infty$ the solution is *conditionally* (nonlinearly) stable. For nonlinear stability problems it is an important goal to derive parameter regions for unconditional nonlinear stability, or at least conditional stability with a finite initial data threshold (i.e. finite, non-vanishing, radius of attraction). It is important to realise that the linearization as in (1.11) and (1.12) can only yield linear *instability*. It tells us nothing whatsoever about stability. There are many equations for which nonlinear solutions will become unstable well before the linear instability analysis predicts this, cf. [253]. Also, when an analysis is performed with $\gamma < \infty$ in (1.14) this yields conditional nonlinear stability, i.e. nonlinear stability for only a restricted class of initial data.

We have only defined stability with respect to the $L^2(\Omega)$ norm in (1.13) and (1.14). However, sometimes it is convenient to use an analogous definition with respect to some other norm or positive-definite solution measure. It will be clear in the text when this is the case. When we refer to continuous dependence on the initial data we mean a phenomenon like (1.13). Thus, a solution (w, r) to equations (1.9) depends continuously on the initial data if a chain of inequalities like (1.13) holds.

Throughout the book we make frequent use of inequalities. In particular, we often use the Cauchy-Schwarz inequality for two functions f and g, i.e.

$$\int_\Omega fg\, dx \le \left(\int_\Omega f^2 dx \right)^{1/2} \left(\int_\Omega g^2 dx \right)^{1/2}, \qquad (1.15)$$

or what is the same in L^2 norm and inner product notation,

$$(f, g) \le \|f\| \, \|g\|. \qquad (1.16)$$

The arithmetic-geometric mean inequality (with a constant weight $\alpha > 0$) is, for $a, b \in \mathbb{R}$,

$$ab \le \frac{1}{2\alpha} a^2 + \frac{\alpha}{2} b^2, \qquad (1.17)$$

and this is easily seen to hold since

$$\left(\frac{a}{\sqrt{\alpha}} - \sqrt{\alpha} b \right)^2 \ge 0.$$

Another inequality we frequently have recourse to is Young's inequality, which for $a, b \in \mathbb{R}$ we may write as

$$ab \leq \frac{|a|^p}{p} + \frac{|b|^q}{q}, \qquad \frac{1}{p} + \frac{1}{q} = 1, \quad p, q \geq 1. \tag{1.18}$$

1.1.3 Overview

The layout of the book is now briefly described. In section 1.2 we introduce the equations for thermal convection in an incompressible, viscous fluid allowing for a Boussinesq approximation. In sections 1.3–1.7 we discuss models for flow in a porous medium based on theories of Darcy, Darcy with anisotropy, Forchheimer, Brinkman, and when inertia is present. Section 1.8 discusses how effects like temperature and a salt field are incorporated into porous media. Section 1.9 discusses various boundary conditions which may be used, including the important boundary conditions where a fluid may slip at the boundary.

Chapter 2 introduces the topic of thermal convection with local thermal non-equilibrium effects, i.e. LTNE, where the solid and fluid temperatures in a saturated porous medium may be different. Chapter 3 continues this development and allows the horizontal layer to be in a state of constant angular rotation. Chapter 4 extends the work of chapter 2 to include diffusion and convection of a salt field in addition to LTNE effects.

Chapter 5 investigates LTNE thermal convection in a vertical layer of saturated porous material when the vertical bounding walls have different temperatures. Chapter 6 incorporates penetrative convection effects into the LTNE model of chapter 2. Chapter 7 looks at thermal convection with LTNE effects in a series of horizontal porous layers separated by finite depth horizontal layers of rigid solids. Chapter 8 includes a description of various thermal convection problems where LTNE and other microfluidic effects play a role.

Chapter 9 investigates thermal convection in a fluid when the boundaries allow for slip. In chapter 10 we investigate recent work involving thermal convection with a single temperature field. The geometry is one with a series of horizontal porous layers separated by rigid solids. Chapter 11 analyses how a baffle, or obstruction, on a wall may strongly increase the critical Rayleigh number for thermal convection and thus be of interest in insulation.

In chapter 12 we study stability when the theory for a saturated porous material incorporates an inertia coefficient which may be strongly anisotropic. Chapter 13 investigates thermal convection in a bidispersive porous material. This is a class of man made materials which have a double porosity structure, one on the usual macro scale, but another one well defined on a micro level.

In the penultimate chapter, chapter 14, we examine how convection may occur simultaneously in different sub-layers in a saturated porous material. This leads to much increased critical Rayleigh numbers and is likely to be of interest in insulation.

Finally, the monograph is completed with chapter 15 where we describe models for thermal convection in a nanofluid. Nanofluids are man made fluids which contain a suspension of metallic or metallic-like particles in a carrier fluid and hence may possess good thermal properties to enhance thermal convection.

One aspect of thermal convection we pay particular attention to is when an effect may substantially change the critical Rayleigh number. While it depends on the definition of Rayleigh number, as a general rule, increasing the critical Rayleigh number means convective motion is less likely to occur. Since thermal convection greatly assists heat transfer, an increase of critical Rayleigh number is likely to be of interest in insulation studies. Conversely, a strong decrease in Rayleigh number will lead more easily to convection and so heat transfer is more likely to occur more readily, and this scenario is of interest to the heat transfer industry.

The critical Rayleigh number is strongly affected in the analyses in chapters 7, 9, 10, and 12. Furthermore, a strong increase in critical Rayleigh number is frequently observed in situations in chapters 11, 13, and 14. On the contrary, a strong decrease in critical Rayleigh number is frequently observed in chapter 15.

1.2 The Navier–Stokes Equations for Incompressible Fluid Flow

For many flows it is sufficient to regard the fluid as incompressible. In this situation, with a constant temperature throughout, the velocity field, $\mathbf{v}$, and pressure, p, are determined from the *Navier–Stokes* equations,

$$v_{i,t} + v_j v_{i,j} = -\frac{1}{\rho} p_{,i} + \nu \Delta v_i + f_i,$$
$$v_{i,i} = 0. \tag{1.19}$$

Here ρ is the (constant) density, ν (positive constant) is the kinematic viscosity, $\mathbf{f}$ is the external body force, and Δ is the Laplacian operator.

We shall study equations (1.19) on a bounded spatial region $\Omega \, (\subset \mathbb{R}^3)$ with $\mathbf{v}$ given on the boundary Γ, i.e.,

$$v_i(\mathbf{x}, t) = \hat{v}_i(\mathbf{x}, t), \qquad \mathbf{x} \in \Gamma, \tag{1.20}$$

where $\hat{v}_i$ is a prescribed function.

In situations where the dimensions are on a microscale, i.e. microfluidics, it may not be adequate to prescribe the velocity field on the boundary of the domain as in (1.20). In this case one may need to allow for the physically real effect of slip on the boundary, and this is discussed at length in chapter 9 of this book, see also section 1.9.1.

In the interests of clarity, and to ensure full understanding of the indicial notation used throughout this book we write out equations (1.19) in full as a system of partial differential equations for the four dependent variables v_1, v_2, v_3 and p. It is convenient to use the notation $\mathbf{v} = (u, v, w)$ and $\mathbf{x} = (x, y, z)$, and then system (1.19) is equivalent to

$$
\begin{aligned}
&\frac{\partial u}{\partial t} + u\frac{\partial u}{\partial x} + v\frac{\partial u}{\partial y} + w\frac{\partial u}{\partial z} = -\frac{1}{\rho}\frac{\partial p}{\partial x} + v\Delta u + f_1, \\
&\frac{\partial v}{\partial t} + u\frac{\partial v}{\partial x} + v\frac{\partial v}{\partial y} + w\frac{\partial v}{\partial z} = -\frac{1}{\rho}\frac{\partial p}{\partial y} + v\Delta v + f_2, \\
&\frac{\partial w}{\partial t} + u\frac{\partial w}{\partial x} + v\frac{\partial w}{\partial y} + w\frac{\partial w}{\partial z} = -\frac{1}{\rho}\frac{\partial p}{\partial z} + v\Delta w + f_3, \\
&\frac{\partial u}{\partial x} + \frac{\partial v}{\partial y} + \frac{\partial w}{\partial z} = 0.
\end{aligned}
\tag{1.21}
$$

It is also worth mentioning that when the viscosity is zero equations (1.19) reduce to Euler's equations for an incompressible fluid, given by

$$
\begin{aligned}
&v_{i,t} + v_j v_{i,j} = -\frac{1}{\rho} p_{,i} + f_i, \\
&v_{i,i} = 0.
\end{aligned}
\tag{1.22}
$$

1.2.1 The Balance of Energy and the Boussinesq Approximation

When we deal with motion of a fluid driven by buoyancy forces, such as those caused by heating the fluid, equations (1.19) are not sufficient and it is necessary to also add an equation for the temperature field, T. This is the equation for the *balance of energy*,

$$
\rho T\dot{\varepsilon} = \sigma_{ij}d_{ij} + \rho r - q_{i,i},
\tag{1.23}
$$

where $\varepsilon, \sigma_{ij}, d_{ij}, r$, and q_i are the internal energy, stress tensor, symmetric part of the velocity gradient, external supply of heat, and heat flux, and

$$
\dot{\varepsilon} = \frac{\partial \varepsilon}{\partial t} + \mathbf{v}.\nabla\varepsilon.
$$

For a linear viscous fluid we take

$$
\sigma_{ij} = -p\delta_{ij} + 2\mu d_{ij},
$$

where μ $(= \rho v)$ is the dynamic viscosity. In general, the fluid is compressible, but for many convective motions the system may be considerably simplified by assuming the motion is isochoric, i.e., essentially incompressible flow, except in the body force term $\mathbf{f}$ in (1.19)$_1$. This approximation has caused a lot of concern in the literature and is known as the *Boussinesq approximation*, associated with the work of [59].

Justifications of the Boussinesq approximation and a presentation of the conditions under which it is likely to be valid have been given by [406, 290] [379, pp. 194-197], [138, 139], pages 72–74, and [184].

We wish to reduce equation (1.23) to a form suitable for mathematical analysis. Thus, we commence with the second law of thermodynamics in the form of the Clausius-Duhem inequality, namely,

$$\rho \dot{\eta} \geq \frac{\rho r}{T} - \left(\frac{q_i}{T}\right)_{,i} \tag{1.24}$$

where η is the entropy. We shall also require the Helmholtz free energy, ψ, which is defined by $\psi = \varepsilon - \eta T$. Employ equation (1.23) to remove the heat supply term r from inequality (1.24) to find

$$-\rho \dot{\psi} - \rho \eta \dot{T} + \sigma_{ij} d_{ij} - \frac{q_i T_{,i}}{T} \geq 0. \tag{1.25}$$

Since we are dealing with a viscous, incompressible fluid the constitutive theory we require is that $\psi = \psi(T, T_{,i}, d_{ij})$. Thus, employing the chain rule in inequality (1.25) we obtain

$$-\rho \left(\frac{\partial \psi}{\partial T} + \eta\right) \dot{T} - \rho \frac{\partial \psi}{\partial T_{,i}} \frac{d}{dt}(T_{,i}) - \frac{\partial \psi}{\partial d_{ij}} \frac{d}{dt}(d_{ij}) + \sigma_{ij} d_{ij} - \frac{q_i T_{,i}}{T} \geq 0, \tag{1.26}$$

where d/dt or a superposed dot denotes the material derivative. We may now argue that since $dT_{,i}/dt$ and $\dot{T}$ appear linearly in inequality (1.26) their coefficients must be zero, since by suitable choice of r the energy balance equation (1.23) may hold for an independent choice of $dT_{,i}/dt$ and $\dot{T}$. Thus, we must have

$$\psi = \psi(T, d_{ij}) \quad \text{and} \quad \eta = -\frac{\partial \psi}{\partial T}. \tag{1.27}$$

The fluid is incompressible and so $d(d_{ij})/dt$ is not independent for all six components. Thus we must introduce the constraint $d_{ii} = 0$ into inequality (1.26) via a Lagrange multiplier λ. Hence what remains of (1.26) may be written as

$$\frac{d}{dt}(d_{ij})\left(\lambda \delta_{ij} - \rho \frac{\partial \psi}{\partial d_{ij}}\right) + \sigma_{ij} d_{ij} - \frac{q_i T_{,i}}{T} \geq 0. \tag{1.28}$$

Now we may argue that all six components of $d(d_{ij})/dt$ are independent because λ is arbitrary, the momentum equation being balanced by a suitable body force f_i. Thus, since $d(d_{ij})/dt$ appears linearly in inequality (1.28),

$$\rho \frac{\partial \psi}{\partial d_{ij}} = \lambda \delta_{ij}. \tag{1.29}$$

Next, ψ is a scalar quantity and so it can only depend on scalar quantities derived from d_{ij}. Thus, by the Cayley–Hamilton theorem ψ will depend on the invariants of

d_{ij}, i.e. on $I_1 = d_{ii}$, $I_2 = \text{tr}\mathbf{d}^2$, $I_3 = D$, where $D = \det\mathbf{d}$. Hence,

$$\psi = \psi(I_1, I_2, I_3).$$

Employ the relation $6D = \varepsilon_{ijk}\varepsilon_{abc}d_{ia}d_{jb}d_{kc}$ and then one finds $\partial I_3/\partial d_{ij} = f_{ij}$ where f_{ij} is a quadratic function in d_{ij}. Hence,

$$\frac{\partial\psi}{\partial d_{ij}} = \frac{\partial\psi}{\partial I_1}\delta_{ij} + 2\frac{\partial\psi}{\partial I_2}d_{ij} + \frac{\partial\psi}{\partial I_3}f_{ij}. \tag{1.30}$$

However, employing (1.29) we must have $\partial\psi/\partial I_2 = 0$, $\partial\psi/\partial I_3 = 0$. Then from (1.29) and (1.30)

$$\frac{\partial\psi}{\partial d_{ii}} = \frac{3\lambda}{\rho} = \frac{\partial\psi}{\partial I_1}$$

and so

$$\psi = \frac{3\lambda}{\rho}I_1 + \psi_0(T).$$

Thus we deduce $\psi = \psi(T)$. Since we know $\eta = -\partial\psi/\partial T$ it follows that $\eta = \eta(T)$. What remains from inequality (1.28) is

$$\sigma_{ij}d_{ij} - \frac{q_i T_{,i}}{T} \geq 0. \tag{1.31}$$

For a linearly viscous fluid obeying Fourier's law we know that $\sigma_{ij} = -p\delta_{ij} + 2\mu d_{ij}$ and $q_i = -kT_{,i}$. These expressions inserted into inequality (1.31) lead to

$$2\mu d_{ij}d_{ij} + k\frac{T_{,i}T_{,i}}{T} \geq 0.$$

Hence, necessarily we obtain the restrictions on the dynamic viscosity and on the thermal conductivity, entirely consistent with the physics, $\mu \geq 0$, $k \geq 0$.

To reduce the energy balance equation (1.23) we note $\varepsilon = \psi + \eta T$ so $\varepsilon = \varepsilon(T)$ and then we obtain

$$\rho T\dot{\eta} = \sigma_{ij}d_{ij} + \rho r - q_{i,i}. \tag{1.32}$$

Take now $r = 0$ and put $\eta = -\partial\psi/\partial T$ and employ the constitutive theory for σ_{ij} and q_i to obtain

$$-\rho T\psi_{TT}\dot{T} = (kT_{,i})_{,i} + 2\mu d_{ij}d_{ij}. \tag{1.33}$$

From thermodynamics we know

$$c_p = \frac{\partial\varepsilon}{\partial T}\bigg|_p + p\frac{\partial V}{\partial T}\bigg|_p$$

where the derivatives are at constant pressure, V denotes the specific volume, and c_p is the specific heat at constant pressure. Thus, in this case equation (1.33) may be written

$$\rho c_p\dot{T} = (kT_{,i})_{,i} + 2\mu d_{ij}d_{ij}. \tag{1.34}$$

It is usually argued that when one is interested in convection problems before the instability onset or not too far from the onset that one may discard the viscous dissipation term $2\mu d_{ij}d_{ij}$. Thus, for k constant we derive the reduced energy balance equation

$$\rho c_p \dot{T} = k\Delta T . \tag{1.35}$$

Of course, this supposes the velocity gradients are small. When one is studying strongly convective motion or certainly thermal turbulence, the $d_{ij}d_{ij}$ term should be retained.

For thermal convection with a linearly viscous incompressible fluid we thus have equation (1.35) governing the temperature field and equation (1.19)$_2$ expressing the solenoidal character of the velocity field. It remains to develop the momentum equation (1.19)$_1$ and this equation we rewrite as

$$\rho(v_{i,t} + v_j v_{i,j}) = -p_{,i} + \mu\Delta v_i + \rho f_i. \tag{1.36}$$

The pressure in (1.36) is regarded as an unknown to be solved for in the problem, and the density is assumed constant except in the body force term $\mathbf{f}$, where

$$\rho\mathbf{f} = -g\mathbf{k}\rho(T), \tag{1.37}$$

assuming the only force acting is gravity, and where $\mathbf{k} = (0,0,1)$. The density in (1.37) is expanded in a Taylor series, and at this stage we consider only the first term, so

$$\rho = \rho_0\left[1 - \alpha(T - T_R)\right], \tag{1.38}$$

where ρ_0 is the density at temperature T_R, and α is the coefficient of thermal expansion given by

$$\alpha = -\frac{1}{\rho}\left(\frac{\partial\rho}{\partial T}\right)_{T_R}. \tag{1.39}$$

From this point we follow the mathematical approach described in [139, pp. 72–74], or what is essentially an equivalent approach, the mathematical procedure presented in [379, pp. 194–197].

Suppose now the fluid is contained in the layer $(x,y) \in \mathbb{R}^2 \times \{z \in (0,d)\}$ and

$$\left|\alpha(T - T_R)\right| \ll 1.$$

(This is not a serious restriction since the temperature field is normally restricted to a small interval and the coefficient of thermal expansion is usually very small.) Write equation (1.38) as $\rho = \rho_0 + \rho_1$ so that

$$\left|\frac{\rho_1}{\rho_0}\right| = \left|\alpha(T - T_R)\right| \ll 1.$$

Non-dimensionalize equation (1.36) with the time scale $\mathcal{T} = d^2\rho_0/\mu$, velocity scale $U = \mu/d\rho_0$, pressure scale $P = \mu^2/\rho_0 d^2$, and with $\mathbf{x} = \mathbf{x}^*d$, $p = Pp^*$, $T = T^\sharp T^*$. Drop the stars in what follows although it is now understood that the variables

v_i, p, T and x_i are non-dimensional. Then equation (1.36) may be rewritten as

$$\left(1 + \frac{\rho_1}{\rho_0}\right)(v_{i,t} + v_j v_{i,j}) = -p_{,i} + \Delta v_i - \frac{gd^3 \rho_0^2}{\mu^2} k_i [1 - \alpha T^\sharp (T - T_R)], \qquad (1.40)$$

where T_R is the non-dimensional form for T_R. Since $|\rho_1/\rho_0| \ll 1$ we discard the $(\rho_1/\rho_0)\dot{v}_i$ term. We cannot similarly discard the term in $g\alpha T^\sharp (T - T_R)$ since this is the driving force for the convective motion. Instead we argue as [139, p. 74], that we retain the matching orders of magnitude of the pressure and buoyancy terms. This means that we balance the buoyancy term involving $g\alpha T^\sharp$ by the pressure gradient.

For fixed upper and lower constant temperatures T_U and T_L we have in non-dimensional form a steady solution (where $\mathbf{v} \equiv 0$) for the temperature field of form $\bar{T} = -(T_L - T_U)z + T_L$. This leads to an ordinary differential equation for the steady pressure field $\bar{p}$,

$$\frac{d\bar{p}}{dz} = -\frac{gd^3 \rho_0^2}{\mu^2}\left(1 - \alpha T^\sharp [-(T_L - T_U)z + T_L - T_R]\right).$$

The next step of the procedure is to perturb the steady solution and now write $p = \bar{p} + \beta_1 \pi$ and $T = \bar{T} + \gamma\theta$, where β_1 and γ are parameters in an expansion for p and T. Then from equation (1.40), for v_i a velocity perturbation, one may derive the following equation

$$v_{i,t} + v_j v_{i,j} = -\beta_1 \pi_{,i} + \frac{gd^3 \rho_0^2 \alpha \gamma}{\mu^2} k_i \theta + \Delta v_i.$$

We pick β_1 to balance the $\pi_{,i}$ and θ terms, namely, $\beta_1 = gd^3 \rho_0^2 \alpha \gamma / \mu^2$. Then pick γ such that $\gamma = \mu^2 / gd^3 \rho_0^2 \alpha$. This leads to the equation

$$v_{i,t} + v_j v_{i,j} = -\pi_{,i} + k_i \theta + \Delta v_i. \qquad (1.41)$$

This is the momentum equation after the Boussinesq approximation has been applied.

Thus, taking account of (1.35), (1.41), and (1.19)$_2$, the equations for a linearly viscous, heat conducting, incompressible fluid (utilizing the Boussinesq approximation) may be written as

$$\begin{aligned} v_{i,t} + v_j v_{i,j} &= -\frac{1}{\rho_0} p_{,i} + \nu \Delta v_i - k_i g\left(1 - \alpha[T - T_R]\right), \\ v_{i,i} &= 0, \\ T_{,t} + v_i T_{,i} &= \kappa \Delta T, \end{aligned} \qquad (1.42)$$

where $\kappa = k/\rho_0 c_p$ is the thermal diffusivity.

For water between 0°C and 100°C (see e.g., [39 p. 596, 597]) c_p varies from 4.27 joule gm^{-1} deg C^{-1} to 4.216 joule gm^{-1} deg C^{-1}, κ varies from 1.33×10^{-3} cm^2 sec^{-1} to 1.66×10^{-3} cm^2 sec^{-1}, so it is certainly reasonable to treat them as constant in many convection studies. The kinematic viscosity v varies from $1.787 \times 10^{-2}\,cm^2\,sec^{-1}$ to $0.295 \times 10^{-2}\,cm^2\,sec^{-1}$ so its variation is greater. We shall primarily look at thermal convection with v constant, but there are practical situations where a strongly varying viscosity with temperature is necessary.

More elaborate and relatively recent treatments of the Boussinesq approximation are available and some of these are discussed in chapter 15 of the book [414], and in [31, 134, 159, 160, 311, 359], and in addition [360, 361]. Without a Boussinesq approximation to yield a simplified system like (1.42) with a solenoidal velocity field, one is left with treating convection in a compressible fluid, and this is a far more complicated issue, cf. e.g. [414], chapter 15.

1.3 The Darcy Model

The celebrated Darcy equation is believed to originate from work of [103], and this equation is discussed in detail in [312], section 1.4. Darcy's law basically states that the flow rate of a fluid in a porous material is proportional to the pressure gradient. In current terminology, if the flow is in the $x-$direction and the speed in that direction is u then this may be represented as

$$\mu u = -k \frac{dp}{dx},\qquad(1.43)$$

where μ is the dynamic viscosity of the fluid, and k denotes the permeability of the porous skeleton, with p being the pressure of the fluid in the porous medium. Despite its apparent simplicity, equation (1.43) has been very successful in providing a theoretical description of flow in porous media.

Equation (1.43) is generalized to three spatial dimensions and we additionally include a term to allow for external forces (such as gravity). Thus, if we denote the velocity field in the porous medium by $\mathbf{v}$, where $\mathbf{v} = (u, v, w)$, and denote the body force by $\mathbf{f}$, we have a three-dimensional version of equation (1.43) which takes form

$$0 = -\frac{\partial p}{\partial x_i} - \frac{\mu}{k} v_i + \rho f_i,\qquad(1.44)$$

where ρ is the density of the fluid, cf. [197], while μ and k continue to be the dynamic viscosity and permeability. In direct vector notation equation (1.44) may be written as

$$0 = -\nabla p - \frac{\mu}{k}\mathbf{v} + \rho \mathbf{f}.\qquad(1.45)$$

Equation (1.44) describes flow of a fluid in a saturated porous medium, provided the flow rate is sufficiently low, which is often the case with a material which has a small porosity. Extensive use has been made of equation (1.44), often coupled to other equations, for the temperature, or salt concentration, for example.

If the fluid in the porous medium is incompressible then we must couple equation (1.44) with the incompressibility condition

$$\frac{\partial v_i}{\partial x_i} = 0. \tag{1.46}$$

Thus, the complete system of equations describing flow of an incompressible fluid in a saturated porous medium of Darcy type consists of equations (1.44) and (1.46), for the variables v_i and p.

To be absolutely clear about the notation in equations (1.44) and (1.46) we write them here in full component form, with $\mathbf{x} = (x, y, z)$,

$$
\begin{aligned}
\frac{\partial p}{\partial x} &= -\frac{\mu}{k} u + \rho f_1, \\
\frac{\partial p}{\partial y} &= -\frac{\mu}{k} v + \rho f_2, \\
\frac{\partial p}{\partial z} &= -\frac{\mu}{k} w + \rho f_3, \\
\frac{\partial u}{\partial x} + \frac{\partial v}{\partial y} + \frac{\partial w}{\partial z} &= 0.
\end{aligned}
\tag{1.47}
$$

Derivations of Darcy's law based on various assumptions may be found in many places in the literature, cf. [312], section 1.4. Very interesting accounts based on the technique of homogenization and on an approach employing asymptotic expansions are provided by [140] and by [158], respectively, see also [478]. The history of porous media model development may be found in [106].

1.3.1 The Porous Medium Equation

In this book we frequently deal with a porous medium described by the equations of Darcy. However, there is an equation which frequently appears in the mathematical analysis literature known as the porous medium equation, cf. [10, 16, 17, 111, 142]. We show how this equation is connected to the equations for a porous medium as described by Darcy's law. To derive "the porous medium equation" from Darcy's law suppose the fluid is compressible, i.e. equation (1.46) does not hold, then the density instead satisfies the equation

$$\frac{\partial \rho}{\partial t} + \frac{\partial}{\partial x_i}(\rho v_i) = 0. \tag{1.48}$$

We couple this equation to equation (1.44). Equation (1.44) with $f_i = 0$ may be rewritten as,

$$v_i = -\frac{k}{\mu} \frac{\partial p}{\partial x_i}. \tag{1.49}$$

Note that p is a function of the density ρ. Now substitute in equation (1.48) the expression (1.49) for v_i. In this way one obtains the equation

$$\frac{\partial \rho}{\partial t} - \frac{k}{\mu} \frac{\partial}{\partial x_i} \left(\rho \frac{\partial p}{\partial x_i} \right) = 0$$

or, expanding $\partial p / \partial x_i$,

$$\frac{\partial \rho}{\partial t} - \frac{k}{\mu} \frac{\partial}{\partial x_i} \left(\rho p'(\rho) \frac{\partial \rho}{\partial x_i} \right) = 0. \tag{1.50}$$

Let $\Phi(\rho)$ be a potential (integral) for the function $(k\rho p'/\mu)(\partial \rho/\partial x_i)$, i.e.

$$\frac{\partial \Phi}{\partial x_i} = \Phi'(\rho) \frac{\partial \rho}{\partial x_i} = \frac{k\rho p'(\rho)}{\mu} \frac{\partial \rho}{\partial x_i}. \tag{1.51}$$

Now substitute in equation (1.50) from equation (1.51) and then equation (1.50) may be rewritten as

$$\frac{\partial \rho}{\partial t} - \Delta \Phi(\rho) = 0. \tag{1.52}$$

Equation (1.52) is that equation often referred to as the porous medium equation.

Some early articles investigating properties of solutions to the porous medium equation are those of [10, 16, 17, 111]. A more recent interesting article dealing with a novel application of the porous medium equation to image contour enhancement in image processing is [28]. Further interesting pointwise stability results and finite time blow-up results for a solution to equations like (1.52) may be found in [141, 143, 145], for various choices of nonlinear function Φ. The work of [378] studies continuous dependence in a weighted L^2 norm for a similar problem and [369] contains some interesting asymptotic results.

1.4 Anisotropic Darcy Model

The models discussed in sections 1.3, 1.5, 1.6 are frequently very useful when dealing with the flow in a porous medium when the situation is isotropic, i.e. the response is the same in all directions. However, many porous media exhibit strongly anisotropic characteristics. For example, wood behaves very differently along the grain to the way it does across the grain, cf. figure 1.8. Layered rock forming strata provides another example of highly anisotropic porous media, as displayed in figure 1.7. If one is interested in modelling flow in anisotropic porous media, it is necessary to modify each of the models in section 1.3, 1.5, 1.6, accordingly. In particular, for flow in a micro dimension situation anisotropy may well be very important, cf. chapter 12.

In this section we indicate how we may modify the model in section 1.3, namely the model employing Darcy's law. Anisotropic modifications for the other models of Brinkman and Forchheimer type follow in a similar manner.

Typically the permeability, i.e. the ease with which the fluid flows, will vary if the solid fraction of the porous medium displays a strong anisotropy. To account for this we replace the permeability k in (1.44) by a tensor K_{ij}. Thus, we may replace equation (1.44) by

$$K_{ij}\frac{\partial p}{\partial x_j} = -\mu v_i + \rho K_{ij} f_j. \tag{1.53}$$

If we introduce a generalized inverse tensor to K_{ij}, say M_{ij}, then we may recast equation (1.53) in a form not dissimilar to (1.44). To do this we suppose

$$\mathbf{MK} = c\mathbf{I}, \quad \text{i.e.} \quad \mathbf{M} = c\mathbf{K}^{-1},$$

for $c > 0$ a constant, and then equation (1.53) is equivalent to

$$0 = -\frac{\mu}{c} M_{ij} v_j - \frac{\partial p}{\partial x_i} + \rho f_i. \tag{1.54}$$

Equation (1.54) is to be coupled with equation (1.46). The precise form for the permeability tensor $\mathbf{K}$ depends on what the structure is of the underlying solid matrix in the porous medium.

Thus, for incompressible flow in a saturated Darcy porous material with anisotropic permeability the governing system of equations consists of the following equations of momentum and continuity,

$$0 = -\frac{\mu}{c} M_{ij} v_j - \frac{\partial p}{\partial x_i} + \rho f_i,$$
$$\frac{\partial v_i}{\partial x_i} = 0. \tag{1.55}$$

Again, in the interests of fully understanding the indicial notation employed we put $\mathbf{v} = (u, v, w)$, $\mathbf{x} = (x, y, z)$ and then equations (1.55) may be written out fully in the form

$$0 = -\frac{\mu}{c} M_{11} u - \frac{\mu}{c} M_{12} v - \frac{\mu}{c} M_{13} w - \frac{\partial p}{\partial x} + \rho f_1,$$
$$0 = -\frac{\mu}{c} M_{21} u - \frac{\mu}{c} M_{22} v - \frac{\mu}{c} M_{23} w - \frac{\partial p}{\partial y} + \rho f_2,$$
$$0 = -\frac{\mu}{c} M_{31} u - \frac{\mu}{c} M_{32} v - \frac{\mu}{c} M_{33} w - \frac{\partial p}{\partial z} + \rho f_3,$$
$$\frac{\partial u}{\partial x} + \frac{\partial v}{\partial y} + \frac{\partial w}{\partial z} = 0. \tag{1.56}$$

Note that equations (1.56) show the effects of cross influence of velocity components due to anisotropy. For example, equation (1.56)$_1$ contains not only u, as in equation (1.47)$_1$, but also terms in v and w, namely the M_{12} and M_{13} terms. These cross effects due to anisotropy make equations (1.56) more difficult to deal with mathematically.

As a specific example we consider the case where the permeability in the vertical direction is different from that in the horizontal directions, cf. the rock strata in picture 1.7. This is a much simplified situation when compared to that of equations (1.56), but it is nevertheless a highly realistic one for geophysical and microfluidic applications. Then, for $k \neq k_3$, $k, k_3 > 0$, the permeability tensor **K** may be written in the following matrix form

$$\mathbf{K} = \begin{pmatrix} k & 0 & 0 \\ 0 & k & 0 \\ 0 & 0 & k_3 \end{pmatrix}$$

To comply with the definition of the inverse permeability tensor **M** we may select $c = k_3$, and then **M** becomes

$$\mathbf{M} = \begin{pmatrix} k_3/k & 0 & 0 \\ 0 & k_3/k & 0 \\ 0 & 0 & 1 \end{pmatrix}$$

A porous convection problem with such a permeability is analysed by [78], by [260], and by [172]. A further example of application to thermal convection in a porous medium with transversely isotropic permeability at an angle oblique to the vertical is given in [437], see also [414, p. 338], and section 4.2.6 of [417]. The permeability for this situation is not simply a diagonal matrix and this severely complicates the analysis and numerical calculations giving rise to complex solutions which are not found with the simple permeability tensor written above.

1.5 The Forchheimer Model

If the flow rate exceeds a certain value then it is believed that the linear relationship of (1.43) or (1.44) will be inadequate to describe the velocity field accurately. [146] (see also [120]) proposed modifying the linear velocity / pressure gradient law and replacing it with a nonlinear one. According to [140, 146] proposes replacing the left hand side of equation (1.43) by one of three formulae,

$$\begin{aligned} \alpha &= au + bu^2, \\ \alpha &= mu^n, \\ \text{or} \quad \alpha &= au + bu^2 + cu^3, \end{aligned} \tag{1.57}$$

where α denotes the left hand side of equation (1.43).

1.5.1 Quadratic Forchheimer Model

If one adopts the first of (1.57) then the generalization of equation (1.44) which arises results in the Forchheimer model

$$0 = -\frac{\partial p}{\partial x_i} - \frac{\mu}{k} v_i - b|\mathbf{v}|v_i + \rho f_i, \tag{1.58}$$

where b is a positive constant and $|\mathbf{v}| = \sqrt{u^2 + v^2 + w^2}$ when $\mathbf{v} = (u, v, w)$. For incompressible flow we couple this with equation (1.46) to yield the system of equations

$$0 = -\frac{\partial p}{\partial x_i} - \frac{\mu}{k} v_i - b|\mathbf{v}|v_i + \rho f_i,$$
$$\frac{\partial v_i}{\partial x_i} = 0, \tag{1.59}$$

which must be solved for the variables v_1, v_2, v_3 and p.

Rigorous justifications of equation (1.58) and generalizations may be found in the papers of [45, 140, 158, 479].

1.5.2 Cubic Forchheimer Model

[305] gives very good reasons for suggesting an alternative to (1.58) or (1.59) in which one essentially employs (1.57)$_3$ with $b = 0$. [3] provides a demonstration based on channel flow in support of the same alternative. In fact, for incompressible flow in a Forchheimer porous material these writers suggest instead of (1.59) one should employ the system

$$0 = -\frac{\partial p}{\partial x_i} - \frac{\mu}{k} v_i - c|\mathbf{v}|^2 v_i + \rho f_i,$$
$$\frac{\partial v_i}{\partial x_i} = 0, \tag{1.60}$$

where again one solves for the variables v_1, v_2, v_3 and p.

We remark that system (1.60) does possess some elegant beneficial mathematical properties for nonlinear stability problems, cf. [335].

1.5.3 Generalized Forchheimer Model

Another generalization of (1.59) which is employed in the literature is to adapt system (1.60) to one of form

$$0 = -\frac{\partial p}{\partial x_i} - \frac{\mu}{k} v_i - c_1 |\mathbf{v}|^\alpha v_i + \rho f_i,$$

$$\frac{\partial v_i}{\partial x_i} = 0,$$

(1.61)

where α is a constant with $1 \leq \alpha \leq 2$.

Structural stability issues for systems of equations of type (1.61) have occupied the attention of various writers, cf. [20, 21, 80, 186–188, 203, 205, 207, 210, 242, 244, 245, 247, 248, 328, 424, 455, 490].

1.5.4 Teng–Zhao Model

[449] considered another generalization of Darcy's law which may apply to non-slow flow in a dense porous medium, i.e. non-Stokes flow. They employ an averaging technique over a sub-volume of a porous body, a so-called representative elementary volume, to derive the following system of equations, when the fluid is incompressible,

$$\rho \left(\frac{\partial v_i}{\partial t} + v_j \frac{\partial v_i}{\partial x_j} \right) + \left(\frac{1 + Re_p}{K} \right) \mu v_i = -\frac{\partial p}{\partial x_i} + \rho f_i,$$

$$\frac{\partial v_i}{\partial x_i} = 0,$$

(1.62)

where $Re_p = \rho U_p \ell_p / \mu$ is a microscopic Reynolds number, U_p and ℓ_p being typical velocity and length scales. The derivation of [449] is based on the representative elementary volume being a micro-flow channel.

[449] discuss how to determine U_p and ℓ_p for a real porous material and thus determine their microscopic Reynolds number. They also use an appropriate non-dimensionalization to show that when $Re_p \to 0$ system (1.62) reduces to the Darcy system (1.47).

Observe that in equation (1.62)$_1$ there appears an acceleration term of form $\rho \partial v_i / \partial t$, which hitherto in this book has been omitted. For many flows in porous media it is believed such a term will be small and may be neglected. Nevertheless, this term is present in [449]. We discuss such an acceleration term, which is an inertia term, in section 1.7 and we point out that such terms may well be important in flows on a micro scale.

1.6 The Brinkman Model

1.6.1 Brinkman's Equation

If the porosity, ε, of the porous medium is close to 1, i.e. the solid skeleton occupies little of the total volume, or if the porous medium is adjacent to a solid wall, then there is a belief that neither of the models of sections 1.3 nor 1.5 will prove sufficient. Indeed, when $\varepsilon \approx 1$, one might expect the higher derivatives of the Laplacian in the Navier–Stokes equations to play a role. In fact, there is evidence that equation (1.44) should be replaced by the following

$$0 = -\frac{\partial p}{\partial x_i} - \frac{\mu}{k} v_i + \lambda \Delta v_i + \rho f_i. \tag{1.63}$$

Equation (1.63) is usually associated with [63], it is known as Brinkman's equation, and this subject is discussed in some detail in [312], section 1.5.3. The coefficient λ is usually referred to as an equivalent viscosity. Observe that the presence of the Laplacian term, $\lambda \Delta v_i$, in (1.63) means that the order of this differential equation is higher than that of the Darcy equation (1.44). Care has thus to be taken when switching between equations (1.44) and (1.63), especially with prescription of correct boundary conditions.

If the fluid in the saturated porous material is incompressible then equation (1.63) must be coupled to equation (1.46) for the balance of mass. This leads to the system of equations

$$0 = -\frac{\partial p}{\partial x_i} - \frac{\mu}{k} v_i + \lambda \Delta v_i + \rho f_i,$$
$$\frac{\partial v_i}{\partial x_i} = 0. \tag{1.64}$$

Let us again note that whereas equations $(1.47)_{1-3}$, $(1.55)_1$, $(1.59)_1$, $(1.60)_1$, $(1.61)_1$, do not contain derivatives of v_i, equations (1.64) contain the term $\lambda \Delta v_i$. This leads to different solution behaviour and one has to adopt different boundary conditions on the velocity, see section 1.9.

1.6.2 Alternative Brinkman Equation

[9] employ a different equation to (1.63) which they refer to as the general equation for steady incompressible flow through a porous medium given by [63], although I can only find equation (1.63) in [63]—his equation (5), which is (1.63) with no body force term. The equation of [9] takes the form

$$\rho v_j \frac{\partial v_i}{\partial x_j} = -\frac{\partial p}{\partial x_i} - \frac{\mu}{k} v_i + \lambda \Delta v_i, \tag{1.65}$$

where ρ is the saturating fluid density, and they refer to λ as an effective viscosity for the porous region.

If one wishes to adopt this equation for thermal convection then the body force must be added and thus one would need to solve the system

$$\rho v_j \frac{\partial v_i}{\partial x_j} = -\frac{\partial p}{\partial x_i} - \frac{\mu}{k} v_i + \lambda \Delta v_i + \rho f_i,$$
$$\frac{\partial v_i}{\partial x_i} = 0. \tag{1.66}$$

This system must be solved over a suitable spatial domain subject to appropriate boundary conditions, to determine the variables v_1, v_2, v_3 and p.

1.6.3 Generalizations

There are, in fact, many generalizations of equations (1.64) or (1.66) which have appeared in the literature. We report on some of the common ones.

Many writers combine the Brinkman and Forchheimer equations in some way. For example, one may combine the models of section 1.5.1 and 1.6.1 to arrive at a system of equations of form

$$b|\mathbf{v}|v_i = -\frac{\partial p}{\partial x_i} - \frac{\mu}{k} v_i + \lambda \Delta v_i + \rho f_i,$$
$$\frac{\partial v_i}{\partial x_i} = 0. \tag{1.67}$$

We refer to this as a Brinkman-Forchheimer model. We observe that this is the steady state version of the time dependent model used by [357].

If one includes an inertia term, acceleration term, as [357] do then (1.67) is modified to

$$\rho \frac{\partial v_i}{\partial t} + b|\mathbf{v}|v_i = -\frac{\partial p}{\partial x_i} - \frac{\mu}{k} v_i + \lambda \Delta v_i + \rho f_i,$$
$$\frac{\partial v_i}{\partial x_i} = 0. \tag{1.68}$$

One has now to solve for v_1, v_2, v_3 and p given suitable boundary and initial data on v_i. We observe that [333] provide several convergence and continuous dependence results for a solution to (1.68). The inclusion of an inertia term, i.e. an acceleration term like $\rho \partial v_i / \partial t$ as on the left hand side of (1.68), is discussed in detail in section 1.7.

[261] employ what they call a Lapwood-Brinkman model and this has some resemblance to the equations of section 1.6.2 with an inertia term in. The equations of [261] are

$$\frac{\rho}{\varepsilon}\frac{\partial v_i}{\partial t} + \frac{\rho}{\varepsilon^2}v_j\frac{\partial v_i}{\partial x_j} = -\frac{\partial p}{\partial x_i} - \frac{\mu}{k}v_i + \lambda \Delta v_i + \rho f_i,$$

$$\frac{\partial v_i}{\partial x_i} = 0,$$

(1.69)

where ε is the porosity.

Another generalization of the Brinkman-Forchheimer type is employed by [348] and by [346]. Their system of equations may be written

$$\frac{\rho}{\varepsilon}\frac{\partial v_i}{\partial t} + \frac{\rho}{\varepsilon^2}v_j\frac{\partial v_i}{\partial x_j} = -\frac{\partial p}{\partial x_i} - \frac{\mu}{k}v_i + \lambda \Delta v_i - b|\mathbf{v}|v_i + \rho f_i,$$

$$\frac{\partial v_i}{\partial x_i} = 0,$$

(1.70)

where their Forchheimer coefficient is proportional to $\rho/\sqrt{k}$, and we observe that (1.70) also contain an inertia (acceleration) term. [348] and [346] investigate in detail thermal convection in a horizontal layer using (1.70) coupled to appropriate equations allowing for LTNE effects in the saturated porous material.

1.7 Inertia Coefficient

When dealing with flow in a porous medium, the equations in section 1.3–1.6 have mostly neglected the acceleration, or inertia, term. However, we believe that in microfluidic situations an inertia term may be relevant. Also, while many studes of thermal convection in porous media neglect inertia effects in the momentum equation for the movement of fluid in the porous skeleton, there are increasing instances of where the inertia term can lead to important new physics. [456] showed that the inertia term leads to oscillatory convection which is not present when the inertia is neglected. Other recent studies also show that inclusion of inertia leads to novel effects, see e.g. [12, 51, 70, 130, 172, 403, 421, 422, 425]. The fact that [456] found that thermal convection in a horizontal layer rotating at constant angular speed may be oscillatory rather than stationary when inertia is present is a very striking result. It certainly shows that mathematically the presence of inertia in this situation results in a major change. Thus, we now present equations for the standard Darcy, Forchheimer, and Brinkman theories also when inertia is present.

1.7.1 Darcy Equations

The Darcy equations (1.47) in the presence of an acceleration term may be written in the form

$$a_0 \frac{\partial v_i}{\partial t} = -\frac{\partial p}{\partial x_i} - \frac{\mu}{k} v_i + \rho f_i,$$

$$\frac{\partial v_i}{\partial x_i} = 0,$$

(1.71)

where a_0 is the inertia coefficient which will, in general, depend on the geometric structure and the physical properties of the material of the porous skeleton. Equations (1.71) should be solved simultaneously with prescribed initial and boundary conditions.

1.7.2 Quadratic Forchheimer Equations

The quadratic Forchheimer equations (1.59) with an acceleration term in the momentum equation become

$$a_0 \frac{\partial v_i}{\partial t} = -\frac{\partial p}{\partial x_i} - \frac{\mu}{k} v_i - b|\mathbf{v}|v_i + \rho f_i,$$

$$\frac{\partial v_i}{\partial x_i} = 0,$$

(1.72)

which are to be solved with appropriate boundary and initial conditions for the variables v_1, v_2, v_3 and p. Let us accentuate the fact that equation (1.72)$_1$ differs from equation (1.71)$_1$ due to the nonlinear term $-b|\mathbf{v}|v_i$. This term may also be regarded as a non-local one since $|\mathbf{v}|$ involves all components of v_i.

1.7.3 Cubic Forchheimer Equations

The cubic Forchheimer equations (1.60) with an acceleration term are

$$a_0 \frac{\partial v_i}{\partial t} = -\frac{\partial p}{\partial x_i} - \frac{\mu}{k} v_i - c|\mathbf{v}|^2 v_i + \rho f_i,$$

$$\frac{\partial v_i}{\partial x_i} = 0,$$

(1.73)

which are to be solved with appropriate boundary and initial conditions for the variables v_1, v_2, v_3 and p.

In the interests of clarity we put $\mathbf{v} = (u, v, w)$, $\mathbf{x} = (x, y, z)$, and we write out equations (1.73) in full as

$$a_0 \frac{\partial u}{\partial t} = -\frac{\partial p}{\partial x} - \frac{\mu}{k} u - c(u^2 + v^2 + w^2)u + \rho f_1,$$

$$a_0 \frac{\partial v}{\partial t} = -\frac{\partial p}{\partial y} - \frac{\mu}{k} v - c(u^2 + v^2 + w^2)v + \rho f_2,$$

$$a_0 \frac{\partial w}{\partial t} = -\frac{\partial p}{\partial z} - \frac{\mu}{k} w - c(u^2 + v^2 + w^2)w + \rho f_3,$$

$$\frac{\partial u}{\partial x} + \frac{\partial v}{\partial y} + \frac{\partial w}{\partial z} = 0. \tag{1.74}$$

1.7.4 Brinkman Equations

With acceleration present the Brinkman equations (1.63) are

$$a_0 \frac{\partial v_i}{\partial t} = -\frac{\partial p}{\partial x_i} - \frac{\mu}{k} v_i + \lambda \Delta v_i + \rho f_i,$$

$$\frac{\partial v_i}{\partial x_i} = 0. \tag{1.75}$$

In general, equations (1.75) are to be solved on $\Omega \times \{t > 0\}$, for a domain Ω in $\mathbb{R}^3$, with suitable conditions on v_i on the boundary Γ, see section 1.9, and with suitable initial data, typically v_i given on Ω at $t = 0$.

Written out in component form equations (1.75) are

$$a_0 \frac{\partial u}{\partial t} = -\frac{\partial p}{\partial x} - \frac{\mu}{k} u + \lambda \Delta u + \rho f_1,$$

$$a_0 \frac{\partial v}{\partial t} = -\frac{\partial p}{\partial y} - \frac{\mu}{k} v + \lambda \Delta v + \rho f_2,$$

$$a_0 \frac{\partial w}{\partial t} = -\frac{\partial p}{\partial z} - \frac{\mu}{k} w + \lambda \Delta w + \rho f_3,$$

$$\frac{\partial u}{\partial x} + \frac{\partial v}{\partial y} + \frac{\partial w}{\partial z} = 0. \tag{1.76}$$

1.7.5 Anisotropic Inertia Coefficient

Nearly all of the studies I have seen of flow in a porous medium employing an acceleration term take the inertia coefficient to be isotropic, in the relevant momentum equation. However, we believe that in a microfluidic situation the inertia coefficient could be highly anisotropic. Certainly [312, p. 8, 9], gives convincing arguments to show that the inertia coefficient should really be, in general, described by a (constant) anisotropic tensor. The example they employ is one where the porous medium is such that the pores are parallel circular tubes of the same radius d. Then, if ρ denotes the fluid velocity, ε the porosity, μ the dynamic viscosity of the fluid, and K the permeability, Darcy's law with an inertia term may be written as

$$\frac{\rho}{\varepsilon}\frac{\partial v_i}{\partial t} = -\frac{\partial p}{\partial x_i} - \frac{\mu}{K}v_i,\tag{1.77}$$

where v_i, p are the velocity and pressure, and x_i, t are the spatial coordinate and time. [312] argues that with a constant pressure gradient in the direction of the circular tubes (1.77) leads to a flow in the direction of the tubes which decays like $\exp\{-(\mu\varepsilon/K\rho)t\}$ whereas the classical solution for flow in a circular tube will decay like $\exp\{-(\lambda_1^2\mu/d^2\rho)t\}$ with $\lambda_1 \approx 2.405$. In general these decay rates will differ. Thus [312] concludes (1.77) should be replaced by

$$\rho c_{ij}\frac{\partial v_j}{\partial t} = -\frac{\partial p}{\partial x_i} - \frac{\mu}{K}v_i,\tag{1.78}$$

where c_{ij} is a constant tensor they call the "acceleration coefficient tensor". The tensor c_{ij} will depend on the geometric structure of the porous skeleton comprising the porous medium. [122] also considers periodic porous media such as where the porous skeleton is composed of parallel tubes, and [382] considers flow in an array of ordered microcylinders, and in such a scenario it may be argued that one should include an anisotropic tensor as the coefficient of the fluid inertia term. Such an inertia term has been shown by [98] to lead to novel effects in wave propagation in a porous medium. We here present suitable Darcy equations for flow with an anisotropic inertia coefficient. Such a situation in thermal convection is analysed in detail in [430], see also chapter 12 of this book.

When the inertia coefficient is anisotropic the Darcy equations (1.71) should be replaced by a system of partial differential equations which contains an anisotropic inertia coefficient, say a_{ij}, and thus they assume the form

$$a_{ij}\frac{\partial v_j}{\partial t} = -\frac{\partial p}{\partial x_i} - \frac{\mu}{k}v_i + \rho f_i,$$
$$\frac{\partial v_i}{\partial x_i} = 0.\tag{1.79}$$

Written out in component form involving $\mathbf{v} = (u, v, w)$, the anisotropic system (1.79) is

$$a_{11}\frac{\partial u}{\partial t} + a_{12}\frac{\partial v}{\partial t} + a_{13}\frac{\partial w}{\partial t} = -\frac{\partial p}{\partial x} - \frac{\mu}{k}u + \rho f_1,$$
$$a_{21}\frac{\partial u}{\partial t} + a_{22}\frac{\partial v}{\partial t} + a_{23}\frac{\partial w}{\partial t} = -\frac{\partial p}{\partial y} - \frac{\mu}{k}v + \rho f_2,$$
$$a_{31}\frac{\partial u}{\partial t} + a_{32}\frac{\partial v}{\partial t} + a_{33}\frac{\partial w}{\partial t} = -\frac{\partial p}{\partial z} - \frac{\mu}{k}w + \rho f_3,\tag{1.80}$$
$$\frac{\partial u}{\partial x} + \frac{\partial v}{\partial y} + \frac{\partial w}{\partial z} = 0.$$

Clearly, the presence of cross terms such as $\partial v/\partial t$, $\partial w/\partial t$ in equation $(1.80)_1$, makes this a more complicated system of equations than the analogous isotropic Darcy ones, equations (1.71).

One could allow the permeability k to also be anisotropic and derive a generalization to the anisotropic Darcy system (1.55) such that (possibly different) anisotropy in both the inertia coefficient and in the permeability coefficient is present.

Generalizations of the isotropic inertia systems of Forchheimer equations (1.72), (1.73), and of the isotropic Brinkman system (1.75) which possess anisotropic inertia coefficients as in (1.79) are clearly also possible.

1.8 Equations for Other Fields

1.8.1 Temperature Equations

In this section we suppose the fluid and solid skeleton may have different temperatures, T_f and T_s, respectively. In a given (small) volume, $\tilde{\Omega}$, containing the point $\mathbf{x}$, we denote the solid (porous matrix) part by s while f denotes the fluid. The individual fluid and solid volumes within $\tilde{\Omega}$ are denoted by Ω_f and Ω_s, respectively. The "small" volume is such that a typical length scale is sufficiently larger than the pore scale of the porous material, but the same length scale is much smaller than the overall flow domain. The thermal conductivities are k_s, k_f, the densities are ρ_{0s} and ρ_{0f}, c_s is the specific heat of the solid, and c_{pf} denotes the specific heat at constant pressure of the fluid. Then, let v_i which appears in equation (1.44) be the average velocity of the fluid at the point $\mathbf{x}$. The quantity v_i is called the seepage velocity and is the average of the real fluid velocity over the whole of $\tilde{\Omega}$. Let also V_i, defined by $v_i = \varepsilon V_i$, be the pore average velocity (i.e. the fluid velocity averaged over Ω_f).

For the separate fluid and solid components we write down equations of energy balance. For the fluid the approach taken in section 1.2.1 leads us to an equation for the fluid temperature of form

$$(\rho_0 c_p)_f \left(\frac{\partial T^f}{\partial t} + V_i \frac{\partial T^f}{\partial x_i} \right) = k_f \Delta T^f, \tag{1.81}$$

cf. equation (1.35), when k_f is constant. Assuming the solid skeleton is stationary the equation for the temperature of the solid is

$$(\rho_0 c)_s \frac{\partial T^s}{\partial t} = k_s \Delta T^s. \tag{1.82}$$

These equations would hold in general, but in a fluid saturated porous situation when the solid and fluid temperatures are different one should add interaction terms allowing for the temperature difference effects. This will lead to equations for T^f and T^s which contain terms of form $h_1(T_s - T_f)$ for some constant h_1. Instead of

(1.81) and (1.82) when in an LTNE scenario the accepted equations arise by multiplying the equation which follows from (1.81) by ε and by multiplying the equation which follows from (1.82) by $1 - \varepsilon$ to obtain

$$
\varepsilon(\rho_0 c_p)_f \frac{\partial T^f}{\partial t} + (\rho_0 c_p)_f v_i \frac{\partial T^f}{\partial x_i} = \varepsilon k_f \Delta T^f + h(T_s - T_f),
$$
$$
(1 - \varepsilon)(\rho_0 c)_s \frac{\partial T^s}{\partial t} = (1 - \varepsilon)k_s \Delta T^s - h(T_s - T_f). \tag{1.83}
$$

To investigate thermal convection problems in a saturated porous medium using these equations one also needs appropriate balance of mass and balance of momentum equations and these are discussed in chapter 2.

1.8.2 Average Temperature

Let $\tilde{\Omega}, \Omega_f$ and Ω_s be as in section 1.8.1. While LTNE is a very important concept for thermal convection in a saturated porous material, there are many situations, including microfluidic ones, where it is adequate to treat the solid and fluid temperatures to be the same. In this situation a convenient derivation of the equation for these fields is given by [197]. We follow his derivation.

For the separate solid and fluid components we have for the temperature field, T,

$$
(\rho_0 c)_s \frac{\partial T}{\partial t} = k_s \Delta T, \tag{1.84}
$$
$$
(\rho_0 c_p)_f \left(\frac{\partial T}{\partial t} + V_i \frac{\partial T}{\partial x_i} \right) = k_f \Delta T. \tag{1.85}
$$

We add $(1 - \varepsilon)(1.84)$ and $\varepsilon(1.85)$ to see that

$$
\left[\varepsilon(\rho_0 c_p)_f + (\rho_0 c)_s (1 - \varepsilon) \right] \frac{\partial T}{\partial t}
$$
$$
+ (\rho_0 c_p)_f \varepsilon V_i \frac{\partial T}{\partial x_i} = \left[k_s(1 - \varepsilon) + k_f \varepsilon \right] \Delta T. \tag{1.86}
$$

Denote by $M = (\rho_0 c_p)_f / (\rho_0 c)_m$ where $(\rho_0 c)_m = \varepsilon(\rho_0 c_p)_f + (\rho_0 c)_s (1 - \varepsilon)$, and by $\kappa = k_m / (\rho_0 c_p)_f$, where $k_m = k_s(1 - \varepsilon) + k_f \varepsilon$. Then, equation (1.86) may be rewritten

$$
\frac{1}{M} \frac{\partial T}{\partial t} + v_i \frac{\partial T}{\partial x_i} = \kappa \Delta T. \tag{1.87}
$$

Equation (1.87) is the equation one employs to govern the temperature field in a porous medium when only a single temperature is used. If we couple it with equations (1.44), (1.46), or those for a Forchheimer or Brinkman material, then we must specify how temperature enters equation (1.44). This is usually done via some equation of state for $\rho(T)$.

Note that if one assumes the solid and fluid temperatures are equal, i.e. one puts $T^f = T^s = T$ in equations (1.83), then addition of those equations leads to equation (1.87), as might be expected.

1.8.3 Darcy Equations for Thermal Convection

When one couples equations (1.47) with equation (1.87) for the temperature field T one needs to also assign the density ρ as a function of T as is done for a fluid in section 1.2.1, and we now assume equation (1.37) holds with the density given by (1.38). Thus, utilizing a Boussinesq approximation, when studying thermal convection in a Darcy porous material with a buoyancy force linear in the temperature field the relevant equations are

$$0 = -\frac{\partial p}{\partial x_i} - \frac{\mu}{k} v_i - g\rho_0 k_i \left[1 - \alpha(T - T_R)\right],$$

$$\frac{\partial v_i}{\partial x_i} = 0, \tag{1.88}$$

$$\frac{1}{M}\frac{\partial T}{\partial t} + v_i \frac{\partial T}{\partial x_i} = \kappa \Delta T.$$

Written out in full, assuming gravity in the negative $z-$direction, and with $\mathbf{v} = (u, v, w)$ and $\mathbf{x} = (x, y, z)$, equations (1.88) are

$$\frac{\partial p}{\partial x} = -\frac{\mu}{k} u,$$

$$\frac{\partial p}{\partial y} = -\frac{\mu}{k} v,$$

$$\frac{\partial p}{\partial z} = -\frac{\mu}{k} w - g\rho_0 \left[1 - \alpha(T - T_R)\right], \tag{1.89}$$

$$\frac{\partial u}{\partial x} + \frac{\partial v}{\partial y} + \frac{\partial w}{\partial z} = 0,$$

$$\frac{1}{M}\frac{\partial T}{\partial t} + u\frac{\partial T}{\partial x} + v\frac{\partial T}{\partial y} + w\frac{\partial T}{\partial z} = \kappa \Delta T.$$

Equations (1.89) represent a nonlinear system of five partial differential equations for the five variables u, v, w, p and T.

1.8.4 Forchheimer Equations, Thermal Convection

When one employs a quadratic Forchheimer law, see section 1.5.1, instead of Darcy theory the relevant equations which replace (1.88) are

$$0 = -\frac{\partial p}{\partial x_i} - \frac{\mu}{k} v_i - b|\mathbf{v}|v_i - g\rho_0 k_i \left[1 - \alpha(T - T_R) \right],$$

$$\frac{\partial v_i}{\partial x_i} = 0, \tag{1.90}$$

$$\frac{1}{M}\frac{\partial T}{\partial t} + v_i \frac{\partial T}{\partial x_i} = \kappa \Delta T.$$

Equations (1.90) are a coupled nonlinear system of five partial differential equations for five variables, namely v_i ($i = 1, 2, 3$), p and T, or alternatively, in the variables u, v, w, p and T.

1.8.5 Cubic Forchheimer Thermal Convection

When one employs a cubic Forchheimer law, as in section 1.5.2, instead of Darcy theory the relevant equations to replace (1.88) are

$$0 = -\frac{\partial p}{\partial x_i} - \frac{\mu}{k} v_i - c|\mathbf{v}|^2 v_i - g\rho_0 k_i \left[1 - \alpha(T - T_R) \right],$$

$$\frac{\partial v_i}{\partial x_i} = 0, \tag{1.91}$$

$$\frac{1}{M}\frac{\partial T}{\partial t} + v_i \frac{\partial T}{\partial x_i} = \kappa \Delta T.$$

The nonlinear system (1.91) involves five partial differential equations which are to be solved for the five variables, v_i, ($i = 1, 2, 3$), p and T, or alternatively, for u, v, w, p and T.

1.8.6 Brinkman Equations for Thermal Convection

When one employs a standard Brinkman law, see section 1.6.1, instead of Darcy theory the relevant equations which arise in place of (1.88) are

$$0 = -\frac{\partial p}{\partial x_i} - \frac{\mu}{k} v_i + \hat{\lambda} \Delta v_i - g\rho_0 k_i \left[1 - \alpha(T - T_R) \right],$$

$$\frac{\partial v_i}{\partial x_i} = 0, \tag{1.92}$$

$$\frac{1}{M}\frac{\partial T}{\partial t} + v_i \frac{\partial T}{\partial x_i} = \kappa \Delta T.$$

It is required to solve the nonlinear system of equations (1.92) for the variables u, v, w, p and T, as functions of $\mathbf{x}$ and t.

1.8.7 Salt Field

The inclusion of a salt field into a thermal convection problem is very important. We firstly consider a clear fluid in which is dissolved a salt with concentration C. If k_c is the diffusion coefficient and v_i denotes the fluid velocity then the equation governing the evolution of C is

$$\frac{\partial C}{\partial t} + v_i \frac{\partial C}{\partial x_i} = k_c \Delta C \tag{1.93}$$

provided k_c is constant.

Suppose we now wish to consider thermal convection in a fluid but when a salt field is simultaneously present. We then use equations (1.42) but with a modification to incorporate the salt field. In general, the density will depend on both temperature and salt concentration, so for a linear relationship in T and C we modify equation (1.38) to

$$\rho = \rho_0 \left[1 - \alpha(T - T_R) + \alpha_C(C - C_R) \right], \tag{1.94}$$

where T_R and C_R are constant reference values, $\alpha > 0$ is the coefficient of thermal expansion, and $\alpha_C > 0$ is constant. Note that the signs in front of α and α_C in (1.94) are different. This is because usually when a fluid is heated the temperature increases and the density decreases, whereas when a salt is added the density increases. Thus, the equations for thermal convection with a salt field (so called double diffusive convection, or alternatively thermohaline convection), in a fluid are

$$\begin{aligned}
v_{i,t} + v_j v_{i,j} &= -\frac{1}{\rho_0} \frac{\partial p}{\partial x_i} + \nu \Delta v_i \\
&\quad - g k_i \left[1 - \alpha(T - T_R) + \alpha_C(C - C_R) \right], \\
v_{i,i} &= 0, \\
T_{,t} + v_i T_{,i} &= \kappa \Delta T, \\
C_{,t} + v_i C_{,i} &= k_c \Delta C.
\end{aligned} \tag{1.95}$$

If we now have a fluid with a salt dissolved in it in a porous medium we may use a similar procedure to derive an equation for the salt concentration, C. Suppose the salt is not absorbed by the solid matrix. Then, in the fluid part, the salt concentration, C, satisfies the differential equation

$$\frac{\partial C}{\partial t} + V_i \frac{\partial C}{\partial x_i} = k_c \Delta C. \tag{1.96}$$

Since $V_i = v_i/\varepsilon$, is the actual velocity in the pore, see section 1.8.1, the equation governing C is

$$\varepsilon \frac{\partial C}{\partial t} + v_i \frac{\partial C}{\partial x_i} = \varepsilon k_c \Delta C. \tag{1.97}$$

If we were to non-dimensionalize with time, $\mathcal{T} = d^2/M\kappa$, velocity, $U = \kappa/d$, d being a length scale, then one may show the appropriate non-dimensional form of (1.97) is

$$M\varepsilon\frac{\partial C}{\partial t} + v_i\frac{\partial C}{\partial x_i} = \frac{1}{Le}\Delta C. \tag{1.98}$$

Here, $Le = \varepsilon k_c/\kappa$ is a Lewis number.

We see that some care must be exercised to derive the correct coefficients in a non-dimensional form of the equations governing non-isothermal flow of a salt laden fluid in a porous medium.

The equations for double diffusive convection in a porous medium depend on which flow law we employ in the porous medium. We must again use an equation of state something like (1.94) and so, for example, with Darcy's law, from equations (1.88) we may derive the following equations for double diffusive convection in a porous medium, assuming the acceleration (inertia) term in the momentum equation may be neglected,

$$
\begin{aligned}
0 &= -\frac{1}{\rho_0}\frac{\partial p}{\partial x_i} - \frac{\mu}{k}v_i - \rho_0 g k_i\big[1 - \alpha(T - T_R) + \alpha_C(C - C_R)\big], \\
\frac{\partial v_i}{\partial x_i} &= 0, \\
\frac{1}{M}\frac{\partial T}{\partial t} + v_i\frac{\partial T}{\partial x_i} &= \kappa\Delta T, \\
\varepsilon\frac{\partial C}{\partial t} + v_i\frac{\partial C}{\partial x_i} &= \varepsilon k_c\Delta C.
\end{aligned}
\tag{1.99}
$$

Equations (1.99) constitute a nonlinear system of six equations for the six variables v_i, p, T and C.

For a quadratic Forchheimer flow law the double diffusive equations are, again neglecting the effect of inertia,

$$
\begin{aligned}
0 &= -\frac{1}{\rho_0}\frac{\partial p}{\partial x_i} - \frac{\mu}{k}v_i - b|\mathbf{v}|v_i \\
&\quad - \rho_0 g k_i\big[1 - \alpha(T - T_R) + \alpha_C(C - C_R)\big], \\
\frac{\partial v_i}{\partial x_i} &= 0, \\
\frac{1}{M}\frac{\partial T}{\partial t} + v_i\frac{\partial T}{\partial x_i} &= \kappa\Delta T, \\
\varepsilon\frac{\partial C}{\partial t} + v_i\frac{\partial C}{\partial x_i} &= \varepsilon k_c\Delta C,
\end{aligned}
\tag{1.100}
$$

whilst with a Brinkman law the resulting equations are, once more supposing we are justified in neglecting the effect of fluid acceleration,

$$0 = -\frac{1}{\rho_0}\frac{\partial p}{\partial x_i} - \frac{\mu}{k}v_i + \hat{\lambda}\Delta v_i$$
$$- \rho_0 g k_i\left[1 - \alpha(T - T_R) + \alpha_C(C - C_R)\right],$$
$$\frac{\partial v_i}{\partial x_i} = 0, \tag{1.101}$$
$$\frac{1}{M}\frac{\partial T}{\partial t} + v_i\frac{\partial T}{\partial x_i} = \kappa\Delta T,$$
$$\varepsilon\frac{\partial C}{\partial t} + v_i\frac{\partial C}{\partial x_i} = \varepsilon k_c\Delta C.$$

It is worth again noting that equations (1.101) represent a nonlinear system of six equations for the six variables v_i, p, T and C.

1.8.8 Soret Effect

In section 1.8.7 we have derived equations governing the evolutionary behaviour of the velocity v_i, pressure p, temperature T, and salt concentration C, for double diffusion in a fluid or in a fluid saturated porous medium. The law for the flux of diffusion $\mathbf{J}_C$ was taken to be Fick's law, namely,

$$\mathbf{J}_C = -k_c\nabla C. \tag{1.102}$$

However, in some circumstances the temperature gradient may have a strong influence upon the diffusion flux $\mathbf{J}_C$. This effect is known as the Soret effect, after [405]. An excellent review of this effect is given by [344]. When one incorporates a Soret effect, [190] and [344] give the diffusion flux in the form

$$\mathbf{J}_C = -k_c\nabla C - k_c S_T\, C(1 - C)\nabla T, \tag{1.103}$$

where S_T is the Soret coefficient. Many articles neglect the C effect in the coefficient in the second term and simply take

$$\mathbf{J}_C = -k_c\nabla C - k_c S_T\,\nabla T. \tag{1.104}$$

Equation (1.93) may be written as

$$\frac{\partial C}{\partial t} + v_i\frac{\partial C}{\partial x_i} = -\frac{\partial J_i^C}{\partial x_i},$$

where $\mathbf{J}^C \equiv \mathbf{J}_C$, and $\mathbf{J}_C$ is given by (1.102). One may employ instead (1.103) or (1.104) to derive the analogous equations for double diffusive convection with a

Soret effect in either a fluid, or in a fluid saturated porous material. Thus, with the Soret term (1.103) one finds for a fluid

$$
v_{i,t} + v_j v_{i,j} = -\frac{1}{\rho_0}\frac{\partial p}{\partial x_i} + \nu \Delta v_i
$$
$$
- g k_i \left[1 - \alpha(T - T_R) + \alpha_C(C - C_R)\right],
$$
$$
v_{i,i} = 0, \tag{1.105}
$$
$$
T_{,t} + v_i T_{,i} = \kappa \Delta T,
$$
$$
C_{,t} + v_i C_{,i} = k_c \Delta C + k_c S_T \left[C(1 - C)T_{,i}\right]_{,i},
$$

whereas the analogous equations for flow in a Darcy porous material are,

$$
0 = -\frac{1}{\rho_0}\frac{\partial p}{\partial x_i} - \frac{\mu}{k} v_i - \rho_0 g k_i \left[1 - \alpha(T - T_R) + \alpha_C(C - C_R)\right],
$$
$$
\frac{\partial v_i}{\partial x_i} = 0,
$$
$$
\frac{1}{M}\frac{\partial T}{\partial t} + v_i \frac{\partial T}{\partial x_i} = \kappa \Delta T, \tag{1.106}
$$
$$
\varepsilon \frac{\partial C}{\partial t} + v_i \frac{\partial C}{\partial x_i} = \varepsilon k_c \Delta C + \varepsilon k_c S_T \left[C(1 - C)T_{,i}\right]_{,i}.
$$

When one employs the simpler, but more often used, constitutive equation (1.104) then the relevant equations for a fluid are

$$
v_{i,t} + v_j v_{i,j} = -\frac{1}{\rho_0}\frac{\partial p}{\partial x_i} + \nu \Delta v_i
$$
$$
- g k_i \left[1 - \alpha(T - T_R) + \alpha_C(C - C_R)\right],
$$
$$
v_{i,i} = 0, \tag{1.107}
$$
$$
T_{,t} + v_i T_{,i} = \kappa \Delta T,
$$
$$
C_{,t} + v_i C_{,i} = k_c \Delta C + k_c S_T \Delta T,
$$

while the analogous equations for double-diffusive Soret flow in a Darcy porous material are,

$$
0 = -\frac{1}{\rho_0}\frac{\partial p}{\partial x_i} - \frac{\mu}{k} v_i - \rho_0 g k_i \left[1 - \alpha(T - T_R) + \alpha_C(C - C_R)\right],
$$
$$
\frac{\partial v_i}{\partial x_i} = 0,
$$
$$
\frac{1}{M}\frac{\partial T}{\partial t} + v_i \frac{\partial T}{\partial x_i} = \kappa \Delta T, \tag{1.108}
$$
$$
\varepsilon \frac{\partial C}{\partial t} + v_i \frac{\partial C}{\partial x_i} = \varepsilon k_c \Delta C + \varepsilon k_c S_T \Delta T.
$$

For microfluidic flows the Soret effect is likely to be important. Problems with cross diffusion effects like that of Soret have been the subject of much recent research, cf. [38, 52, 62, 256, 375, 391, 392], and the references therein.

1.8.9 Many Salt Fields

Thermal convection in the presence of a salt field was presented in section 1.8.7 and we now consider a generalization of this which has many practical applications. Namely, thermal convection when there are two or more different salt fields. This is an area of much interest in the microfluidic field, for example, involving DNA dissociation, see e.g. [11, 104].

The derivation of the concentration equations proceeds as in section 1.8.7. Let us suppose there are N different salt fields, $N \in \mathbb{N}, N \geq 2$, and denote their respective concentrations by $C_1, \ldots, C_N$. Then, each C_α, $\alpha = 1, \ldots, N$, satisfies an evolution equation like (1.93), and the equation of state (1.94) must be modified to include all components C_α, e.g.

$$\rho = \rho_0 \left[1 - \alpha(T - T_R) + \sum_{n=1}^{N} \alpha_n (C_n - C_R) \right]. \tag{1.109}$$

In this way we may arrive at the equations governing the evolutionary behaviour of thermal convection with N salt fields in a fluid (multi-component diffusion), so that the equations are

$$v_{i,t} + v_j v_{i,j} = -\frac{1}{\rho_0} \frac{\partial p}{\partial x_i} + \nu \Delta v_i$$

$$- g k_i \left[1 - \alpha(T - T_R) + \sum_{n=1}^{N} \alpha_n (C_n - C_R) \right], \tag{1.110}$$

$$v_{i,i} = 0,$$
$$T_{,t} + v_i T_{,i} = \kappa \Delta T,$$
$$C_{,t}^\alpha + v_i C_{,i}^\alpha = k_\alpha \Delta C^\alpha, \qquad \alpha = 1, \ldots, N,$$

where k_α represents the diffusion coefficient for the C^α species of salt field dissolved in the fluid.

To be specific consider the case of $N = 2$ so there are two separate salts with concentrations C_1 and C_2. In this case (1.110) become the equations for so-called triply - diffusive convection (i.e. convective motion when three fields T, C_1 and C_2

are present). The relevant equations are

$$v_{i,t} + v_j v_{i,j} = -\frac{1}{\rho_0}\frac{\partial p}{\partial x_i} + \nu \Delta v_i$$
$$\quad - gk_i\left[1 - \alpha(T - T_R) + \alpha_1(C_1 - C_R) + \alpha_2(C_2 - C_R)\right],$$
$$v_{i,i} = 0,$$
$$T_{,t} + v_i T_{,i} = \kappa \Delta T,$$
$$C_{,t}^1 + v_i C_{,i}^1 = k_1 \Delta C^1,$$
$$C_{,t}^2 + v_i C_{,i}^2 = k_2 \Delta C^2.$$

$$(1.111)$$

A similar procedure to that used in section 1.8.7 may be used to derive the analogous system of equations to those of (1.110) but for multi-component diffusion in a porous layer. For example, in a Darcy porous medium the equations are,

$$0 = -\frac{1}{\rho_0}\frac{\partial p}{\partial x_i} - \frac{\mu}{k} v_i$$
$$\quad - \rho_0 gk_i\left[1 - \alpha(T - T_R) + \sum_{n=1}^{N} \alpha_n(C_n - C_R)\right],$$
$$\frac{\partial v_l}{\partial x_i} = 0,$$
$$\frac{1}{M}\frac{\partial T}{\partial t} + v_i\frac{\partial T}{\partial x_i} = \kappa \Delta T,$$
$$\varepsilon\frac{\partial C^\alpha}{\partial t} + v_i\frac{\partial C^\alpha}{\partial x_i} = \varepsilon k_\alpha \Delta C^\alpha, \quad \alpha = 1,\ldots,N.$$

$$(1.112)$$

Triple diffusive convection in a fluid layer was analysed by [337] and including penetrative convection effects by [438]. [436] examined triple diffusive convection in a finite box. There are many recent articles dealing with triple diffusive convection, especially from a nonlinear stability point of view, see e.g. [68, 69, 84, 371–374, 376, 401], and the references therein.

1.9 Boundary Conditions

For the Darcy model of section 1.3 or the Forchheimer model of section 1.5 we do not have derivatives present in the v_i term. Thus, we need to prescribe the normal component of v_i on the boundary, Γ, of a volume Ω, i.e. we prescribe $v_i n_i$, n_i being the unit outward normal to Γ. However, when dealing with the Brinkman model of section 1.6 the presence of the Δv_i term means we need to prescribe v_i on the whole of Γ. Thus, if we consider the Darcy model for thermal convection of section 1.8.3

we employ equations (1.88) in a domain $\Omega \times \{t > 0\}$ with on the boundary Γ of Ω,

$$v_i n_i = v_B(\mathbf{x}, t), \qquad \text{on } \Gamma \times \{t > 0\}, \tag{1.113}$$

where v_B is a prescribed function. If, in particular, we consider thermal convection in a horizontal layer $\mathbb{R}^2 \times \{z \in (0, d)\}$, then the domain Ω is a periodicity cell V, typically of hexagonal shape, cf. [414, p. 51]. In that case since $n_i = k_i = (0, 0, 1)$ on $z = d$ and $n_i = -k_i$ on $z = 0$, the boundary conditions on v_i on the boundary of V are periodicity on the lateral walls of V and

$$v_i n_i \equiv w = 0, \qquad \text{on } z = 0, d, \tag{1.114}$$

where $w = v_3$.

The same boundary conditions apply for Forchheimer models, (1.90) and (1.91), if no Brinkman term is present.

When we consider the Brinkman equations for thermal convection, equations (1.92), then instead of (1.113) we must prescribe all of v_i, i.e.

$$v_i(\mathbf{x}, t) = v_i^B(\mathbf{x}, t), \qquad \text{on } \Gamma \times \{t > 0\}, \tag{1.115}$$

where v_i^B is a prescribed function. If Ω is a periodic cell V then instead of (1.114) we have periodicity in the x, y directions and

$$v_i = 0, \qquad \text{on } z = 0, d. \tag{1.116}$$

In addition to boundary data, if we are considering the boundary - initial value problem for equations (1.88), (1.90), (1.91) or (1.92) then one needs to prescribe initial data on the temperature field T. If one considers the equivalent models with inertia, cf. section 1.7, then initial data must be prescribed on both v_i and T.

For the temperature field, T, we may prescribe T on Γ if the temperature is measurable there. If, on the other hand, we can measure the heat flux on the boundary Γ, then since the heat flux $\mathbf{q}$ is usually given by $q_i = -kT_{,i}$, $k > 0$, we can prescribe $q_i n_i$ on Γ. In other words, we may prescribe $\partial T / \partial n$ on Γ. Thus, on Γ one may prescribe

$$T(\mathbf{x}, t) = T_B(\mathbf{x}, t) \qquad \text{on } \Gamma \times \{t > 0\}, \tag{1.117}$$

or

$$\frac{\partial T}{\partial n}(\mathbf{x}, t) = \hat{T}_B(\mathbf{x}, t) \qquad \text{on } \Gamma \times \{t > 0\}, \tag{1.118}$$

for given functions T_B or $\hat{T}_B$, or one could have a combination of both, in the sense that (1.117) holds on a part of Γ, say Γ_1, while (1.118) holds on the rest of Γ, say Γ_2, where $\Gamma_1 \cup \Gamma_2 = \Gamma$. In addition a combination boundary condition may also be appropriate, i.e. one of form

$$\frac{\partial T}{\partial n} + \gamma T = a(\mathbf{x}, t), \qquad \text{on } \Gamma \times \{t > 0\},$$

where a is a prescribed function and γ is known. If radiation heating is the dominant effect, e.g. a surface directly in sunlight, then γ is likely to be small. If, however, there is little radiant heating, it may be more appropriate to assign T on Γ. The combination boundary condition is between both extremes and may hold for several real problems.

For a thermal convection problem in a horizontal layer $\mathbb{R}^2 \times \{z \in (0,d)\}$ we would have periodicity in T in the x,y directions and T or $q_i n_i$ given on $z = 0, d$, e.g.

$$T = T_L \quad \text{on } z = 0, \qquad T = T_U \quad \text{on } z = d, \tag{1.119}$$

for T_L, T_U constants.

For a salt field, if there is zero flux of the salt out of the boundary, we may assume $\partial C / \partial n = 0$ there. However, there are some instances where it is possible to control the concentration field on Γ, cf. [217] and in this instance we would have a boundary condition of form

$$C = C_G, \qquad \mathbf{x} \in \Gamma,$$

where C_G is a known function. For double diffusive convection in a horizontal layer $\mathbb{R}^2 \times \{z \in (0,d)\}$ we would often have periodicity in x,y and

$$C = C_L \quad \text{on } z = 0, \qquad C = C_U \quad \text{on } z = d, \tag{1.120}$$

where C_L and C_U are constants.

1.9.1 Slip Boundary Conditions

There is very strong evidence that flows in a microfluidic situation are inadequately described by conditions of no slip at the boundary. In fact, boundary conditions which incorporate slip have been suggested for a long time starting with the early work of [304] and [278]. They are the subject of intense recent work, especially in microfluidic and nanofluidic situations, as witnessed by [14, 24, 64, 118, 119, 233, 295, 350, 382, 383, 489, 494, 496], and many references including historical ones are given in these articles. A lucid historical account of the origin of slip boundary conditions in also given in chapter 1 of Webber [474]. Slip boundary conditions also lead to some extremely interesting mathematical problems regarding such issues as convergence, see e.g. [42, 43, 408, 472], and the references therein.

The standard slip boundary condition was proposed by [304], and this relates the velocity of the fluid at the boundary to the shear rate. While one may derive slip boundary conditions for a general shaped domain it is easier to consider a domain which is an infinite horizontal layer between two planes, $z = d_1$ and $z = d_2$, with $d_1 < d_2$. Suppose these planes remain horizontal but they may move and between them is a viscous, incompressible fluid. Then, if v_i denotes the velocity field in the domain, the stress tensor t_{ij} is given by

$$t_{ij} = -p\delta_{ij} + 2\mu d_{ij},\qquad(1.121)$$

where p is the fluid pressure, μ is the dynamic viscosity, and $d_{ij} = (v_{i,j} + v_{j,i})/2$. If V_i denotes the velocity at a point on the planes $z = d_1$ or $z = d_2$, the [304] slip conditions may be written as

$$
\begin{aligned}
v_i n_i &= V_i n_i, && \text{on } z = d_1 \text{ or } z = d_2,\\
v_i t_i^\alpha &= (V_i - \hat{\lambda} t_{ij} n_j) t_i^\alpha, && \text{on } z = d_1 \text{ or } z = d_2,
\end{aligned}
\qquad(1.122)
$$

where $\hat{\lambda}$ is a positive constant, t_i^α, $\alpha = 1$ or 2, are tangential vectors to $z = d_1, d_2$ and n_i is the unit outward normal to $z = d_1, d_2$.

Equation $(1.122)_1$ represents the fact that there is no flux of fluid through the planes $z = d_1$ or $z = d_2$. Equations $(1.122)_2$ shows that the fluid does not "stick" to the planes in the horizontal directions but may slip. If we write $\mathbf{V} = (U, V, W)$ and $\mathbf{v} = (u, v, w)$ then we have $\mathbf{n} = (0, 0, -1)$ on $z = d_1$ and $\mathbf{n} = (0, 0, 1)$ on $z = d_2$, and we may take $\mathbf{t}^1 = (1, 0, 0)$, $\mathbf{t}^2 = (0, 1, 0)$ on $z = d_1, d_2$. In this case equations $(1.122)_2$ may be written as

$$
\begin{aligned}
v_i t_i^\alpha &= V_i t_i^\alpha - \lambda (v_{i,3} + w_{,i}) t_i^\alpha && \text{on } z = d_2,\\
v_i t_i^\alpha &= V_i t_i^\alpha + \lambda (v_{i,3} + w_{,i}) t_i^\alpha && \text{on } z = d_1,
\end{aligned}
\qquad(1.123)
$$

where $\lambda = \mu \hat{\lambda}$ and λ has the dimension of length. In fact, λ is sometimes referred to as a "slip length".

For many problems we keep the planes fixed in the vertical directions so that $w = 0$ and $W = 0$ there, so then the slip boundary conditions (1.123) may be rewritten as

$$
\begin{aligned}
u &= U - \lambda u_{,z}, & v &= V - \lambda v_{,z}, && \text{on } z = d_2,\\
u &= U + \lambda u_{,z}, & v &= V + \lambda v_{,z}, && \text{on } z = d_1.
\end{aligned}
\qquad(1.124)
$$

If we further restrict attention to fixed planes, as is usually the case with thermal convection, then (1.124) become

$$
\begin{aligned}
u &= -\lambda u_{,z}, & v &= -\lambda v_{,z}, && \text{on } z = d_2,\\
u &= \lambda u_{,z}, & v &= \lambda v_{,z}, && \text{on } z = d_1.
\end{aligned}
\qquad(1.125)
$$

One could have different slip conditions on $z = d_1$ and on $z = d_2$, which would result in different values of λ in the conditions of (1.125) on $z = d_1$ and on $z = d_2$.

For an incompressible fluid we have the mass balance equation

$$u_{,x} + v_{,y} + w_{,z} = 0$$

holding in the whole domain. We may differentiate the u parts of (1.125) with respect to x and differentiate the v parts of (1.125) with respect to y and then employ the mass balance equation to see that

$$
\begin{aligned}
w_{,z} &= -\lambda w_{,zz}, && \text{on } z = d_2, \\
w_{,z} &= \lambda w_{,zz}, && \text{on } z = d_1.
\end{aligned}
\tag{1.126}
$$

The equations (1.126) are very useful in thermal convection studies.

Much more detail about fluid mechanics problems involving slip boundary conditions is available in the thesis of [474]. We also study thermal convection and related problems which are appropriate for microfluidic situations in chapter 9.

Chapter 2
Thermal Convection with LTNE

2.1 Stability and Symmetry

The object of this chapter is to present the equations of thermal convection with local thermal non-equilibrium (LTNE) effects and analyse stability properties of their solutions. The linear operator which arises in the LTNE equations often belongs to a special class of operators, known as symmetric operators, and this class of linear operator has special mathematical properties in the context of stability. Before we briefly discuss an important general nonlinear stability result we include an example of a symmetric linear operator which occurs in the classical theory of thermal convection in a fluid, a phenomenon usually referred to as Bénard convection.

2.1.1 Classical Bénard Convection

Let us consider a linear viscous fluid governed by equations (1.42). The fluid occupies the domain contained between the horizontal planes $z = 0$ and $z = d$ with $(x, y) \in \mathbb{R}^2$. The upper plane is held at a fixed constant temperature T_U, while the lower plane is held at a fixed constant temperature T_L, with $T_L > T_U$.

In the above situation equations (1.42) possess the motionless conduction solution

$$\bar{v}_i \equiv 0, \qquad \bar{T} = \beta z + T_L, \tag{2.1}$$

where $\beta = (T_L - T_U)/d > 0$ and $\bar{p}(z)$ is determined up to a constant from equation (1.42)$_1$. To study the stability of solution (2.1) we introduce perturbations u_i, θ, π by $v_i = \bar{v}_i + u_i$, $T = \bar{T} + \theta$, $p = \bar{p} + \pi$ and we non-dimensionalize the resulting perturbation equations with the scalings of pressure $P = \rho_0 \nu U / d$, time $\mathscr{T} = d^2/\nu$, length d, and velocity $U = \nu/d$. Introduce the Prandtl number $Pr = \nu/\kappa$ and the parameter $R = \sqrt{\alpha g \beta d^4 / \kappa \nu}$ where $Ra = R^2$ is the Rayleigh number. The

© Springer International Publishing Switzerland 2015
B. Straughan, *Convection with Local Thermal Non-Equilibrium and Microfluidic Effects*,
Advances in Mechanics and Mathematics 32, DOI 10.1007/978-3-319-13530-4_2

non-dimensional perturbation equations may then be shown to be

$$\frac{\partial u_i}{\partial t} + u_j \frac{\partial u_i}{\partial x_j} = -\frac{\partial \pi}{\partial x_i} + \Delta u_i + R\theta k_i,$$

$$\frac{\partial u_i}{\partial x_i} = 0, \tag{2.2}$$

$$Pr\left(\frac{\partial \theta}{\partial t} + u_i \frac{\partial \theta}{\partial x_i}\right) = Rw + \Delta\theta,$$

for $(x, y) \in \mathbb{R}^2$ and $z \in (0, d)$, and where $w = u_3$. The boundary conditions to be satisfied are those of no slip on the velocity field together with constant temperatures on the upper and lower horizontal bounding planes of the fluid. Hence, we require the perturbation variables u_i and θ to be such that

$$u_i = 0, \qquad \theta = 0, \qquad \text{on } z = 0, 1, \tag{2.3}$$

together with u_i, π, θ satisfying a plane tiling periodicity in the (x, y) plane. This is consistent with the formation of hexagonal thermal convection cells in a horizontal strip of fluid.

Our intention here is not to give a full stability analysis of the steady solution (2.1) via equations (2.2). Instead, we wish to emphasize the special structure of system (2.2). Equations (2.2) may be written in the form

$$Au_t = Lu + N(u)$$

where $u = (u_1, u_2, u_3, \theta)$, A is the operator

$$A = \begin{pmatrix} 1 & 0 & 0 & 0 \\ 0 & 1 & 0 & 0 \\ 0 & 0 & 1 & 0 \\ 0 & 0 & 0 & Pr \end{pmatrix}$$

while L is the linear operator

$$L = \begin{pmatrix} \Delta & 0 & 0 & 0 \\ 0 & \Delta & 0 & 0 \\ 0 & 0 & \Delta & R \\ 0 & 0 & R & \Delta \end{pmatrix}$$

Let $(\cdot, \cdot)$ denote the inner product on the Hilbert space $(L^2(V))^4$ where V is a periodic cell for the solution (u_i, θ, π). Suppose $u = (u_1, u_2, u_3, \theta)$ and $v = (v_1, v_2, v_3, \phi)$ are functions which satisfy (2.2) and (2.3). Then it is straightforward to demonstrate that

$$(u, Lv) = (v, Lu). \tag{2.4}$$

Since (2.4) holds the operator L is symmetric. Furthermore, we may here show $(u, N(u)) = 0$, where $N(u)$ refers to the convective nonlinearities in (2.2). To be precise

$$
\begin{aligned}
(u, N(u)) &= \int_V u_i u_j u_{i,j} dV + Pr \int_V u_i \theta \theta_{,i} dV \\
&= \frac{1}{2} \int_V u_j (u_i u_i)_{,j} dV + \frac{Pr}{2} \int_V u_i \theta_{,i}^2 dV \\
&= -\frac{1}{2} \int_V u_{j,j} u_i u_i dV + \frac{1}{2} \oint_{\partial V} n_j u_j u_i u_i dS \\
&\quad - \frac{Pr}{2} \int_V u_{i,i} \theta^2 dV + \frac{Pr}{2} \oint_{\partial V} n_i u_i \theta^2 dS \\
&= 0
\end{aligned}
$$

where we have integrated by parts and used the boundary conditions (2.3). In this case since L is symmetric one has the very important result that the linear instability boundary governed by equations (2.2) is exactly the same as the global nonlinear stability boundary governed by the same equations. This result follows from theorem 2 of [150].

Our purpose here is to present a simple example showing that the case of a symmetric linear operator is a very special one. However, the class of symmetric linear operators is one which occurs frequently in thermal convection in a porous medium taking into account LTNE effects.

We now briefly consider the question of linear instability versus nonlinear stability in a general setting before embarking on describing various thermal convection problems in porous media in the presence of LTNE effects.

2.1.2 Symmetric Operators

In general, the equations governing problems in hydrodynamic stability (including those in porous media) are typically of the form

$$
Au_t = L_S u + L_A u + N(u), \tag{2.5}
$$

where u is a Hilbert space valued function, u_t is its time derivative, A is a bounded linear operator (typically a matrix with constant entries), $L = L_S + L_A$ is an unbounded, sectorial linear operator, and $N(u)$ represents the nonlinear terms. The operator L_S is the symmetric part of L while L_A denotes the anti-symmetric part. (Very roughly, a sectorial operator is one where the eigenvalues all lie in a sector in the complex plane. Detailed accounts of sectorial operators may be found in the books by [170] and by [484], chapter 2, and a very readable account may be found in [252]. In this book the linear operators all consist of terms like the identity or the Laplacian operator and under the boundary conditions we employ they are sectorial.)

The classical theory of linear instability writes

$$u = e^{\sigma t} \phi$$

and discards the $N(u)$ term in (2.5). One is then faced with solving the eigenvalue problem

$$\sigma A \phi = L_S \phi + L_A \phi, \tag{2.6}$$

where σ is the eigenvalue and ϕ the eigenfunction.

It is important to note that equation (2.5) involves both the skew-symmetric operator L_A and the symmetric operator L_S. In general, σ is complex, and one looks for the eigenvalue with largest real part to become positive for instability.

A classical nonlinear energy stability analysis, on the other hand, commences by forming the inner product of u with (2.5). If $(\cdot, \cdot)$ denotes the inner product on the Hilbert space in question then one finds

$$\frac{d}{dt} \frac{1}{2} (u, Au) = (u, L_S u) + (u, N(u)) \tag{2.7}$$

since $(u, L_A u) = 0$. Nonlinear energy stability follows from (2.7) and it is very important to note that in this way the nonlinear stability boundary does not involve the skew part of L, L_A. Thus, one may expect, in general, that the linear instability and nonlinear stability boundaries are very different. Details of how nonlinear stability follows from (2.7) may be found in section 4.3 of [414], or from the paper of [150].

In fact, the reason why the nonlinear energy stability analyses discussed in the present chapter give optimal results is due to the fact that the associated operator L is symmetric.

There are two fundamental problems arising from (2.7) when one is faced with deriving *unconditional* nonlinear stability results. These are

(a) the effect of L_A on the nonlinear stability boundary;

(b) what does one do when $(u, N(u)) \ngtr 0$?

When the operator L is far from symmetric traditional energy stability arguments can break down completely, or yield very poor results for certain classes of problem. For example, in parallel shear flows progress is very difficult, as explained in chapter 8 of [410]. In this regard though, an interested reader may wish to consider the articles of [113, 201] and [202]. Certain classes of viscoelastic flows prove severely problematic to tackle via energy methods, as is shown in the interesting paper of [114].

Due to the failure of the classical energy method to yield sharp, or at least useful, nonlinear stability thresholds in problems such as shear flows, much research effort has recently been directed toward this area and a variety of novel approaches involving clever choices of Lyapunov functional have been suggested, cf. [68–70, 74, 131, 132, 182, 250, 271, 298, 299, 336, 342, 372, 374].

2.2 Darcy Theory

We begin with a description of the problem of a horizontal layer of porous material saturated with an incompressible fluid and heated below, allowing the fluid and solid temperatures to be different, LTNE. Our first step is to employ Darcy theory, see sections 1.3 and 1.8.3 but additionally utilize the LTNE equations for the temperature fields T^f and T^s as given in equations (1.83). Thus, we begin with the problem considered by [27]. The paper of [27] deals with the onset of thermal convection when the porous medium is modelled using Darcy's law. It is worth noting that [363] is a useful article which lucidly shows how one may compare convection in theories of Darcy and of Brinkman with an average temperature. His asymptotic estimates are very useful, and he also shows how the Forchheimer theory enters the picture.

In figure 2.1 we show a possible scenario for LTNE in a porous medium composed of spherical beads. Note that spherical beads are often used in a laboratory to perform experiments of thermal convection in a fluid saturated porous material, cf. [86]. For a real LTNE porous material we would expect to have many more beads than in figure 2.1 and the beads would be very small in keeping with a microfluidic setting. The continuum approximation assumes T_f and T_s are both defined at all points $\mathbf{x}$ for time t and we likewise have a seepage velocity $v_i = \varepsilon V_i$ at all points $\mathbf{x}$. The seepage velocity is defined in section 1.8.1. For the configuration of figure 2.1 if we have a similar picture in the direction orthogonal to that of the figure then we have a three-dimensional box containing 320 beads, i.e. 8×5 in each orthogonal projection direction. If the beads have diameters D then the porosity may be calculated as $\varepsilon = 320D^3(1 - 4\pi/24)/320D^3 = 1 - \pi/6 \approx 0.4767$. The beads in figure 2.1 are not positioned in a close packing format. For a close packing format the porosity would be smaller, the lowest value of the porosity being $\varepsilon = 0.2595$.

Consider now a layer of porous material saturated with fluid and contained between the planes $z = 0$ and $z = d$. The temperatures of the solid, T_s, and fluid, T_f, are maintained at constants on the planes $z = 0$ and $z = d$ with

$$T_s = T_f = T_L, \quad z = 0; \qquad T_s = T_f = T_U, \quad z = d; \tag{2.8}$$

where $T_L > T_U$. (If the layer is heated above, i.e. $T_U \geq T_L$, then one may demonstrate global nonlinear stability always holds.) The equations are formed by essentially a combination of equations (1.88) and (1.83), cf. Banu and Rees [27], and are

$$v_i = -\frac{K}{\mu}p_{,i} + \frac{\rho_f g \alpha K}{\mu}T_f k_i, \tag{2.9}$$

$$v_{i,i} = 0, \tag{2.10}$$

$$\varepsilon(\rho c)_f T_{,t}^f + (\rho c)_f v_i T_{,i}^f = \varepsilon k_f \Delta T_f + h(T_s - T_f), \tag{2.11}$$

$$(1 - \varepsilon)(\rho c)_s T_{,t}^s = (1 - \varepsilon)k_s \Delta T_s - h(T_s - T_f). \tag{2.12}$$

Note that in deriving equation (2.9) one makes the assumption that the density in the body force term in equation (1.88)$_1$ is linear in the fluid temperature T^f, i.e.

$$\rho = \rho_f \left[1 - \alpha(T^f - T_0)\right] \tag{2.13}$$

for a reference temperature T_0. Equations (2.9)–(2.12) hold in the domain $\mathbb{R}^2 \times \{z \in (0,d)\} \times \{t > 0\}$, $\mathbf{k} = (0,0,1)$, and Δ is the three-dimensional Laplacian. The variables v_i, p, T_f and T_s are the velocity, pressure and fluid and solid temperatures, respectively. The constants $K, \mu, g, \alpha, \varepsilon, \rho_\alpha, c_\alpha, k_\alpha$ ($\alpha = f,s$), appearing in equations (2.9)–(2.13), are permeability, dynamic viscosity, gravity, thermal expansion coefficient, porosity, density, specific heat, thermal diffusion coefficient (where $\alpha = f,s$, denotes fluid or solid), $(\rho c)_\alpha = \rho_\alpha c_\alpha$, $\alpha = f,s$, and h is an interaction coefficient.

The steady solution whose stability is under investigation is

$$\begin{aligned}
&\bar{\mathbf{v}} \equiv 0, \\
&\bar{T}_f = \bar{T}_s = -\beta z + T_L, \qquad z \in (0,d),
\end{aligned} \tag{2.14}$$

where β is the temperature gradient given by

$$\beta = \frac{T_L - T_U}{d} \tag{2.15}$$

and the steady pressure $\bar{p}(z)$ is a quadratic function determined from (2.9) (up to an arbitrary constant which defines the pressure scale).

2.2.1 Linear Instability

To investigate stability we introduce perturbations u_i, π, θ, ϕ to $\bar{v}_i, \bar{p}, \bar{T}_f$ and $\bar{T}_s$ by

$$v_i = u_i + \bar{v}_i, \quad p = \pi + \bar{p}, \quad T_f = \theta + \bar{T}_f, \quad T_s = \phi + \bar{T}_s. \tag{2.16}$$

The perturbation equations are derived from (2.9) to (2.12) and are non-dimensionalized with velocity, pressure, temperature, time and length scales of $U = \varepsilon k_f / (\rho c)_f d$, $P = \mu dU/K$, $T^\sharp = Ud\sqrt{\mu \beta c_f / \varepsilon k_f g \alpha K}$, $\mathscr{T} = (\rho c)_f d^2 / k_f$, $L = d$. The Rayleigh number Ra is defined by

$$Ra = R^2 = d^2 \rho_f^2 \sqrt{\frac{\beta c_f g \alpha K}{\varepsilon k_f \mu}}.$$

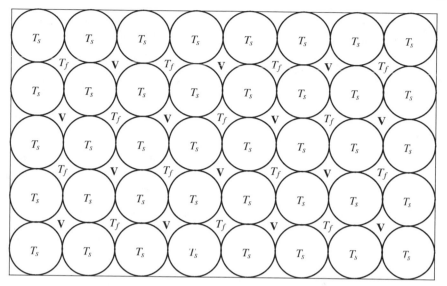

Fig. 2.1 An example of a local thermal non-equilibrium porous medium composed of spherical beads. The solid is fixed so the velocity is zero in the skeleton. In the fluid the velocity is the pore average velocity **V**, cf. section 1.8.1. This is a horizontal projection from the side of a three-dimensional body comprised of 320 beads.

In addition $H = hd^2/\varepsilon k_f$ and $\gamma = \varepsilon k_f/(1-\varepsilon)k_s$ are the non-dimensional coefficients introduced by [27]. It then follows that the non-dimensional perturbation equations have form

$$u_i = -\pi_{,i} + R\theta k_i, \tag{2.17}$$

$$u_{i,i} = 0, \tag{2.18}$$

$$\theta_{,t} + u_i\theta_{,i} = Rw + \Delta\theta + H(\phi - \theta), \tag{2.19}$$

$$A\phi_{,t} = \Delta\phi - H\gamma(\phi - \theta), \tag{2.20}$$

where these equations hold on $\mathbb{R}^2 \times \{z \in (0,1)\} \times \{t > 0\}$, $w = u_3$, and $A = \rho_s c_s k_f/k_s\rho_f c_f$ is a non-dimensional thermal inertia coefficient. Observe that equations (2.17)–(2.20) are still nonlinear due to the presence of the $u_i\theta_{,i}$ term in (2.19). Also, the form of the equations is different to that of [27] because we employ R rather than Ra (where $Ra = R^2$), although equations (2.17)–(2.20) are easily transformed to the equations of [27] and they are equivalent to them.

The boundary conditions to be satisfied are

$$u_i n_i = 0, \quad \theta = 0, \quad \phi = 0, \quad \text{on } z = 0,1, \tag{2.21}$$

where n_i denotes the unit outward normal, together with u_i, π, θ, ϕ satisfying a plane tiling periodicity in x, y. Such forms are discussed in e.g. [85], p. 43, [414], p. 51, where especially the hexagonal planform of [91] is described.

One linearizes (2.19) and then we put $u_i = e^{\sigma t} u_i(\mathbf{x})$ with a similar representation for π, θ and ϕ to derive from (2.17) to (2.20) the linearized instability equations

$$
\begin{aligned}
u_i &= -\pi_{,i} + R\theta k_i, \\
u_{i,i} &= 0, \\
\sigma\theta &= Rw + \Delta\theta + H(\phi - \theta), \\
\sigma A\phi &= \Delta\phi - H\gamma(\phi - \theta).
\end{aligned}
\tag{2.22}
$$

If we put $u = (u_1, u_2, u_3, \theta, \phi)^T$ and consider the linear operator, L, as defined in the abstract equation (2.5), from equations (2.22) but with (2.22)$_4$ multiplied by γ^{-1}, we see that

$$
L = \begin{pmatrix}
-1 & 0 & 0 & 0 & 0 \\
0 & -1 & 0 & 0 & 0 \\
0 & 0 & -1 & R & 0 \\
0 & 0 & R & \Delta - H & H \\
0 & 0 & 0 & H & \frac{1}{\gamma}\Delta - H
\end{pmatrix}
\tag{2.23}
$$

In this form the symmetry of L, taking account of boundary conditions (2.21) is evident.

Exchange of stabilities follows immediately from the symmetry of L, although it is easily proved directly from (2.22). To see this let V be a period cell for the solution and then multiply (2.22)$_1$ by u_i^*, (2.22)$_3$ by θ^*, (2.22)$_4$ by ϕ^*/γ where the $*$ denotes complex conjugate. The resulting equations are integrated over V and added, using the boundary conditions (2.21) to see that

$$
\begin{aligned}
\frac{\sigma}{2}\left(\|\theta\|^2 + \frac{A}{\gamma}\|\phi\|^2\right) &= -\|\mathbf{u}\|^2 - \|\nabla\theta\|^2 - \frac{1}{\gamma}\|\nabla\phi\|^2 \\
&\quad - H\|\theta - \phi\|^2 + R[(\theta, w^*) + (w, \theta^*)],
\end{aligned}
\tag{2.24}
$$

where $\|\cdot\|$ and $(\cdot, \cdot)$ momentarily denote the norm and inner product on the complex Hilbert space $L^2(V)$. Put $\sigma = \sigma_r + i\sigma_1$ and then take the imaginary part of (2.24) to obtain

$$
\frac{\sigma_1}{2}\left(\|\theta\|^2 + \frac{A}{\gamma}\|\phi\|^2\right) = 0.
$$

For a non-zero solution we must have $\sigma_1 = 0$ and so exchange of stabilities holds, as observed by [27].

One may now set $\sigma = 0$ in (2.22) and solve these equations for the Rayleigh number Ra. [27] show that

$$
Ra = \frac{\Lambda^2}{a^2}\left(\frac{\Lambda + H(1 + \gamma)}{\Lambda + \gamma H}\right)
\tag{2.25}
$$

where a is the wavenumber and $\Lambda = \pi^2 + a^2$. [27] minimize Ra in (2.25) over the wavenumber to find the critical Rayleigh number for instability, Ra_c, for many values of γ and H. They find that increasing γ and H increases Ra_c and so stabilizes the solution. In addition they provide many useful asymptotic results for Ra_c for small H and γ.

At this point it is worth pointing out that values of the parameter γ are certainly available for real materials. However, values for the thermal interaction coefficient, H, are more elusive. In this regard the papers of [364, 365] are extremely important. He presents very interesting analyses where he produces a possible way to calculate H for a variety of porous media. The porous media he uses consist of one-dimensional stripes of fluid between the solid skeleton, randomly striped one-dimensional porous media, two-dimensional porous media where the fluid occupies a checkerboard pattern, box type configurations, and random networks. The calculations of [364, 365] indicate there is a strong correlation between the porosity and the thermal conductivities of the fluid and solid components, but there is also a major effect due to a geometrical factor. This is a very interesting calculation and will be very useful when dealing with a known geometrical pattern of porous media and given solid and fluid components.

2.2.2 Nonlinear Stability

[416] demonstrated that the results of [27] are optimal in that their linearized instability results yield exactly the same Rayleigh number threshold as one obtains with a global (for all initial data) nonlinear stability analysis. This means that the results of [27] are particularly useful because they show that the linearized theory has captured the physics of the onset of convection. One may deduce the equivalence between the linear instability boundary and the nonlinear stability one by writing the problem (2.17)–(2.21) as an abstract system of partial differential equations in a Hilbert space and then verifying that appropriate conditions hold, namely that L given by (2.23) is symmetric and $(u, N(u)) \geq 0$, where N denotes the nonlinear terms as indicated in (2.5), see also (2.7). In fact, for equations (2.17)–(2.21), it is straightforward to show $(u, N(u)) = 0$.

It is instructive to include here a direct proof of the equivalence of the linear and nonlinear stability boundaries, as is done in [416]. Let V be a three-dimensional period cell for the solution to (2.17)–(2.21) and let $(\cdot, \cdot)$ and $\| \cdot \|$ denote the inner product and norm on $L^2(V)$. The idea is to construct "energy identities" by multiplying (2.17) by u_i, (2.19) by θ, and (2.20) by ϕ/γ to obtain after integration by parts and use of (2.18),

$$0 = -\|\mathbf{u}\|^2 + R(\theta, w), \tag{2.26}$$

and

$$\frac{d}{dt} \frac{1}{2} \|\theta\|^2 = R(w, \theta) - \|\nabla\theta\|^2 - H(\theta, \theta - \phi), \tag{2.27}$$

and

$$\frac{d}{dt} \frac{A}{2\gamma} \|\phi\|^2 = -\frac{1}{\gamma} \|\nabla\phi\|^2 - H(\phi, \phi - \theta). \tag{2.28}$$

Define an energy function E, an indefinite production term I, and the dissipation D by

$$E(t) = \frac{1}{2}\|\theta\|^2 + \frac{A}{2\gamma}\|\phi\|^2,$$

$$I = 2(\theta, w), \tag{2.29}$$

$$D = \|\mathbf{u}\|^2 + \|\nabla\theta\|^2 + \frac{1}{\gamma}\|\nabla\phi\|^2 + H\|\theta - \phi\|^2.$$

Now add equations (2.26), (2.27) and (2.28) to arrive at the energy identity

$$\frac{dE}{dt} = RI - D. \tag{2.30}$$

From this equation one may deduce

$$\frac{dE}{dt} \le -D\left(1 - \frac{R}{R_E}\right), \tag{2.31}$$

where the number R_E is defined by the relation

$$R_E^{-1} = \max_{\mathcal{H}} \frac{I}{D} \tag{2.32}$$

with $\mathcal{H}$ being the space of admissible solutions, namely, $\mathcal{H} = \{(\mathbf{u}, \theta, \phi) \mid u_i \in L^2(V), \theta, \phi \in H^1(V), u_{i,i} = 0, u_i, \theta, \phi, \pi$ are periodic over a plane tiling domain in x and $y\}$. By use of Poincaré's inequality we may show $D \ge 2\pi^2 \zeta_1 E$, where $\zeta_1 = \min\{1, A^{-1}\}$. Then for $R < R_E$, put $c = 1 - R/R_E > 0$, and from (2.31) we may deduce $dE/dt \le -kE$, where $k = 2\pi^2 \zeta_1 c$. This inequality integrates to see that

$$E(t) \le \exp(-kt)E(0). \tag{2.33}$$

From inequality (2.33) it follows that $E \to 0$ at least exponentially in time.

The exponential decay of E guarantees exponential decay of θ and ϕ (in $L^2(V)$ norm). To obtain decay of $\mathbf{u}$ we observe that from (2.26) we may deduce

$$\|\mathbf{u}\|^2 = R(\theta, w)$$

$$\le \frac{R^2}{2}\|\theta\|^2 + \frac{1}{2}\|w\|^2,$$

where in the last line the arithmetic-geometric mean inequality has been employed. In this manner we see that

$$\|\mathbf{u}\|^2 \le R^2\|\theta\|^2. \tag{2.34}$$

This shows that the condition $R < R_E$ also guarantees exponential decay of $\|\mathbf{u}\|$.

The value of R_E thus represents a global (i.e. for all initial data) nonlinear stability threshold. The number R_E is calculated from the Euler-Lagrange equations which arise from the maximum in (2.32), and these are

$$R_E \theta k_i - u_i = \omega_{,i},$$
$$u_{i,i} = 0,$$
$$R_E w + \Delta \theta - H\theta + H\phi = 0, \tag{2.35}$$
$$\frac{1}{\gamma}\Delta\phi + H\theta - H\phi = 0,$$

where ω is a Lagrange multiplier.

Since we have shown that the strong form of the principle of exchange of stabilities holds for the linearised version of (2.17)–(2.21) one may take the growth rate σ equal to zero. Now observe that the system of equations (2.35) is exactly the same eigenvalue problem as the linearised one from equations (2.17)–(2.21) when the growth rate σ is set equal to zero. Hence, the linear instability eigenvalues of [27], R_L^2, are the same as the ones for global nonlinear stability, R_E^2. This means that if $R^2 > R_L^2$ there is instability of solution (2.14). When, instead, $R^2 < R_E^2 \equiv R_L^2$ there is definitely global nonlinear asymptotic stability of solution (2.14). Because we have shown $R_L = R_E$ then no subcritical instabilities can arise.

2.3 Anisotropy

[260] extend the work of [27] to the case where the permeability in the Darcy theory is an anisotropic tensor, and they also allow the thermal conductivity to be anisotropic. Thus, the starting point for [260] is the anisotropic LTNE Darcy equations

$$\mu v_i = -K_{ij}p_{,j} - g\rho(T^f)K_{ij}k_j,$$
$$v_{i,i} = 0,$$
$$\varepsilon(\rho c)_f T^f_{,t} + (\rho c)_f v_i T^f_{,i} = \varepsilon k^f_{ij} T^f_{,ij} + h(T_s - T_f), \tag{2.36}$$
$$(1 - \varepsilon)(\rho c)_s T^s_{,t} = (1 - \varepsilon)k^s_{ij} T^s_{,ij} - h(T_s - T_f),$$

with the density in the buoyancy force given as

$$\rho = \rho_0 \left[1 - \alpha(T^f - T_U)\right], \tag{2.37}$$

where α is the coefficient of thermal expansion and T_U is the constant upper temperature of the horizontal layer, T_L being the temperature of the lower boundary. Here [260] take

$$K_{ij} = \text{diag}\,(K_h, K_h, K_z),$$
$$k^f_{ij} = \text{diag}\,(k^f_h, k^f_h, k^f_z),$$
$$k^s_{ij} = \text{diag}\,(k^s_h, k^s_h, k^s_z),$$

with $K_h, K_z, k_h^f, k_z^f, k_h^s, k_z^s$ being constants, where diag indicates diagonal as defined in section 1.1.2, equation (1.6). Thus, the permeability and thermal conductivities are transversely isotropic with isotropy in the horizontal directions. The rest of the notation used is as in section 2.2.

The steady state of [260] is again linear in z like (2.14). If we let $\mathbf{L}$ be the inverse tensor to $\mathbf{K}$ then $\mathbf{L} = \text{diag}\,(K_h^{-1}, K_h^{-1}, K_z^{-1})$ and equation (2.36)$_1$ may be rewritten

$$\mu L_{ij} v_j + \rho_0 \alpha g T^f k_i = -p_{,i}, \tag{2.38}$$

where constant terms arising from (2.37) have been absorbed in p. [260] introduce perturbations u_i, π, θ and ϕ as in section 2.2.1 and they derive their linearized equations. They use a Rayleigh number attached to the buoyancy term, but one may write their non-dimensional perturbation equations in a similar manner to equations (2.17)–(2.20), in the form

$$\begin{aligned}
\mathscr{L}_{ij} u_j &= -\pi_{,i} + R\theta k_i, \\
u_{i,i} &= 0, \\
\theta_{,t} + u_i \theta_{,i} &= Rw + \ell_{ij} \theta_{,ij} + H(\phi - \theta), \\
A\phi_{,t} &= q_{ij} \phi_{,ij} - H\gamma(\phi - \theta),
\end{aligned} \tag{2.39}$$

where $\mathscr{L}_{ij}, \ell_{ij}, q_{ij}$ are diagonal matrices with constant entries such that the 11 terms (one-one terms) equal the 22 terms. In fact, $\mathscr{L}_{ij}, \ell_{ij}, q_{ij}$ are simply non-dimensional forms of the matrices L_{ij}, k_{ij}^f and k_{ij}^s. Equations (2.39) hold on $\mathbb{R}^2 \times \{z \in (0,1)\} \times \{t > 0\}$. The boundary conditions to be satisfied are

$$u_i n_i = u_3 = w = 0, \quad \theta = 0, \quad \phi = 0, \qquad \text{on } z = 0, 1, \tag{2.40}$$

together with u_i, π, θ, ϕ satisfying a plane tiling periodicity in x, y.

The Rayleigh number of [260] is

$$Ra = R^2 = \frac{\rho_0 g \alpha (T_L - T_U) K_z d}{\varepsilon \mu \kappa_{fz}}.$$

[260] show that the Rayleigh number is determined by the expression

$$Ra = \left(\frac{\pi^2 + \xi a^2}{\xi a^2} \right) \left[\frac{\Lambda_f \Lambda_s + H\{\gamma(\Lambda_f + \Lambda_s)\}}{\Lambda_s + \gamma H} \right] \tag{2.41}$$

where $\Lambda_f = \pi^2 + \eta_f a^2$, $\Lambda_s = \pi^2 + \eta_s a^2$, $\eta_f = k_h^f / k_z^f$, $\eta_s = k_h^s / k_z^s$ and $\xi = K_h / K_z$. The analytical expression for the Rayleigh number (2.41) should be compared with the analogous expression derived by [27]. The effect of anisotropy in (2.41) is immediately evident through the presence of the ξ, η_f and η_s terms. [260] minimize Ra over a^2 numerically to obtain the critical threshold for thermal convection and they provide many numerical results assessing the effect of the anisotropy in the permeability and the thermal conductivities. They also provide many asymptotic results.

As [260] note exchange of stabilities holds for system (2.39) together with the boundary conditions (2.40). In fact, this is due to the symmetry of the linear operator in (2.39). To see that the linear operator is symmetric we write it as

$$
L = \begin{pmatrix}
-\mathscr{L}_{11} & 0 & 0 & 0 & 0 \\
0 & -\mathscr{L}_{22} & 0 & 0 & 0 \\
0 & 0 & -\mathscr{L}_{33} & R & 0 \\
0 & 0 & R & A_{44} - H & H \\
0 & 0 & 0 & H & A_{55} - H
\end{pmatrix}
$$

where A_{44} and A_{55} are the differential operators

$$
A_{44} = \ell_{11}\left(\frac{\partial^2}{\partial x^2} + \frac{\partial^2}{\partial y^2}\right) + \ell_{33}\frac{\partial^2}{\partial z^2},
$$

$$
A_{55} = \frac{1}{\gamma}\left[q_{11}\left(\frac{\partial^2}{\partial x^2} + \frac{\partial^2}{\partial y^2}\right) + q_{33}\frac{\partial^2}{\partial z^2}\right].
$$

In the present scenario one may easily show $(u, N(u)) = 0$ where $N(u)$ is the non-linearity, and then from the symmetry result of section 2.1 we may again deduce that the linear instability Rayleigh number threshold is identical to the global non-linear stability one. Thus, the instability results of [260] represent not only linear instability ones but yield also the global nonlinear stability boundary.

2.4 Forchheimer Theory

We presently wish to consider the problem addressed in section 2.2, but now to incorporate Forchheimer effects. Hence, consider now a layer of quadratic Forchheimer porous material saturated with fluid and contained between the planes $z = 0$ and $z = d$. The temperatures of the solid, T_s, and fluid, T_f, are maintained at constants on the planes $z = 0$ and $z = d$ with

$$
T_s = T_f = T_L, \quad z = 0; \qquad T_s = T_f = T_U, \quad z = d; \tag{2.42}
$$

where $T_L > T_U$. The relevant LTNE Forchheimer equations are

$$
v_i = -\frac{K}{\mu}p_{,i} + \frac{\rho_f g\alpha K}{\mu}T_f k_i - \gamma_1|\mathbf{v}|v_i, \tag{2.43}
$$

$$
v_{i,i} = 0, \tag{2.44}
$$

$$
\varepsilon(\rho c)_f T_{,t}^f + (\rho c)_f v_i T_{,i}^f = \varepsilon k_f \Delta T_f + h(T_s - T_f), \tag{2.45}
$$

$$
(1-\varepsilon)(\rho c)_s T_{,t}^s = (1-\varepsilon)k_s \Delta T_s - h(T_s - T_f). \tag{2.46}
$$

These equations hold in the domain $\mathbb{R}^2 \times \{z \in (0,d)\} \times \{t > 0\}$. The notation is as in section 2.2 except we now have a Forchheimer term with coeffcient γ_1.

The steady solution whose stability is under investigation is

$$
\bar{\mathbf{v}} \equiv 0, \qquad \bar{T}_f = \bar{T}_s = -\beta z + T_L, \tag{2.47}
$$

where

$$\beta = \frac{T_L - T_U}{d} \tag{2.48}$$

is the temperature gradient and $\bar{p}(z)$ is a quadratic function found from equation (2.43).

To investigate stability we introduce perturbations u_i, π, θ, ϕ to $\bar{v}_i, \bar{p}, \bar{T}_f$ and $\bar{T}_s$ by

$$v_i = u_i + \bar{v}_i, \quad p = \pi + \bar{p}, \quad T_f = \theta + \bar{T}_f, \quad T_s = \phi + \bar{T}_s. \tag{2.49}$$

The perturbation equations are derived from system (2.43)–(2.46) and are non-dimensionalized with velocity, pressure, temperature, time and length scales of $U = \varepsilon k_f/(\rho c)_f d$, $P = \mu d U/K$, $T^\sharp = U d \sqrt{\mu \beta c_f/\varepsilon k_f g \alpha K}$, $\mathscr{T} = (\rho c)_f d^2/k_f$, $L = d$. The Rayleigh number, Ra, is defined by

$$Ra = R^2 = d^2 \rho_f^2 \sqrt{\frac{\beta c_f g \alpha K}{\varepsilon k_f \mu}} .$$

The non-dimensional Forchheimer coefficient is denoted by F, where $F = \gamma_1 U$, and as in section 2.2, $H = h d^2/\varepsilon k_f$ and $\gamma = \varepsilon k_f/(1 - \varepsilon)k_s$ are again the non-dimensional coefficients introduced by Banu and Rees [27]. One may show that the non-dimensional perturbation equations have form

$$u_i = -\pi_{,i} + R\theta k_i - F|\mathbf{u}|u_i, \tag{2.50}$$

$$u_{i,i} = 0, \tag{2.51}$$

$$\theta_{,t} + u_i \theta_{,i} = Rw + \Delta\theta + H(\phi - \theta), \tag{2.52}$$

$$A\phi_{,t} = \Delta\phi - H\gamma(\phi - \theta), \tag{2.53}$$

where these equations hold on $\mathbb{R}^2 \times \{z \in (0,1)\} \times \{t > 0\}$, $w = u_3$, and $A = \rho_s c_s k_f/k_s \rho_f c_f$ is a non-dimensional thermal inertia coefficient.

The boundary conditions to be satisfied are

$$u_i n_i = w = 0, \quad \theta = 0, \quad \phi = 0, \quad \text{on } z = 0, 1, \tag{2.54}$$

and u_i, π, θ, ϕ satisfy a plane tiling periodicity in x, y.

[27] observe that in a linearized instability analysis the Forchheimer term drops out and thus the numerical results of [27] apply also to the Forchheimer theory (2.50)–(2.54).

2.4.1 Global Nonlinear Stability

[416] shows that one may deduce the equivalence between the linear instability boundary and the nonlinear stability one and hence the linear results of [27] are very strong capturing the instability threshold completely.

To investigate nonlinear stability directly let V be a three-dimensional period cell for the solution to (2.50)–(2.54). Construct identities by multiplying equation (2.50) by u_i, equation (2.52) by θ, and equation (2.53) by ϕ/γ and integrate each over V. Use integration by parts and employ equation (2.51). From equation (2.50) one finds

$$0 = -\|\mathbf{u}\|^2 - F\|\mathbf{u}\|_3^3 + R(\theta, w), \tag{2.55}$$

whereas from equation (2.52) we obtain,

$$\frac{d}{dt}\frac{1}{2}\|\theta\|^2 = R(w, \theta) - \|\nabla\theta\|^2 - H(\theta, \theta - \phi), \tag{2.56}$$

whilst from equation (2.53) one may deduce

$$\frac{d}{dt}\frac{A}{2\gamma}\|\phi\|^2 = -\frac{1}{\gamma}\|\nabla\phi\|^2 - H(\phi, \phi - \theta), \tag{2.57}$$

where in equation (2.55) $\|\cdot\|_3$ denotes the $L^3(V)$ norm. Define the energy function E, the production term I, and the dissipation D, by

$$E(t) = \frac{1}{2}\|\theta\|^2 + \frac{A}{2\gamma}\|\phi\|^2,$$

$$I = 2(\theta, w), \tag{2.58}$$

$$D = \|\mathbf{u}\|^2 + \|\nabla\theta\|^2 + \frac{1}{\gamma}\|\nabla\phi\|^2 + H\|0 - \phi\|^2.$$

Add equations (2.55), (2.56) and (2.57) to obtain

$$\frac{dE}{dt} = RI - D - F\|\mathbf{u}\|_3^3$$

and then drop the Forchheimer term to see that

$$\frac{dE}{dt} \le RI - D.$$

From this inequality we obtain

$$\frac{dE}{dt} \le -D\left(1 - \frac{R}{R_E}\right), \tag{2.59}$$

where we have defined R_E by

$$R_E^{-1} = \max_{\mathscr{H}} \frac{I}{D} \tag{2.60}$$

with $\mathscr{H}$ being the space of admissible solutions, namely, $\mathscr{H} = \{(\mathbf{u}, \theta, \phi) \,|\, u_i \in L^2(V), \theta, \phi \in H^1(V), u_{i,i} = 0, u_i, \theta, \phi, \pi$ are periodic over a plane tiling domain in x and $y\}$. As in section 2.2.2 we may show from inequality (2.59) that if $R < R_E$ then $E \to 0$ at least exponentially in time.

The exponential decay of E also guarantees exponential decay of θ and ϕ (in $L^2(V)$ norm). Decay of $\mathbf{u}$ follows from equation (2.55) which is

$$\|\mathbf{u}\|^2 + F\|\mathbf{u}\|_3^3 = R(\theta, w).$$

Use of the arithmetic-geometric mean inequality on the right hand side allows one to deduce

$$R(\theta, w) \leq \frac{R^2}{2}\|\theta\|^2 + \frac{1}{2}\|w\|^2 \leq \frac{R^2}{2}\|\theta\|^2 + \frac{1}{2}\|\mathbf{u}\|^2.$$

Combining the two previous expressions we may then show that

$$\|\mathbf{u}\|^2 + 2F\|\mathbf{u}\|_3^3 \leq R^2\|\theta\|^2. \tag{2.61}$$

Therefore, $R < R_E$ also guarantees exponential decay of $\|\mathbf{u}\|$ and of $\|\mathbf{u}\|_3$, and so decay in a stronger norm than that of Darcy theory follows.

The quantity R_E represents a global (i.e. for all initial data) nonlinear stability threshold. This quantity is calculated from the Euler-Lagrange equations which from the maximum problem (2.60) are found to be

$$
\begin{aligned}
& R_E \theta k_i - u_i = \zeta_{,i}, \\
& u_{i,i} = 0, \\
& R_E w + \Delta\theta - H\theta + H\phi = 0, \\
& \frac{1}{\gamma}\Delta\phi + H\theta - H\phi = 0,
\end{aligned}
\tag{2.62}
$$

where ζ is a Lagrange multiplier.

As remarked in section 2.2 [27] show that the strong form of the principle of exchange of stabilities holds for the linearised version of system (2.50)–(2.54), i.e. they show one may take the growth rate σ equal to zero. Therefore, the Euler-Lagrange equations (2.62) define an eigenvalue problem which is identically the same as the linearised one from (2.50) to (2.54) with the growth rate σ equal to zero.

2.5 Brinkman Theory

LTNE thermal convection with a Brinkman term present was investigated by [348], by [261], and by [346]. These writers all included inertia terms and convective non-linearities, with [348] and [346] also including a Forchheimer effect. In this section we describe the simplest LTNE thermal convection scenario with Brinkman theory, without involving inertia or any other nonlinearities. The critical Rayleigh number thresholds are, however, the same as those of [261, 348] and [346]. More information on these works is given at the end of this section. We consider again a horizontal layer of porous material saturated with fluid and contained between the planes $z = 0$

and $z = d$, but now the porous material is considered to be of Brinkman type. The temperatures of the solid, T_s, and fluid, T_f, are maintained at constants on the planes $z = 0$ and $z = d$ so that

$$T_s = T_f = T_L, \quad z = 0; \qquad T_s = T_f = T_U, \quad z = d; \tag{2.63}$$

where $T_L > T_U$. The govering equations for the fluid saturated porous material are

$$v_i = -\frac{K}{\mu} p_{,i} + \frac{\rho_f g \alpha K}{\mu} T_f k_i + \hat{\lambda} \Delta v_i, \tag{2.64}$$

$$v_{i,i} = 0, \tag{2.65}$$

$$\varepsilon(\rho c)_f T_{,t}^f + (\rho c)_f v_i T_{,i}^f = \varepsilon k_f \Delta T_f + h(T_s - T_f), \tag{2.66}$$

$$(1 - \varepsilon)(\rho c)_s T_{,t}^s = (1 - \varepsilon) k_s \Delta T_s - h(T_s - T_f). \tag{2.67}$$

These equations hold in the domain $\mathbb{R}^2 \times \{z \in (0, d)\} \times \{t > 0\}$. The notation is as in section 2.2 excepting we now have present the Brinkman term $\hat{\lambda} \Delta v_i$ where $\hat{\lambda}$ is the effective viscosity.

The steady solution whose stability is under investigation is

$$\bar{\mathbf{v}} \equiv 0, \qquad \bar{T}_f = \bar{T}_s = -\beta z + T_L, \tag{2.68}$$

where

$$\beta = \frac{T_L - T_U}{d} \tag{2.69}$$

is the temperature gradient and $\bar{p}(z)$ is a quadratic function found from (2.64).

To investigate stability we introduce perturbations u_i, π, θ, ϕ to $\bar{v}_i, \bar{p}, \bar{T}_f$ and $\bar{T}_s$ by

$$v_i = u_i + \bar{v}_i, \quad p = \pi + \bar{p}, \quad T_f = \theta + \bar{T}_f, \quad T_s = \phi + \bar{T}_s. \tag{2.70}$$

The perturbation equations are derived from (2.64) to (2.67) and are non-dimensionalized with velocity, pressure, temperature, time and length scales of $U = \varepsilon k_f / (\rho c)_f d$, $P = \mu d U / K$, $T^\sharp = U d \sqrt{\mu \beta c_f / \varepsilon k_f g \alpha K}$, $\mathcal{T} = (\rho c)_f d^2 / k_f$, $L = d$. The Rayleigh number is defined by

$$Ra = R^2 = d^2 \rho_f^2 \sqrt{\frac{\beta c_f g \alpha K}{\varepsilon k_f \mu}}.$$

The non-dimensional Brinkman coefficient is denoted by λ and is defined by $\lambda = \hat{\lambda}/d^2$. Again, $H = h d^2 / \varepsilon k_f$ and $\gamma = \varepsilon k_f / (1 - \varepsilon) k_s$ are the non-dimensional coefficients introduced by [27]. One may then derive the non-dimensional perturbation equations in the form

$$u_i = -\pi_{,i} + R\theta k_i + \lambda \Delta u_i, \tag{2.71}$$

$$u_{i,i} = 0, \tag{2.72}$$

$$\theta_{,t} + u_i \theta_{,i} = Rw + \Delta \theta + H(\phi - \theta), \tag{2.73}$$

$$A\phi_{,t} = \Delta \phi - H\gamma(\phi - \theta), \tag{2.74}$$

where these equations hold on $\mathbb{R}^2 \times \{z \in (0,1)\} \times \{t > 0\}$, $w = u_3$, and $A = \rho_s c_s k_f / k_s \rho_f c_f$ is a non-dimensional thermal inertia coefficient.

The boundary conditions to be satisfied are in this case

$$u_i = 0, \quad \theta = 0, \quad \phi = 0, \quad \text{on } z = 0, 1, \tag{2.75}$$

together with u_i, π, θ, ϕ satisfying a plane tiling periodicity in x, y.

For stress free surfaces [261, 348] show that the Rayleigh number may be calculated to be

$$Ra = \frac{\Lambda^2}{a^2}(1 + Da\Lambda)\left(1 + \frac{H}{\Lambda + \gamma H}\right)$$

where $\Lambda = \pi^2 + a^2$, a is the wavenumber, and $Da = \hat{\lambda}K/\mu d^2$ is the Darcy number. Both sets of writers minimize the Rayleigh number expression over a^2 to find the critical Rayleigh number threshold and they provide extensive numerical results, and additionally they establish a variety of asymptotic results. The critical Rayleigh number for the fixed surface problem is found by a Galerkin numerical technique by [346] who also includes a substantial number of numerical results.

2.5.1 Global Nonlinear Stability

One may again deduce the equivalence between the linear instability boundary and the nonlinear stability one by appealing to the symmetry of the linear operator involved with equations (2.71)–(2.75). This means that the linear results of [261, 348] and [346] are very powerful and actually yield a global nonlinear stability threshold. This may be proved directly and to do this let V be a three-dimensional period cell for the solution to (2.71)–(2.75). Energy identities may be constructed by multiplying (2.71) by u_i, (2.73) by θ, and (2.74) by ϕ/γ to obtain after integration by parts and use of equations (2.72) and (2.75),

$$0 = -\|\mathbf{u}\|^2 - \lambda\|\nabla\mathbf{u}\|^2 + R(\theta, w), \tag{2.76}$$

and

$$\frac{d}{dt}\frac{1}{2}\|\theta\|^2 = R(w, \theta) - \|\nabla\theta\|^2 - H(\theta, \theta - \phi), \tag{2.77}$$

together with

$$\frac{d}{dt}\frac{A}{2\gamma}\|\phi\|^2 = -\frac{1}{\gamma}\|\nabla\phi\|^2 - H(\phi, \phi - \theta). \tag{2.78}$$

The energy E, production term I, and the dissipation D, in this section are

$$E(t) = \frac{1}{2}\|\theta\|^2 + \frac{A}{2\gamma}\|\phi\|^2,$$

$$I = 2(\theta, w), \tag{2.79}$$

$$D = \|\mathbf{u}\|^2 + \lambda\|\nabla\mathbf{u}\|^2 + \|\nabla\theta\|^2 + \frac{1}{\gamma}\|\nabla\phi\|^2 + H\|\theta - \phi\|^2.$$

One adds equations (2.76), (2.77) and (2.78) to derive the energy equation

$$\frac{dE}{dt} = RI - D.$$

From this equation one obtains

$$\frac{dE}{dt} \leq -D\left(1 - \frac{R}{R_E}\right), \tag{2.80}$$

where R_E is found from the expression

$$R_E^{-1} = \max_{\mathscr{H}} \frac{I}{D} \tag{2.81}$$

with $\mathscr{H}$ being the space of admissible solutions, namely, $\mathscr{H} = \{(\mathbf{u}, \theta, \phi) \mid u_i \in L^2(V), \theta, \phi \in H^1(V), u_{i,i} = 0, u_i, \theta, \phi, \pi$ are periodic over a plane tiling domain in x and $y\}$. As in section 2.2.2, one may show from inequality (2.80) that if $R < R_E$ then $E \to 0$ exponentially in time.

The exponential decay of E guarantees exponential decay of θ and ϕ (in $L^2(V)$ norm). To obtain decay of $\mathbf{u}$ rearrange equation (2.76) to find

$$\|\mathbf{u}\|^2 + \lambda \|\nabla \mathbf{u}\|^2 - R(\theta, w)$$

$$\leq \frac{R^2}{2} \|\theta\|^2 + \frac{1}{2} \|w\|^2,$$

from which

$$\|\mathbf{u}\|^2 + 2\lambda \|\nabla \mathbf{u}\|^2 \leq R^2 \|\theta\|^2. \tag{2.82}$$

This shows $R < R_E$ also guarantees exponential decay of $\|\mathbf{u}\|$ and of $\|\nabla \mathbf{u}\|$ when the Brinkman model is employed.

The term R_E represents a global (i.e. for all initial data) nonlinear stability threshold and may be calculated from the Euler-Lagrange equations which arise from (2.81), namely

$$R_E \theta k_i - u_i + \lambda \Delta u_i = \omega_{,i},$$
$$u_{i,i} = 0,$$
$$R_E w + \Delta \theta - H\theta + H\phi = 0, \tag{2.83}$$
$$\frac{1}{\gamma} \Delta \phi + H\theta - H\phi = 0,$$

where ω is a Lagrange multiplier.

Equations (2.83) are the same as the corresponding ones of linear instability theory, since exchange of stabilities holds, see e.g. [261, 348], and [346]. This then leads to coincidence of the linear and nonlinear stability thresholds. We observe that the linear instability and global nonlinear stability thresholds one obtains from the equations of this section are the same as those of [261, 348], and [346], since inertia does not play a pivotal role due to exchange of stabilities, and as [348] note, the extra nonlinear velocity terms in their work have no effect on stability since the basic state is one of zero flow.

For completeness we record the momentum equations of [261] which are, in our notation,

$$\frac{\rho}{\varepsilon}\frac{\partial v_i}{\partial t} + \frac{\rho}{\varepsilon^2}v_j\frac{\partial v_i}{\partial x_j} = -\frac{\partial p}{\partial x_i} - \frac{\mu}{K}v_i + \hat{\lambda}\Delta v_i + \rho_0 g\alpha T^f k_i,$$

while those of [348] and [346] have form

$$\frac{\rho}{\varepsilon}\frac{\partial v_i}{\partial t} + \frac{\rho}{\varepsilon^2}v_j\frac{\partial v_i}{\partial x_j} = -\frac{\partial p}{\partial x_i} - \frac{\mu}{K}v_i + \hat{\lambda}\Delta v_i - \frac{\rho b}{\sqrt{K}}|\mathbf{v}|v_i + \rho_0 g\alpha T^f k_i.$$

Chapter 3
Rotating Convection with LTNE

3.1 Rotating Thermal Convection, Single Temperature

The aim of this chapter is to describe work analogous to that of chapter 2 but when the porous layer is rotating about an axis orthogonal to the layer and in the opposite direction to gravity. Before we consider the LTNE case we briefly describe thermal convection in a rotating layer with a single temperature. We begin with a layer of viscous, incompressible fluid contained in the horizontal layer between the planes $z = 0$ and $z = d$, such that the layer is in rotation with constant angular velocity, about an axis in the vertical direction. The lower and upper planes are held at constant temperatures T_L and T_U, respectively, with $T_L > T_U$. This is the classical Bénard problem in a rotating layer, the linear instability theory of which is described in detail by [85]. In his book [85] shows that in addition to stationary convection, oscillatory convection is possible, where the growth rate σ is complex at the critical transistion to thermal convection. He also studies the situation where the axis of rotation may be in a different direction to that of the gravity field.

The steady state is one where the velocity is zero apart from the rigid rotation, and the temperature field is linear in z. Then the non-dimensional equations for a perturbation to the steady state for thermal convection in an incompressible, viscous fluid in a rotating layer as derived by [85] may be written in the form, cf. [149],

$$\frac{\partial u_i}{\partial t} + u_j \frac{\partial u_i}{\partial x_j} = -\frac{\partial p}{\partial x_i} + R\theta k_i + \Delta u_i + \varepsilon_{ijk} T u_j \delta_{k3},$$

$$\frac{\partial u_i}{\partial x_i} = 0, \tag{3.1}$$

$$Pr\left(\frac{\partial \theta}{\partial t} + u_i \frac{\partial \theta}{\partial x_i}\right) = Rw + \Delta\theta,$$

where $w = u_3$. In equations (3.1) the spatial region is the layer $\{(x,y) \in \mathbb{R}^2\} \times \{z \in (0,1)\}$, and the term $\varepsilon_{ijk} T u_j \delta_{k3}$ arises from the Coriolis force and is the component form of $T(\mathbf{u} \times \mathbf{k})$ with $\mathbf{k} = (0,0,1)$. Here u_i, π, θ are perturbations to velocity, pres-

© Springer International Publishing Switzerland 2015
B. Straughan, *Convection with Local Thermal Non-Equilibrium and Microfluidic Effects*,
Advances in Mechanics and Mathematics 32, DOI 10.1007/978-3-319-13530-4_3

sure, temperature, $Ra = R^2$ is the Rayleigh number, $\mathcal{T} = T^2$ is the Taylor number, and Pr is the Prandtl number. The boundary conditions adopted in [149] are that

$$u_i, \theta, p \text{ are periodic in } x, y, \text{ with periods } 2a_1, 2a_2, \text{ respectively,} \qquad (3.2)$$

and the planes $z = 0, 1$ are *free surfaces* on which no tangential stresses act (cf. [85], pp. 21–22), so that

$$\frac{\partial u}{\partial z} = \frac{\partial v}{\partial z} = w = \theta = 0, \qquad \text{on} \quad z = 0, 1, \qquad (3.3)$$

where $\mathbf{u} = (u, v, w)$.

It is necessary to exclude rigid motions and in order to achieve this we assume the mean values of u and v are zero, i.e., $< u > = < v > = 0$, where $< \cdot >$ denotes integration over a periodicity cell V. For the boundary conditions (3.2) one may take $V = [0, 2a_1) \times [0, 2a_2) \times (0, 1)$. When one is dealing with rotating thermal convection in a clear fluid the assumption of free surfaces is not overly unrealistic. For high rotation rates Ekman layers form and the convecting layer is bounded by these boundary layers which may act like free surfaces. If one has fixed bounding surfaces then equations (3.3) are replaced by

$$u = v = w = \theta = 0, \qquad \text{on} \quad z = 0, 1. \qquad (3.4)$$

[85] demonstrates that increasing the Taylor number strongly increases the critical Rayleigh number at which thermal convection will commence. One important thing which is immediately evident is that any standard type of energy stability analysis cannot capture this inhibiting effect of rotation on convection. To understand this mutiply equation $(3.1)_1$ by u_i and integrate over V. Whatever boundary conditions one employs the effect of rotation immediately disappears in the energy stability analysis since

$$T \int_V (\mathbf{u} \times \mathbf{k}) \cdot \mathbf{u} \, dx = 0.$$

This was overcome by [149] by modifying the standard Lyapunov functional. They let $\omega = \text{curl} \, \mathbf{u}$, $\omega = \omega_3$, and then introduced a variable F by $F = \omega - (T/R)\xi Pr\theta_{,z}$ for a coupling parameter ξ to be opportunely selected. To obtain a stabilizing effect of rotation they employed a generalized energy of form

$$E(t) = \frac{1}{2} \left\{ \left(\lambda \|\mathbf{u}\|^2 + Pr\|\theta\|^2 \right) + c_1 Pr\|\theta_{,z}\|^2 + c_2 \|F\|^2 + \|\nabla w\|^2 \right\}$$

where λ, c_1, c_2 are further coupling parameters to be selected and $\| \cdot \|$ denotes the norm on $L^2(V)$. [149] did achieve the stabilizing effect of rotation nonlinearly for two stress free surfaces, but their initial energy $E(0)$ had to be restricted, i.e. the nonlinear stability was conditional.

For a porous medium the results for the analogous class of thermal convection problem in a rotating horizontal layer are somewhat different. A fundamental piece of work in rotating porous convection in a Darcy porous medium is that of [456]. Let us consider only one (average) temperature field and let the porous medium occupy a horizontal layer, where gravity acts in the negative z−direction, and the layer rotates about the z−axis. The nonlinear, non-dimensionalized perturbation equations for thermal convection in a saturated Darcy porous medium are, cf. [456]

$$\frac{1}{Va}\frac{\partial u_i}{\partial t} = -\frac{\partial \pi}{\partial x_i} + R\theta k_i - T(\mathbf{k} \times \mathbf{u})_i - u_i,$$

$$\frac{\partial u_i}{\partial x_i} = 0, \tag{3.5}$$

$$\frac{\partial \theta}{\partial t} + u_i\frac{\partial \theta}{\partial x_i} = Rw + \Delta\theta.$$

The domain for equations (3.5) is $\{(x,y) \in \mathbb{R}^2\} \times \{z \in (0,1)\} \times \{t > 0\}$. The non-dimensional numbers R, T and Va are introduced where R^2 is the Rayleigh number, T^2 is the Taylor number, and $Va = \varepsilon Pr/Da$ is the Vadasz number. The coefficients ε, Pr and $Da = K/d^2$, are the porosity, Prandtl number and Darcy number, where K is the permeability and d is the depth of the layer.

In [456] Vadasz established the striking result that when $Va \neq \infty$, oscillatory convection is possible. For many situations there is no loss in ignoring the acceleration term in (3.5)₁, i.e. let $Va \to \infty$. This becomes the classical theory of Darcy with no inertia, and is the case covered by [411]. The boundary conditions considered in connection with (3.5) are

$$w(\mathbf{x},t) = 0, \quad \theta(\mathbf{x},t) = 0, \qquad z = 0, 1, \tag{3.6}$$

with u_i, θ, π satisfing a plane tiling periodic boundary condition in x and y.

In the case where $Va \to \infty$ in (3.5)₁ one may show, cf. [411], that the growth rate σ for linear instability is real, i.e. $\sigma \in \mathbb{R}$. In fact, [411] demonstrated that with $Va = \infty$, i.e. no inertia in (3.5), one has coincidence of the linear instability and global (for all initial data) nonlinear stability boundaries for thermal convection. Thus, with zero inertia no sub-critical instabilities may arise.

This work was extended by [249] who include Brinkman effects. They also take inertia to be zero and consider stress free boundaries. Their perturbation equations are

$$0 = -\frac{\partial \pi}{\partial x_i} + R\theta k_i - T(\mathbf{k} \times \mathbf{u})_i - u_i + Da\Delta u_i,$$

$$\frac{\partial u_i}{\partial x_i} = 0, \tag{3.7}$$

$$\frac{\partial \theta}{\partial t} + u_i\frac{\partial \theta}{\partial x_i} = Rw + \Delta\theta,$$

where Da is the Darcy number. [249] show that one may also prove the optimal result that the linear instability boundary is the same as the nonlinear stability one. The same writers further consider the analogous rotating double diffusive problem when a salt field is also present for which the non-dimensional perturbation equations are

$$0 = -\frac{\partial \pi}{\partial x_i} + (R\theta - R_c\phi)k_i - T(\mathbf{k} \times \mathbf{u})_i - u_i + Da\,\Delta u_i,$$

$$\frac{\partial u_i}{\partial x_i} = 0,$$

$$\frac{\partial \theta}{\partial t} + u_i\frac{\partial \theta}{\partial x_i} = Rw + \Delta\theta,$$

$$\varepsilon_1 \frac{\partial \phi}{\partial t} + u_i\frac{\partial \phi}{\partial x_i} = R_c w + \frac{1}{Le}\Delta\phi,$$

(3.8)

where R_c^2 is a salt Rayleigh number, ϕ is the salt concentration perturbation, and ε_1, Le are constants. [249] provide linear instability results for this more complicated problem.

Further interesting recent results for rotating convection problems with one temperature field may be found in [161–163], and in [70], and the references therein. Our main aim in this chapter is to study the problem of thermal convection in a rotating porous layer but when the solid and fluid temperatures may be unequal, i.e. with LTNE effects, and so we now turn our attention to this goal.

3.2 Rotation with Darcy Theory

We now consider further the problem of convection in a horizontal porous layer which is rotating with a constant angular speed Ω about a vertical axis in the z−direction. This section describes work of [411] and so we restrict attention to Darcy's law with zero inertia, but now we allow for local thermal non-equilibrium effects. The situation is pictured in the diagram 3.1.

The equations for a perturbation about the steady solution allowing for the local thermal non-equilibrium effect are described by [411] and are

$$u_i + T(\mathbf{k} \times \mathbf{u})_i = -\pi_{,i} + R\theta k_i,$$

$$u_{i,i} = 0,$$

$$\theta_{,t} + u_i\theta_{,i} = Rw + \Delta\theta + H(\phi - \theta),$$

$$A\phi_{,t} = \Delta\phi - H\gamma(\phi - \theta).$$

(3.9)

Thus, equations (3.9) are the analogue of (2.17)–(2.20), but allowing for the Coriolis force, i.e. the T term.

3.2.1 Linear Instability

Consider now the linearised theory in three space dimensions. Put $u_i = e^{\sigma t}u_i(\mathbf{x})$, $\pi = e^{\sigma t}\pi(\mathbf{x})$, $\theta = e^{\sigma t}\theta(\mathbf{x})$, and $\phi = e^{\sigma t}\phi(\mathbf{x})$, and then equations $(3.9)_{3,4}$ after lin-

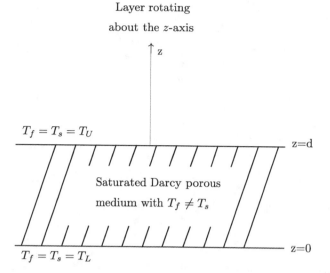

Layer rotating
about the z-axis

Fig. 3.1 LTNE convection in a horizontal porous layer $\{(x,y) \in \mathbb{R}^2\} \times (0,d)$ rotating with constant angular speed Ω around the vertical axis

earization give

$$\sigma\theta = Rw + \Delta\theta + H(\phi - \theta),$$
$$A\sigma\phi = \Delta\phi - H\gamma(\phi - \theta). \tag{3.10}$$

We wish to demonstrate that the strong form of the principle of exchange of stabilities holds, i.e. $\sigma \in \mathbb{R}$. Thus we treat the variables in equations (3.10) as complex from the outset, eg. $\theta = \theta_r + i\theta_i$, $\sigma = \sigma_r + i\sigma_i$, etc. Denote by Δ^* the horizontal Laplacian $\Delta^* = \partial^2/\partial x^2 + \partial^2/\partial y^2$. Multiply (3.10)$_1$ by $-\Delta^*\theta^*$, where θ^* denotes the complex conjugate of θ, i.e. $\theta^* = \theta_r - i\theta_i$. Integrate the result over V with some integration by parts. Then multiply (3.10)$_2$ by $-\Delta^*\phi^*/\gamma$ and likewise integrate over V, together with some integration by parts. One may show,

$$\sigma\|\nabla^*\theta\|^2 = R(\nabla^*w, \nabla^*\theta^*) - \|\nabla^*\nabla\theta\|^2 + H(\nabla^*\theta^*, \nabla^*\phi) - H\|\nabla^*\theta\|^2, \tag{3.11}$$

and

$$\frac{A\sigma}{\gamma}\|\nabla^*\phi\|^2 = -\frac{1}{\gamma}\|\nabla^*\nabla\phi\|^2 + H(\nabla^*\phi^*, \nabla^*\theta) - H\|\nabla^*\phi\|^2, \tag{3.12}$$

where $\nabla^* \equiv (\partial/\partial x, \partial/\partial y, 0)$, and where the norm and inner product are momentarily on the complex space $L^2(V)$.

Define $\omega = \text{curl}\,\mathbf{u}$ to be the vorticity, and then curl curl of equation (3.9)$_1$ and the third components of the resulting equations yield, cf. the calculations for (1.3) and (1.5),

$$\omega_3 = T\frac{\partial w}{\partial z}, \tag{3.13}$$

and

$$\Delta w + T \frac{\partial \omega_3}{\partial z} = R \Delta^* \theta. \tag{3.14}$$

Multiply (3.14) by w^* (the complex conjugate of w) and then integrate over V to find after integration by parts and use of (3.13),

$$0 = R(\nabla^*\theta, \nabla^*w^*) - \|\nabla w\|^2 - T^2\|w_{,z}\|^2. \tag{3.15}$$

After addition of equations (3.11), (3.12) and (3.15) we may arrive at

$$\sigma\left(\|\nabla^*\theta\|^2 + \frac{A}{\gamma}\|\nabla^*\phi\|^2\right) = R[(\nabla^*\theta, \nabla^*w^*) + (\nabla^*w, \nabla^*\theta^*)]$$
$$- \|\nabla w\|^2 - T^2\|w_{,z}\|^2 - \|\nabla^*\nabla\theta\|^2 - \frac{1}{\gamma}\|\nabla^*\nabla\phi\|^2 \tag{3.16}$$
$$- H(\|\nabla^*\theta\|^2 + \|\nabla^*\phi\|^2) + H[(\nabla^*\theta^*, \nabla^*\phi) + (\nabla^*\phi^*, \nabla^*\theta)].$$

Upon inspection one sees that the right hand side of equation (3.16) is real. Thus, taking the imaginary part of equation (3.16) leads to

$$\sigma_i\left(\|\nabla^*\theta\|^2 + \frac{A}{\gamma}\|\nabla^*\phi\|^2\right) = 0. \tag{3.17}$$

Therefore, for a non-zero solution $\sigma_i = 0$ and $\sigma \in \mathbb{R}$.

The linear instability eigenvalue problem is then governed by the linearised equations (3.9), but with the growth rate, σ, set equal to zero. Hence, one may replace equation (3.9)$_1$ by (3.14) and the relevant equations governing the linear instability problem are

$$\Delta w + T^2 w_{,zz} = R \Delta^* \theta, \tag{3.18}$$

and

$$0 = Rw + \Delta\theta + H(\phi - \theta), \tag{3.19}$$

together with

$$0 = \Delta\phi - H\gamma(\phi - \theta), \tag{3.20}$$

where $\Delta^* = \partial^2/\partial x^2 + \partial^2/\partial y^2$. We seek a solution of form $w = W(z)f(x,y)$, $\theta = \Theta(z)f(x,y)$, $\phi = \Phi(z)f(x,y)$, where f is a planform satisfying $\Delta^* f = -a^2 f$, a being a wavenumber. One may eliminate Θ and Φ to derive the equation, cf. [411],

$$L_1 L_2 L W = R^2 a^2 L_2 W + H^2 \gamma L W, \tag{3.21}$$

where $L \equiv (1+T^2)D^2 - a^2$, $L_1 \equiv D^2 - a^2 - H$, $L_2 \equiv D^2 - a^2 - \gamma H$, with $D = d/dz$. We take $W = \sin n\pi z$ and then we find

$$R^2 = \frac{([1+T^2]n^2\pi^2 + a^2)(n^2\pi^2 + a^2)(n^2\pi^2 + a^2 + [1+\gamma]H)}{a^2(n^2\pi^2 + a^2 + \gamma H)}. \tag{3.22}$$

One differentiates (3.22) with respect to n^2 to see that $n = 1$ minimizes R^2 as a function of n, and so we may restrict attention to the case

$$R^2 = \frac{([1+T^2]\pi^2 + a^2)(\pi^2 + a^2)(\pi^2 + a^2 + [1+\gamma]H)}{a^2(\pi^2 + a^2 + \gamma H)}. \qquad (3.23)$$

To determine the critical linear Rayleigh number R_L^2 one minimises (3.23) in a^2. Numerical values for the critical Rayleigh number may be found in the stationary convection results of [263], in tables 1–3 and in figure 1.

3.2.2 Nonlinear Stability

We now multiply equation $(3.9)_3$ by θ and equation $(3.9)_4$ by ϕ and integrate over V with some integration by parts. After addition of the results we may derive an energy identity of form

$$\frac{dE}{dt} = I - D, \qquad (3.24)$$

where the energy function E, the production term I, and the dissipation D, are given by

$$E(t) = \frac{1}{2}\|\theta\|^2 + \frac{A}{2\gamma}\|\phi\|^2, \qquad (3.25)$$

$$I(t) = R(w, \theta), \qquad (3.26)$$

and

$$D(t) = \|\nabla\theta\|^2 + \frac{1}{\gamma}\|\nabla\phi\|^2 + H\|\theta - \phi\|^2. \qquad (3.27)$$

From equation (3.24) we may thus deduce

$$\frac{dE}{dt} \le -D\left(1 - \frac{1}{\Lambda}\right), \qquad (3.28)$$

where Λ is defined by

$$\frac{1}{\Lambda} = \max_{\mathcal{H}} \frac{I}{D}. \qquad (3.29)$$

If $\Lambda^{-1} < 1$, then using the Poincaré inequality one shows $D \ge \pi^2\|\theta\|^2 + (\pi^2/\gamma)\|\phi\|^2$ and so from (3.28) one may deduce that $E(t) \to 0$ at least exponentially. To deduce decay of u_i one multiplies equation $(3.9)_1$ by u_i and integrates over V to find

$$\|\mathbf{u}\|^2 = R(\theta, w) \le \frac{R^2}{2}\|\theta\|^2 + \frac{1}{2}\|w\|^2.$$

Hence, it follows that

$$\|\mathbf{u}\| \le R\|\theta\|. \qquad (3.30)$$

Therefore, the condition $\Lambda^{-1} < 1$ also guarantees exponential decay of $\|\mathbf{u}(t)\|$. One may show that $\Lambda^{-1} < 1$ is equivalent to $R < R_E$, where R_E is the value of R for which $\Lambda = 1$. This value of R_E is the nonlinear stability threshold.

To determine Λ and hence R_E one needs to solve the maximum problem (3.29). Details of this are given by [411] who regards (3.18) as a constraint, and solves the maximum problem

$$\frac{1}{\Lambda} = \max_{\mathcal{H}} \frac{R(w,\theta) + \int_V \ell(\Delta w + T^2 w_{,zz} - R\Delta^* \theta)\, dx}{D}, \qquad (3.31)$$

where ℓ is a Lagrange multiplier. We do not describe details of this procedure but we observe that [411] shows that the critical value of R^2 for global nonlinear stability, R_E^2, is the same as the linear instability one R_L^2. Thus, even for thermal convection LTNE with rotation one has the coincidence of the linear instability and nonlinear stability thresholds and one can be assured that the linearized theory has completely captured the physics of the onset of convection.

3.3 Rotation with Vadasz–Darcy Theory

[263] extend the work described in section 3.2 by allowing for the effect of fluid acceleration, or inertia. This is very important and does potentially change the nature of convection allowing for the possibility of a complex growth rate and hence oscillatory convection.

The equations for a perturbation about the steady state are like those of (3.9) but include an inertia term and so may be written [[263] use a different non-dimensionalization so their equations appear different but are *entirely* equivalent]

$$\begin{aligned}
\frac{1}{Va}\frac{\partial u_i}{\partial t} &= -\frac{\partial \pi}{\partial x_i} + R\theta k_i - u_i - T(\mathbf{k} \times \mathbf{u})_i, \\
\frac{\partial u_i}{\partial x_i} &= 0, \\
\frac{\partial \theta}{\partial t} + u_i \frac{\partial \theta}{\partial x_i} &= Rw + \Delta\theta + H(\phi - \theta), \\
A\frac{\partial \phi}{\partial t} &= \Delta\phi - H\gamma(\phi - \theta),
\end{aligned} \qquad (3.32)$$

where Va is the Vadasz number and $Ra = R^2$ and $\mathcal{T} = T^2$ are again the Rayleigh and Taylor numbers.

[263] perform a linearized instability analysis and also a weakly nonlinear analysis for system (3.32) subject to the boundary conditions

$$w(\mathbf{x},t) = 0, \quad \theta(\mathbf{x},t) = 0, \quad \phi(\mathbf{x},t) = 0, \qquad z = 0, 1, \qquad (3.33)$$

with u_i, θ, π, ϕ satisfing a plane tiling periodic boundary condition in x and y. For the stationary convection problem in linear instability theory, i.e. where $\sigma = 0$, they find the Rayleigh number is given by equation (3.23). However, due to the presence of the Vadasz inertia term $Va^{-1}\partial u_i/\partial t$, $Va < \infty$, exchange of stabilities does not hold and [263] derive conditions for oscillatory convection, Hopf bifurcation, to occur. They derive a complicated but analytical expression for the critical Rayleigh number of oscillatory convection, Ra_{osc}, and this is given by their equation (47).

[263] report extensive tables and graphs for critical Rayleigh and wave numbers for various values of $\mathscr{T}, \gamma, H$, for both stationary and oscillatory convection. They find that if the Taylor number $\mathscr{T}$ is large enough then oscillatory convection is the preferred mode of instability. For example, when $H = 100, \gamma = 0.5, A = 0.01$, and $\varepsilon Pr/Da = 10$, [263] find stationary convection for $\mathscr{T} < 4.483$ whereas there is oscillatory convection when $\mathscr{T} > 4.483$.

As far as I know there are no results for global nonlinear stability for a solution to equations (3.32) and (3.33). We remark that [70] recently address this question for the Vadasz model, equations (3.5).

3.4 Rotation with Vadasz–Brinkman Theory

[259] extend their work described in section 3.3 by also including the higher derivatives Brinkman term. They again include inertia and so one may also expect oscillatory convection.

The equations for a perturbation about the steady solution may be derived as those of (3.32) and with the Brinkman term they may be written

$$\frac{1}{Va}\frac{\partial u_i}{\partial t} = -\frac{\partial \pi}{\partial x_i} + R\theta k_i - u_i - T(\mathbf{k} \times \mathbf{u})_i + Da\Delta u_i,$$

$$\frac{\partial u_i}{\partial x_i} = 0,$$

$$\frac{\partial \theta}{\partial t} + u_i\frac{\partial \theta}{\partial x_i} = Rw + \Delta\theta + H(\phi - \theta),$$

$$A\frac{\partial \phi}{\partial t} = \Delta\phi - H\gamma(\phi - \theta),$$

(3.34)

where now Da is the Darcy number as mentioned in section 3.1. Now the boundary conditions chosen by [259] are those for two stress free surfaces and so are

$$\frac{\partial u}{\partial z} = \frac{\partial v}{\partial z} = w = \theta = 0, \qquad \text{on} \quad z = 0, 1,$$

(3.35)

where $\mathbf{u} = (u, v, w)$, with u_i, π, θ, ϕ periodic in the x, y directions.

[259] perform a detailed linear instability analysis and a weakly nonlinear analysis for (3.34) and (3.35). For the stationary convection boundary they find that the

Rayleigh number is given by

$$Ra_{stat} = \frac{\Lambda^2}{a^2}\left[\frac{\Lambda + H(1+\gamma)}{\Lambda + \gamma H}\right]\left\{\Lambda(\Lambda Da + 1) + \frac{\pi^2 \mathscr{T}}{(\Lambda Da + 1)}\right\}.$$

They also derive an analytical expression for the oscillatory convection Rayleigh number, Ra_{osc}, which is their equation (3.18). They again find that when the Taylor number is large enough, oscillatory convection is the preferred mode of instability.

Apart from the weakly nonlinear stability results of [259] we are unaware of any nonlinear stability results for system (3.34) and (3.35).

Chapter 4
Double Diffusive Convection with LTNE

In the standard Bénard problem described in section 2.1.1 the instability is driven by a density difference caused by a temperature difference between the upper and lower planes bounding the fluid. If the fluid layer additionally has salt dissolved in it then there are potentially two destabilizing sources for the density difference, the temperature field and the salt field. A similar scenario could be witnessed in isothermal conditions but with two dissolved salts such as sodium and potassium chloride. When there are two effects such as this the phenomenon of convection which arises is called *double diffusive convection*. For the specific case involving a temperature field and sodium chloride it is frequently referred to as *thermohaline convection*. There are many recent studies involving three or more fields, such as temperature and two salts, for example, NaCl, KCl. For the three or greater field case we shall refer to this as *multi-component convection*.

The driving force for the interest in double diffusive or multi-component convection is largely physical applications. One such important application is modelling geothermal reservoirs. For example, [192] study double diffusive convection in a brine saturated Darcy porous medium modelling a saturated geothermal reservoir in connection with the storage of CO_2 to control carbon emissions. This paper contains many references to modelling geothermal reservoirs and applications.

Other important applications of double diffusive or multi component convection include the oceans, cf. [288, 338], analysis of planetary cores, e.g. [268, 345, 448], and drainage in a mangrove system, [466].

Solar ponds are a particularly promising means of harnessing energy from the Sun by preventing convective overturning in a thermohaline system by salting from below, cf. [180, 181]. Salinization, where convective motion can actually transport salt to the Earth's surface is likewise a very interesting example of double diffusive convection, cf. [41, 157, 334, 467].

Within microfluidic situations diffusion and flows in porous media and clear fluids have attracted much attention, see e.g. [14, 151, 209, 329, 330, 450, 482], and the review of numerical techniques by [481]. We should also point out extensive recent activity in a very closely related field namely that of triply diffusive convection, involving a temperature field and two different salt fields, see

© Springer International Publishing Switzerland 2015
B. Straughan, *Convection with Local Thermal Non-Equilibrium and Microfluidic Effects*,
Advances in Mechanics and Mathematics 32, DOI 10.1007/978-3-319-13530-4_4

[68, 69, 84, 372, 374, 401], and the references therein. Furthermore, within microfluidic situations multi-component diffusive convection is a very live field, especially involving DNA dissociation; see [11, 104], and the work of [14] shows that slip boundary conditions may be important in this context.

The really interesting situation from both a geophysical and a mathematical viewpoint arises when the layer is simultaneously heated from below and salted from below. In this situation heating expands the fluid at the bottom of the layer and this in turn wants to rise thereby encouraging motion due to thermal convection. On the other hand, the heavier salt at the lower part of the layer has exactly the opposite effect and this acts to prevent motion through convective overturning. Thus, these two physical effects are competing against each other. Due to this competition, it means that the linear theory of instability does not always capture the physics of instability completely and (sub-critical) instabilities may arise before the linear threshold is reached, cf. [175, 352, 469]. Due to the possibility of sub-critical instabilities occurring, it is very important to obtain (unconditional) nonlinear stability thresholds which guarantee bounds below which convective overturning will not occur. The mathematical theory for the topic of thermal convection in a fluid layer which also contains the presence of a salt field was analysed employing nonlinear energy stability methods by [396]. [196] addressed the tricky heated and salted below problem by introducing a generalized energy functional and he was thus able to produce a nonlinear stability threshold which improves upon that which one finds with a classical energy theory. From a mathematical point of view the paper of [196] is important in that it has instigated much research into the construction of novel energy functionals, or Lyapunov functionals. [297] continued the development of [196], introducing, in some sense, a natural Lyapunov functional. We should mention that among the fundamental early contributions to stability in thermohaline convection in a fluid were the papers of [25, 196, 396], while fundamental work in the analogous problem in saturated porous media is due to [306].

We now describe some work on double diffusive convection in a fluid saturated porous medium taking into account LTNE effects.

4.1 Darcy Theory with Inertia

The problem of double diffusive convection in a Darcy porous material with LTNE effects was studied by [264]. These writers considered the inertia effect and employed the configuration of a horizontal layer bounded by the planes $z = 0$ and $z = d$ with gravity in the negative $z-$direction. They suppose the upper plane is kept at constant temperature T_U with the lower one at constant temperature T_L, while the salt concentration is kept at the constant value S_U at $z = d$ and is kept at the constant value S_L at $z = 0$. They define $\Delta T = T_L - T_U$, $\Delta S = S_L - S_U$, and they assume $\Delta T > 0$ and $\Delta S > 0$ and so they are considering the problem when the buoyancy effects due to temperature difference and due to salt diffusion are opposing each other. The configuration is pictured in figure 4.1.

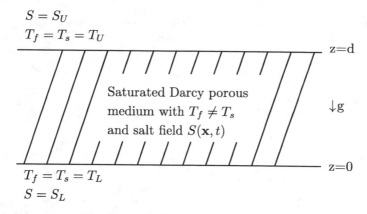

$$S = S_U$$
$$T_f = T_s = T_U$$

z=d

Saturated Darcy porous medium with $T_f \neq T_s$ and salt field $S(\mathbf{x}, t)$

$\downarrow$g

$$T_f = T_s = T_L$$
$$S = S_L$$

z=0

Fig. 4.1 Double diffusive LTNE convection in a horizontal porous layer. Temperatures are given by T_f and T_s, concentration of salt field is S. Gravity is in the negative $z-$ direction.

In terms of the velocity field $v_i(\mathbf{x},t)$, pressure field $p(\mathbf{x},t)$, fluid temperature $T^f(\mathbf{x},t)$, porous solid skeleton temperature $T^s(\mathbf{x},t)$, and the salt concentration $S(\mathbf{x},t)$, the model of [264] consists of the following system of nonlinear partial differential equations

$$\frac{1}{\varepsilon}\frac{\partial v_i}{\partial t} = -\frac{1}{\rho_f}\frac{\partial p}{\partial x_i} - \frac{v}{K}v_i - k_i\big[1 - \alpha(T^f - T_L) + \alpha_c(S - S_L)\big],$$

$$\frac{\partial v_i}{\partial x_i} = 0,$$

$$\varepsilon(\rho c)_f\frac{\partial T^f}{\partial t} + (\rho c)_f v_i\frac{\partial T^f}{\partial x_i} = \varepsilon k_f \Delta T^f + h(T^s - T^f),$$

$$(1-\varepsilon)(\rho c)_s\frac{\partial T^s}{\partial t} = (1-\varepsilon)k_s \Delta T^s - h(T^s - T^f),$$

$$\frac{\partial S}{\partial t} + \frac{1}{\varepsilon}v_i\frac{\partial S}{\partial x_i} = \kappa_c \Delta S,$$

(4.1)

where the notation is as in chapter 2 with specifically ε, v, K being the porosity, kinematic viscosity of the fluid, permeability, and κ_c is the salt diffusion coefficient.

[264] study the instability of the steady state solution with $v_i \equiv 0$ and the temperatures and the salt concentration given by

$$T^f = -\frac{\Delta T}{d}z + T_L = T^s,$$

$$S = -\frac{\Delta S}{d}z + S_L.$$

(4.2)

They restrict attention to two-dimensional disturbances and show that a perturbation to the steady solution (4.2) satisfies the non-dimensional perturbation equations

$$
\left(\frac{1}{Va}\frac{\partial}{\partial t}+1\right)\Delta\psi = -Ra\frac{\partial\theta}{\partial x}+Ra_s\frac{\partial\phi}{\partial x},
$$
$$
\left(\frac{\partial}{\partial t}-\Delta\right)\theta - \frac{\partial(\psi,\theta)}{\partial(x,z)}+\frac{\partial\psi}{\partial x} = H(\phi-\theta),
$$
$$
\left(\alpha_1\frac{\partial}{\partial t}-\Delta\right)\phi = -\gamma H(\phi-\theta),
$$
$$
\left(\frac{\partial}{\partial t}-\frac{1}{Le}\Delta\right)\mathscr{S} - \frac{\partial(\psi,\mathscr{S})}{\partial(x,z)}+\frac{\partial\psi}{\partial x} = 0,
$$

(4.3)

where ψ,θ,ϕ are the perturbation streamfunction, perturbation fluid temperature field, perturbation solid temperature field, and $\mathscr{S}$ is the perturbation salt field. The quantity Va is the Vadasz number, Le is the Lewis number, α_1 is the ratio of the fluid to solid thermal diffusivity, and $\partial(\psi,\theta)/\partial(x,z)$ denotes the Jacobian. The Rayleigh number and salt Rayleigh number are given by

$$
Ra = \frac{\alpha g\Delta T\, dK(\rho c)_f}{\varepsilon v k_f}, \qquad Ra_s = \frac{\alpha_c g\Delta S\, dK(\rho c)_f}{\varepsilon v k_f}.
$$

The boundary conditions for (4.3) are that $\psi,\theta,\phi,\mathscr{S}$ are zero on $z = 0,1$, and periodicity in the horizontal coordinate x.

[264] study both stationary and oscillatory convection, with the latter definitely occurring in this case. They also develop an asymptotic analysis for small H. They obtain the analytical expression

$$
Ra_{stat} = \frac{[\Lambda + H(\gamma+1)][\Lambda^2 + a^2 LeRa_s]}{a^2(\Lambda + \gamma H)}
$$

where a is the wavenumber and $\Lambda = \pi^s + a^2$, for the stationary convection Rayleigh number, while the oscillatory convection Rayleigh number expression is complicated and needs numerical evaluation. [264] report extensive numerical results and they also analyse weakly nonlinear finite amplitude solutions. They show that when the Vadasz number is sufficiently large then oscillatory convection will occur. The interaction coefficient H essentially stabilizes the steady solution, although oscillatory convection may still occur.

4.2 Brinkman Theory with Inertia

[258] analyze the analogous instability problem to that of section 4.1 but they employ the theory of Brinkman rather than that of Darcy. They again analyse the heated-salted below case. The equations they employ are

$$\frac{1}{\varepsilon}\frac{\partial v_i}{\partial t} = -\frac{1}{\rho_f}\frac{\partial p}{\partial x_i} - \frac{v}{K}v_i + v\Delta v_i$$

$$- k_i\left[1 - \alpha(T^f - T_L) + \alpha_c(S - S_L)\right],$$

$$\frac{\partial v_i}{\partial x_i} = 0,$$

$$\varepsilon(\rho c)_f\frac{\partial T^f}{\partial t} + (\rho c)_f v_i\frac{\partial T^f}{\partial x_i} = \varepsilon k_f\Delta T^f + h(T^s - T^f),$$ (4.4)

$$(1 - \varepsilon)(\rho c)_s\frac{\partial T^s}{\partial t} = (1 - \varepsilon)k_s\Delta T^s - h(T^s - T^f),$$

$$\frac{\partial S}{\partial t} + \frac{1}{\varepsilon}v_i\frac{\partial S}{\partial x_i} = \kappa_c\Delta S.$$

[258] again find (4.2) as the steady solution and analyse two-dimensional perturbations. The non-dimensional perturbation equations are

$$\left(\frac{1}{Va}\frac{\partial}{\partial t} + 1\right)\Delta\psi - Da\Delta^2\psi = -Ra\frac{\partial\theta}{\partial x} + Ra_s\frac{\partial\phi}{\partial x},$$

$$\left(\frac{\partial}{\partial t} - \Delta\right)\theta - \frac{\partial(\psi,\theta)}{\partial(x,z)} + \frac{\partial\psi}{\partial x} = H(\phi - \theta),$$

$$\left(\alpha_1\frac{\partial}{\partial t} - \Delta\right)\phi = -\gamma H(\phi - \theta),$$ (4.5)

$$\left(\frac{\partial}{\partial t} - \frac{1}{Le}\Delta\right)\mathscr{S} - \frac{\partial(\psi,\mathscr{S})}{\partial(x,z)} + \frac{\partial\psi}{\partial x} = 0,$$

where Da is the Darcy number, but now they employ stress free boundary conditions so that

$$\psi = \frac{\partial^2\psi}{\partial z^2} = \theta = \phi = \mathscr{S} = 0, \qquad \text{on } z = 0, 1,$$

together with periodicity in the x-direction.

[258] develop analyses of stationary convection, oscillatory convection, an asymptotic analysis for H small, and a weakly nonlinear analysis. Their findings are not dissimilar to those reported in section 4.1 but now the Darcy number also stabilizes.

4.3 Double Diffusion with Rotation

The problem of instability in double diffusion with rotation in a Darcy theory was tackled by [257]. This is effectively analysing the joint effects presented in sections 4.1 and 3.3. The analogous problem of double diffusion with rotation in a Brinkman theory was analysed by [265]. This thus combines the effects of section 4.2 and 3.4. Both sets of writers [257] and [265] included inertia and studied the heated-salted below problem when the porous layer is undergoing a rotation about the vertical axis with constant angular rotation.

The equations studied by [257] are

$$\frac{1}{\varepsilon}\frac{\partial v_i}{\partial t} = -\frac{1}{\rho_f}\frac{\partial p}{\partial x_i} - \frac{v}{K}v_i - \frac{2}{\varepsilon}(\Omega \times \mathbf{v})_i$$

$$- k_i[1 - \alpha(T^f - T_L) + \alpha_c(S - S_L)],$$

$$\frac{\partial v_i}{\partial x_i} = 0,$$

$$\varepsilon(\rho c)_f\frac{\partial T^f}{\partial t} + (\rho c)_f v_i\frac{\partial T^f}{\partial x_i} = \varepsilon k_f \Delta T^f + h(T^s - T^f),$$

$$(1 - \varepsilon)(\rho c)_s\frac{\partial T^s}{\partial t} = (1 - \varepsilon)k_s\Delta T^s - h(T^s - T^f),$$

$$\frac{\partial S}{\partial t} + \frac{1}{\varepsilon}v_i\frac{\partial S}{\partial x_i} = \kappa_c\Delta S,$$

(4.6)

where $\Omega = (0, 0, \Omega)$ is the angular velocity of the layer.

The equations studied by [265] are

$$\frac{1}{\varepsilon}\frac{\partial v_i}{\partial t} = -\frac{1}{\rho_f}\frac{\partial p}{\partial x_i} - \frac{v}{K}v_i + v\Delta v_i - \frac{2}{\varepsilon}(\Omega \times \mathbf{v})_i$$

$$- k_i[1 - \alpha(T^f - T_L) + \alpha_c(S - S_L)],$$

$$\frac{\partial v_i}{\partial x_i} = 0,$$

$$\varepsilon(\rho c)_f\frac{\partial T^f}{\partial t} + (\rho c)_f v_i\frac{\partial T^f}{\partial x_i} = \varepsilon k_f \Delta T^f + h(T^s - T^f),$$

$$(1 - \varepsilon)(\rho c)_s\frac{\partial T^s}{\partial t} = (1 - \varepsilon)k_s\Delta T^s - h(T^s - T^f),$$

$$\frac{\partial S}{\partial t} + \frac{1}{\varepsilon}v_i\frac{\partial S}{\partial x_i} = \kappa_c\Delta S.$$

(4.7)

The boundary conditions employed by [265] are those appropriate to two stress free surfaces.

[257] and [265] analyse stationary convection, oscillatory convection, and a weakly nonlinear instability, in much detail. The effect of rotation is to stabilize. There are many parameters to consider, in particular the effects of H, Va, Ra_s, Da and rotation, on the critical Rayleigh number of instability. Many such numerical results depicting the effect of these parameters on the critical instability threshold are discussed in detail in [257] and [265].

The problem of double diffusive convection in a viscoelastic fluid of Oldroyd B type, taking into account LTNE effects, is considered by [266]. This work is discussed in chapter 8 along with other viscoelastic fluid work.

4.4 Double Diffusion with Reaction

[90] study a double diffusion problem in a Darcy porous material with LTNE effects. While they employ zero inertia the key new element to their work is the inclusion of a chemical reaction term in the equation for the solute concentration. The analogous problem without LTNE effects was analysed by [351].

The geometric configuration is again one of a Darcy porous material confined in a horizontal layer between the two planes $z = 0$ and $z = d$ with gravity in the negative $z-$direction. We adopt notation in keeping with the rest of this chapter rather than that of [90] and in our notation their equations are

$$v_i = -\frac{K}{\mu}\frac{\partial p}{\partial x_i} - \frac{Kg}{\mu}k_i\big[\rho_0 - \alpha(T^f - T_L) + \alpha_c(C - C_L)\big],$$

$$\frac{\partial u}{\partial x} + \frac{\partial w}{\partial z} = 0,$$

$$\varepsilon(\rho c)_f\frac{\partial T^f}{\partial t} + (\rho c)_f v_i \frac{\partial T^f}{\partial x_i} = \varepsilon k_f \Delta T^f + h(T^s - T^f), \qquad (4.8)$$

$$(1-\varepsilon)(\rho c)_s\frac{\partial T^s}{\partial t} = (1-\varepsilon)k_s\Delta T^s - h(T^s - T^f),$$

$$\varepsilon\frac{\partial C}{\partial t} + v_i\frac{\partial C}{\partial x_i} = \varepsilon\kappa_c\Delta C + \bar{k}\big[C_{eq}(T^f) - C\big],$$

where $C(\mathbf{x}, t)$ is the solute concentration and [90] work in two spatial dimensions (x, z) so that $\mathbf{v} = (u, w)$. The key term in equations (4.8) is the chemical reaction term $\bar{k}\big[C_{eq}(T^f) - C\big]$. The values of C and T on the boundaries are $C = C_L, T = T_L, z = 0$, and $C = C_U, T - T_U, z = d$, for constants C_L, T_L, C_U and T_U.

A key assumption in [90] is that the chemical equilibrium term $C_{eq}(T^f)$ is linear in T^f and in the steady state equation (4.8)$_5$ reduces to

$$C_{eq}(\bar{T}_f) = \bar{C},$$

where $\bar{C}$ and $\bar{T}^f$ denote the solution in the steady state. This allows [90] to have the steady solution $v_i = 0$, together with

$$\bar{T}^f = \bar{T}^s = T_L + \frac{(T_U - T_L)}{d}z, \qquad \bar{C} = C_L + \frac{(C_U - C_L)}{d}z. \qquad (4.9)$$

If the above assumptions were not made then the temperature fields could still be linear in z but one would be faced with solving the following equation for $\bar{C}(z)$ in the steady state,

$$\frac{d^2\bar{C}}{dz^2} = \frac{\bar{k}}{\varepsilon\kappa_c}\bar{C} + az + b,$$

for constants a and b. This would yield a steady state for $\bar{C}$ which is exponentially varying in z and not the linear solution as in (4.9).

[90] derive non-dimensional perturbation equations for a two space dimensional perturbation to (4.9) and these are

$$
\Delta \psi = -Ra \frac{\partial \theta}{\partial x} + Ra_c \frac{\partial \mathscr{C}}{\partial x},
$$
$$
\frac{\partial \theta}{\partial t} + \frac{\partial \psi}{\partial x} + \frac{\partial (\psi, \theta)}{\partial (x, z)} = \Delta \theta + H(\phi - \theta),
$$
$$
\alpha_1 \frac{\partial \phi}{\partial t} = \Delta \phi - \gamma H(\phi - \theta), \qquad (4.10)
$$
$$
\frac{\partial \mathscr{C}}{\partial t} + \frac{\partial \psi}{\partial x} + \frac{\partial (\psi, \mathscr{C})}{\partial (x, z)} = \frac{1}{Le} \Delta \mathscr{C} + k(\theta - \mathscr{C}).
$$

Here ψ is the streamfunction for the velocity perturbation, θ, ϕ and $\mathscr{C}$ are the perturbations to T^f, T^s and C, $\Delta = \partial^2 / \partial x^2 + \partial^2 / \partial z^2$, and Ra and Ra_c are the Rayleigh and solute Rayleigh numbers given by

$$
Ra = \frac{\alpha (T_L - T_U) g d K}{\mu \varepsilon \kappa_f}, \qquad Ra_c = \frac{\alpha_c (C_L - C_U) g d K}{\mu \varepsilon \kappa_f}.
$$

[90] analyse the linear instability from system (4.10) in some detail. They obtain analytical expressions for both stationary and oscillatory convection which allow them to determine the critical Rayleigh number for linear instability. They analyse carefully the effect of the non-dimensional reaction term k. [90] also provide asymptotic results for small and large values of both H and k, and perform a weakly nonlinear stability analysis. Various numerical results are presented and for the parameter values they present the presence of the reaction destabilizes in the sense that as k decreases the critical Rayleigh number decreases tending to a constant for large values of k.

Chapter 5
Vertical Porous Convection with LTNE

In this chapter we describe work of [366] and of [393] on thermal convection in a Darcy porous material in a vertical layer subject to different temperatures on the vertical walls.

For the case of a single temperature the problem where the porous medium occupies a vertical layer subject to differential heating from one side to the other has attracted much attention. This problem has much application to insulation, such as in the area of building design, and is of importance in window double glazing. For the single temperature situation the first proof that a vertical porous slab of Darcy type which is held at fixed but different temperatures on the vertical walls is stable to perturbations from the equilibrium state is due to [156]. His analysis is based on the linear theory and analyses the two-dimensional problem. [409] further analysed this problem but in three-dimensions and he treated the completely nonlinear situation by employing energy-like integral methods. In particular, [409] produced a threshold for the Rayleigh number which guarantees global nonlinear stability, i.e. regardless of how large the initial perturbation may be. He also employed a generalized energy method to demonstrate that the result of [156] is true in the fully three-dimensional case although the initial data must be restricted. Articles dealing with other interesting aspects of convection in a vertical porous slab or pipe involving effects like double diffusion, and LTNE, include the papers by [32, 33, 36, 46, 211, 220, 221, 329, 330, 362] and [460]. The methods of energy stability techniques applied specifically to convection in a vertical porous slab are addressed in [144, 229] and [356], and these are discussed in detail in the book by [414, pp. 126–134].

The analogous problem in LTNE to that of [156] was first addressed by [366]. Specifically [366] dealt with the [156] problem of thermal convection in a vertical porous medium, but he employed the theory of local thermal non-equilibrium, allowing for different fluid and solid temperatures. The interesting work of [366] demonstrates that the vertical configuration is always stable according to linear theory, even with the much more complicated LTNE theory. [393] continued the problem of [366] but they analysed the complete nonlinear three-dimensional situation.

© Springer International Publishing Switzerland 2015
B. Straughan, *Convection with Local Thermal Non-Equilibrium and Microfluidic Effects*,
Advances in Mechanics and Mathematics 32, DOI 10.1007/978-3-319-13530-4_5

It is very important to do this since the linear theory only delivers information about instability and cannot address the problem of stability at all.

5.1 Vertical Layer

The theory of [366] and [393] begins with a vertical channel filled with a porous medium which is saturated by a viscous incompressible fluid. The layer extends infinitely in the vertical $z-$direction and is bounded by the walls at $x = \pm L/2$, see figure 5.1. The Boussinesq approximation is adopted as is Darcy's law, although the fluid and solid skeleton may have different temperatures. On the vertical walls $x = \pm L/2$ the horizontal speed is zero, i.e. $u = 0$, where the velocity field $\mathbf{v}$ is written as $\mathbf{v} = (u, v, w)$. The temperature T is kept constant at the value T_h on $x = L/2$ whereas it is kept at the constant value T_c on $x = -L/2$, for $T_h > T_c$. As elsewhere in this book sub or superscript f and s denote the fluid and solid components of the porous medium. The full three-dimensional equations with zero fluid inertia are, cf. equations (2.9)–(2.12),

$$
v_i = -\frac{K}{\mu}\frac{\partial p}{\partial x_i} + k_i \frac{\rho^f g \beta K}{\mu}(T^f - T_0),
$$

$$
\frac{\partial v_i}{\partial x_i} = 0,
$$

$$
\varepsilon(\rho c)_f \frac{\partial T^f}{\partial t} + (\rho c)_f v_i \frac{\partial T^f}{\partial x_i} = \varepsilon k^f \Delta T^f + h(T^s - T^f),
$$

$$
(1 - \varepsilon)(\rho c)_s \frac{\partial T^s}{\partial t} = (1 - \varepsilon)k^s \Delta T^s - h(T^s - T^f),
$$

(5.1)

where $\mathbf{v} = (u, v, w)$ is the velocity of the fluid, K is the permeability of the porous medium, β is the thermal expansion coefficient of the fluid, μ is the dynamic viscosity of the fluid, p is the pressure, ρ and c denote the density and the specific heat at constant pressure, k is the thermal conductivity, $T_0 = (T_h + T_c)/2$ is the reference temperature, and h is the interfacial heat transfer coefficient.

Equations (5.1) are written in non-dimensional form by introducing the following set of variables

$$
x_i = L x_i^*, \quad u_i = U u_i^*, \quad U = \frac{\varepsilon k^f}{(\rho c)^f L}, \quad p = p^* \frac{U L \mu}{K}, \quad \Delta T = T_h - T_c,
$$

$$
T^f = \theta \Delta T + T_0, \quad T^s = \phi \Delta T + T_0, \quad t = \frac{(\rho c)^f L^2}{k^f} t^*,
$$

$$
\alpha = \frac{(\rho c)^s}{(\rho c)^f}\frac{k^f}{k^s}, \quad \gamma = \frac{\varepsilon k^f}{(1 - \varepsilon)k^s}, \quad H = \frac{L^2 h}{\varepsilon k^f},
$$

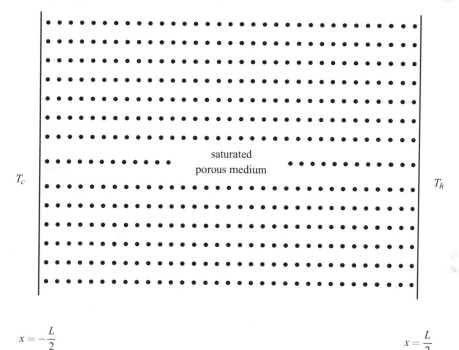

$$x = -\frac{L}{2} \qquad\qquad\qquad\qquad\qquad x = \frac{L}{2}$$

Fig. 5.1 Configuration for convection in a vertical slab of porous material with different temperatures on the vertical walls. The z−axis points upward while the x−axis points to the right. Gravity acts in the negative z−direction.

and the Rayleigh number, Ra, is introduced as

$$Ra = \frac{\rho^f g \beta K L (\rho c)^f \Delta T}{\mu \varepsilon k^f}. \qquad (5.2)$$

Note that θ and ϕ denote the temperature fields in the fluid and solid, respectively. Dropping stars one finds the following system of non-dimensional equations

$$\begin{aligned}
v_i &= -\frac{\partial p}{\partial x_i} + k_i Ra\theta, \\
\frac{\partial v_i}{\partial x_i} &= 0, \\
\frac{\partial \theta}{\partial t} + v_i \frac{\partial \theta}{\partial x_i} &= \Delta\theta + H(\phi - \theta), \\
\alpha \frac{\partial \phi}{\partial t} &= \Delta\phi - H\gamma(\phi - \theta).
\end{aligned} \qquad (5.3)$$

The boundary conditions to be satisfied are

$$u = 0, \qquad \theta = \pm\frac{1}{2}, \qquad \phi = \pm\frac{1}{2}, \qquad \text{on } x = \pm\frac{1}{2}. \qquad (5.4)$$

The steady solution to equations (5.3) and boundary conditions (5.4) which is analysed by [366] and by [393] is

$$\bar{u} = \bar{v} = 0, \qquad \bar{w} = Rax, \qquad \bar{\theta} = x, \qquad \bar{\phi} = x, \qquad (5.5)$$

where an overbar indicates the steady state. The equations governing the perturbation variables to this steady state solution are found to be

$$
\begin{aligned}
&u_i = -\frac{\partial \pi}{\partial x_i} + k_i Ra\theta, \\
&\frac{\partial u_i}{\partial x_i} = 0, \\
&\frac{\partial \theta}{\partial t} + u_i \frac{\partial \theta}{\partial x_i} + u + Rax\frac{\partial \theta}{\partial z} = \Delta\theta + H(\phi - \theta), \\
&\alpha \frac{\partial \phi}{\partial t} = \Delta\phi - H\gamma(\phi - \theta),
\end{aligned}
\qquad (5.6)
$$

and the boundary conditions for this system are

$$u = \theta = \phi = 0 \qquad \text{on } x = \pm\frac{1}{2}. \qquad (5.7)$$

[366] demonstrates from equations (5.6) that the steady state (5.5) is always stable according to linear theory. We here concentrate on nonlinear stability results, but we note that the result of [366] also follows from the nonlinear result in section 5.1.2.

5.1.1 Global Stability

[393] define a convection cell, V, in which $x \in (-1/2, 1/2)$ and in which the solution to equations (5.6) and (5.7) is periodic in y and z. They then let $\|\cdot\|$ and $(\cdot, \cdot)$ be the norm and inner product on $L^2(V)$. The idea is to construct energy relations by multiplying equation (5.6)$_1$ by u_i, equation (5.6)$_3$ by θ, and equation (5.6)$_4$ by ϕ, and they then integrate each over the cell V. Using integration by parts, the boundary conditions and periodicity one may then show that

$$
\begin{aligned}
&0 = -\|\mathbf{u}\|^2 + Ra(\theta, w), \\
&\frac{1}{2}\frac{d}{dt}\|\theta\|^2 = -\|\nabla\theta\|^2 + H(\phi - \theta, \theta) - (u, \theta), \\
&\frac{\alpha}{2}\frac{d}{dt}\|\phi\|^2 = -\|\nabla\phi\|^2 - \gamma H(\phi - \theta, \phi).
\end{aligned}
\qquad (5.8)
$$

[393] let $\lambda > 0$ be a coupling parameter to be chosen judiciously and they form the combination $\lambda(5.8)_1 + \gamma(5.8)_2 + (5.8)_3$ to derive the single equation

$$\frac{1}{2}\frac{d}{dt}\left(\gamma\|\theta\|^2 + \alpha\|\phi\|^2\right) = -\lambda\|\mathbf{u}\|^2 - \gamma\|\nabla\theta\|^2 - \|\nabla\phi\|^2 + Ra\lambda(\theta, w)$$
$$- \gamma(u, \theta) - \gamma H\|\phi - \theta\|^2. \tag{5.9}$$

They then employ the arithmetic-geometric mean inequality together with Poincaré's inequality to deduce from equation (5.9)

$$\frac{1}{2}\frac{d}{dt}\left(\gamma\|\theta\|^2 + \alpha\|\phi\|^2\right) \le -\left(\lambda - \frac{\gamma\xi}{2}\right)\|u\|^2 - \lambda\|v\|^2 - \pi^2\|\phi\|^2$$
$$- \left(\gamma\pi^2 - \frac{Ra\lambda}{2\eta} - \frac{\gamma}{2\xi}\right)\|\theta\|^2$$
$$- \left(\lambda - \frac{Ra\lambda\eta}{2}\right)\|w\|^2$$
$$- \gamma H\|\phi - \theta\|^2. \tag{5.10}$$

[393] choose $\xi = 2\lambda/\gamma$ and $\eta = 2/Ra$ and they require λ such that

$$0 < \lambda_- < \lambda < \lambda_+$$

where λ_- and λ_+ are given by

$$\lambda_\pm = \frac{2\gamma\pi^2}{Ra^2} \pm \frac{\gamma}{Ra^2}\sqrt{4\pi^2 - Ra^2},$$

which in turn requires $Ra < 2\pi$.

The energy function selected by [393] is

$$E(t) = \frac{\gamma}{2}\|\theta\|^2 + \frac{\alpha}{2}\|\phi\|^2,$$

and they show that this function satisfies the inequality

$$\frac{dE}{dt} \le -cE, \tag{5.11}$$

where c is the constant, $c = \min\{2\pi^2/\alpha, 2\omega/\alpha\}$, with ω being defined by $\omega = \gamma\pi^2 - (Ra\lambda/2\eta) - (\gamma/2\xi)$. Inequality (5.11) is integrated to yield

$$E(t) \le e^{-ct}E(0). \tag{5.12}$$

Thus [393] deduce that both θ and ϕ decay at least exponentially in L^2 measure. They further use the Cauchy-Schwarz inequality on equation (5.8)$_1$ to show that

$$\|\mathbf{u}\|^2 \le Ra^2\|\theta\|^2.$$

Hence they deduce that u_i likewise decays at least exponentially in L^2 norm. Therefore, [393] conclude that if $R < 2\pi^2$ then the steady solution (5.5) is stable to perturbations of any amplitude.

5.1.2 Nonlinear Stability for All Rayleigh Numbers

To establish the analog of the result of [393, 409] employed a generalized "energy" function. They take curl curl of equation $(5.6)_1$ and retain only the first component which yields

$$-\Delta u = Ra\,\frac{\partial^2 \theta}{\partial x \partial z}.$$
(5.13)

[393] employ equations $(5.6)_{3,4}$ to derive the identities

$$\frac{1}{2}\frac{d}{dt}\|\nabla\theta\|^2 = -\|\Delta\theta\|^2 + \int_V u_i\theta_{,i}\Delta\theta\,dV - H(\phi-\theta,\Delta\theta)$$
(5.14)

and

$$\frac{\alpha}{2}\frac{d}{dt}\|\nabla\phi\|^2 = -\|\Delta\phi\|^2 + \gamma H(\phi-\theta,\Delta\phi).$$
(5.15)

The same writers bound the nonlinear cubic term in equation (5.14) as follows

$$\int_V u_i\theta_{,i}\Delta\theta\,dV \le \sup_V |\mathbf{u}|\,\|\nabla\theta\|\,\|\Delta\theta\|,$$

and they then form the combination $\gamma(5.14)+(5.15)$. In this way they are able to deduce that

$$\frac{1}{2}\frac{dF}{dt} \le -\gamma(1-CRa\|\nabla\theta\|)\|\Delta\theta\|^2 - \|\Delta\phi\|^2,$$
(5.16)

where the function F is defined by

$$F = \gamma\|\nabla\theta\|^2 + \alpha\|\nabla\phi\|^2,$$

and C is a constant which arises through use of an embedding inequality. [393] estimate $\|\nabla\theta\|$ as

$$\|\nabla\theta\| \le \frac{1}{\sqrt{\gamma}}F^{1/2},$$

and then use this in inequality (5.16) to obtain

$$\frac{1}{2}\frac{dF}{dt} \le -\gamma\Big(1-\frac{CR}{\sqrt{\gamma}}F^{1/2}\Big)\|\Delta\theta\|^2 - \|\Delta\phi\|^2.$$
(5.17)

At this point [393] consider two cases, namely $\alpha \le 1$ and $\alpha > 1$. In the case $\alpha \le 1$ they suppose that

$$\sqrt{F(0)} < \frac{\sqrt{\gamma}}{RaC},$$
(5.18)

and they show that from inequality (5.17) one may deduce that F further satisfies the inequality

$$\frac{1}{2}\frac{dF}{dt} \leq -bF \qquad (5.19)$$

where the constant b is given by $b = 1 - RaCF^{1/2}(0)/\gamma^{1/2} > 0$. Thus, F satisfies

$$F(t) \leq F(0)\exp(-2bt).$$

Then [393] show that provided the initial data restriction (5.18) holds, $\|\nabla\theta\|$, $\|\nabla\phi\|$ and $\|\mathbf{u}\|$ decay at least exponentially.

[393] establish a similar result when $\alpha > 1$ except that they then require the initial data restriction

$$\sqrt{F(0)} < \frac{\sqrt{\gamma}}{\alpha RaC}. \qquad (5.20)$$

These writers present values for a typical laboratory situation to argue that one might expect $\alpha < 1$.

Two other analyses of LTNE thermal convection in a vertical channel are certainly worth highlighting. These are by [460] and by [211].

The paper by [460] studies instability of the situation of a vertical channel in the $z-$direction, of width L filled with a porous medium of Darcy type, with the layer contained between $x = 0$ and $x = L$. [460] has a constant temperature boundary condition on the vertical wall $x = L$. However, on the wall $x = 0$ he prescribes values of the partial heat fluxes in that he selects $-k_s\partial T^s/\partial x = q_0$ and $-k_f\partial T^f/\partial x = q_0$, for some value q_0. In particular, [460] shows that in this case one may have different functions for T^s and T^f even in the steady state, and these may be exponential as opposed to linear when constant temperatures are prescribed on both walls.

[211] analyse a vertical channel but they have a flow in the upward vertical direction. They allow the temperatures at the vertical walls to increase linearly with vertical height. They employ the LTNE equations with Brinkman and nonlinear terms present in the momentum equation. This situation is not obviously amenable to an analytic solution and so [211] employ a numerical solution method. Many numerical results are displayed in [211].

Chapter 6
Penetrative Convection with LTNE

A pioneering piece of work on penetrative convection is the beautiful paper of [468]. Our initial description of penetrative convection relies much on his paper.

In many natural phenomena the process of thermal convection involves a penetrative motion into a stably stratified fluid. Examples of penetrative convection are to be found in several areas of geo- and astrophysical fluid dynamics. For example, the Earth's atmosphere is bounded below by the ground or ocean; this bounding surface is heated by solar radiation, and the air close to the surface then becomes warmer than the upper air, and so a gravitationally unstable system results. When convection occurs, the warm air rises and penetrates into regions that are stably stratified.

In the oceans evaporation is the main cause of gravitational instability near the sea surface. As the cool surface water is carried downward, it too enters regions that are stably stratified. In a star the surface layer is stable. [468] explains that in the interior of a star the temperature gradient may rise to a value which is greater than the adiabatic gradient at locations which are well below the surface. The reason for the rise in the temperature gradient is adiabatic compression which causes negative hydrogen to form. The larger temperature gradient gives rise to an unstable region. The superadiabatic gradient so formed may extend deeply into the star's interior leading to a very high temperature. This high temperature may in turn cause the gas to be ionized which then results in loss of the superadiabatic gradient. The end result is an unstable layer with stable fluid both above and below.

To study penetrative convection on a laboratory scale [468] proposed a simplified model. He suggested one should consider a layer of water with the bottom maintained at 0°C and the top at a temperature greater than 4°C. Because the density of water below 4°C is a decreasing function of temperature, the above situation results in a gravitationally unstable layer of fluid lying below a stably stratified one. When convection occurs in the lower layer the motions will penetrate into the upper layer.

The phenomenon of penetrative convection is lucidly described by [468] who analysed in detail the linearized system and developed a weakly nonlinear finite amplitude analysis for two stress free boundaries. [468] finds that penetrative convection differs markedly from the classical case of Bénard convection for which sub-critical instabilities are not possible and he shows that a finite amplitude solu-

© Springer International Publishing Switzerland 2015

B. Straughan, *Convection with Local Thermal Non-Equilibrium and Microfluidic Effects*, Advances in Mechanics and Mathematics 32, DOI 10.1007/978-3-319-13530-4_6

tion exists for sub-critical Rayleigh numbers. He observes that a finite amplitude instability may commence for a Rayleigh number which is below that given by the linear instability threshold.

We discuss penetrative convection here in an LTNE setting employing a Veronis quadratic temperature in the buoyancy force. However, another mechanism to produce penetrative convection is via employment of a suitable internal heat source or heat sink.

Thermal convection in a plane layer in the presence of an internal heat source or sink is also a well studied phenomenon. In particular, the presence of a heat source or sink may give rise to the idea of penetrative convection where part of the layer has a tendency to move whereas the remainder of the layer will remain motionless until a certain point when movement in the rest of the layer "penetrates" into the stable layer and a resultant motion then ensues, cf. [48, 180, 181, 191, 216], and the references therein. [414], section 17.2, pp. 316–318, shows that the penetrative convection model with an internal heat sink is mathematically equivalent to the adjoint of the model with no heat sink but with a quadratic dependence on temperature in the buoyancy force in the momentum equation. However, the mathematical equations are in some ways simpler since no nonlinearities arise due to the nonlinear density in the buoyancy force.

The subject of penetrative convection is unquestionably one with immense application to problems in the real world at both micro and macro levels. For example, [414], chapter 17, discusses applications in geophysics, [180, 181] develop an application in solar pond design, [217, 231, 232] and [48] analyse applications in atmospheric physics, [289] consider applications in building design, [451] shows penetrative convection occurs in the Sun, [204] show that penetrative convection may be responsible for assisting the rise of volcanic plumes into the Earth's atmosphere, [216] discuss penetrative convection in the Laptev Sea coastal pycnocline layer, [465] investigate the deep mantle of exosolar planets while [191] analyse penetration in clouds surrounding Venus, [254] analyse penetrative convective clouds in connection with cloud to ground electrical discharges, and [353] study internal heat source convection when the heat source occurs in a porous biological material due to organic decay involving microbial activity.

As we mention above in connection with the work of [468] one way to describe penetrative convection mathematically is via using a density relation in the buoyancy force in the momentum equation, (2.9) where the density is not a linear function of the fluid temperature, T_f, as in equation (2.13), but is nonlinear such that the density may achieve a maximum value at an appropriate temperature. [386] has studied thermal convection in a porous medium using an LTNE model when the density is a nonlinear function of the fluid temperature, although he does not concentrate specifically on the aspect of fluid penetrating from one part of the layer into another. [386] also allows heat sources to be present in the temperature equations for T_f and T_s. [386] employs a Brinkman model with zero inertia and the density in the buoyancy force has form

$$\rho = \rho_0\left[1 - \beta_1(T_f - T_U) - \beta_2(T_f - T_U)^2\right]$$

for positive constants β_1, β_2. The fluid layer is held with constant temparatures on the upper and lower surfaces of a horizontal layer, the temperatures being T_U at the upper surface, and T_L at the lower, with $T_U < T_L$. [386] reports many numerical results for the linear instability problem, employing a Galerkin method in the numerical resolution of the eigenvalue problem.

6.1 Quadratic Density Model

We here consider the LTNE analogue of the classical [468] model for penetrative convection in a fluid, cf. the non-LTNE versions in chapter 17 of [414]. Due to the fact that we believe the LTNE effect may be greater with a larger porosity we here employ a Brinkman model.

To commence we employ the LTNE equations (2.9)–(2.12) for the fluid velocity, v_i, pressure, p, fluid temperature, T_f, and solid temperature, T_s, in the form

$$v_i - \hat{\lambda}\Delta v_i = -\frac{K}{\mu}p_{,i} - \frac{gk_i}{\mu}\rho(T_f),$$

$$v_{i,i} = 0,$$

$$\varepsilon(\rho c)_f T_{,t}^f + (\rho c)_f v_i T_{,i}^f = \varepsilon k_f \Delta T_f + h(T_s - T_f),$$

$$(1-\varepsilon)(\rho c)_s T_{,t}^s = (1-\varepsilon)k_s \Delta T_s - h(T_s - T_f),$$

(6.1)

where the notation is as in section 2.2 but now the density ρ is given by the expression

$$\rho = \rho_0\left[1 - \alpha(T_f - T_m)^2\right].$$

(6.2)

Here ρ_0, α are constants and T_m is the temperature at the maximum density. We now restrict attention to when the saturating fluid is water so that the maximum density occurs when $T_m \approx 4°C$.

The saturated porous medium is supposed contained in the horizontal layer between the planes $z = 0$ and $z = d$. The temperatures on the boundaries $z = 0, d$ are

$$T_f = T_s = T_U \quad \text{at } z = d, \qquad T_f = T_s = 0°C \quad \text{at } z = 0,$$

(6.3)

where the constant T_U is such that $T_U \geq T_m$. The boundary conditions on v_i are that

$$v_i = 0 \quad \text{at } z = 0 \quad \text{and at} \quad z = d.$$

(6.4)

For equations (6.1) subject to boundary conditions (6.3) and (6.4) we are interested in the stability of the steady solution

$$\bar{v}_i \equiv 0, \qquad \bar{T}_f(z) = \bar{T}_s(z) = \frac{T_U}{d}z,$$

(6.5)

where the steady pressure $\bar{p}(z)$ is found from (6.1)$_1$.

To examine the stability of this solution we introduce perturbations u_i, π, θ and ϕ such that

$$v_i = \bar{v}_i + u_i, \quad p = \bar{p} + \pi, \quad T_f = \bar{T}_f + \theta, \quad T_s = \bar{T}_s + \phi$$

and then one derives equations for the perturbations from (6.1). The resulting non-linear system of perturbation equations is now non-dimensionalized with the following set of variables

$$x_i = x_i^* d, \qquad u_i = u_i^* U, \qquad \pi = \pi^* P, \qquad \theta = \theta^* T^\sharp, \qquad \phi = \phi^* T^\sharp,$$

$$\lambda = \frac{T_m}{T_U}, \qquad U = \frac{\varepsilon k_f}{(\rho c)_f d}, \qquad t = t^* \mathcal{T}, \qquad \mathcal{T} = \frac{\rho_f c_f d^2}{k_f},$$

$$H = \frac{h d^2}{\varepsilon k_f}, \qquad \lambda_B = \frac{\hat{\lambda}}{d^2}, \qquad P = \frac{\mu d U}{K}, \qquad \varepsilon_1 = \frac{(\rho c)_s k_f}{(\rho c)_f k_s},$$

and with the temperature scale $T^\sharp$ and the Rayleigh number Ra defined by

$$T^\sharp = U \sqrt{\frac{(\rho c)_f d \mu}{\varepsilon k_f g \rho_0 \alpha}}$$

and

$$Ra = R^2 = T_U^2 \frac{(\rho c)_f d g \rho_0 \alpha}{\mu k_f \varepsilon}.$$

Dropping all stars, one may now show the non-dimensional variables u_i, π, θ and ϕ satisfy the equations

$$\begin{aligned}
&u_i - \lambda_B \Delta u_i = -\pi_{,i} - 2R(z - \lambda)\theta k_i + \theta^2 k_i, \\
&u_{i,i} = 0, \\
&\theta_{,t} + u_i \theta_{,i} = Rw + \Delta \theta + H(\phi - \theta), \\
&\varepsilon_1 \phi_{,t} = \Delta \phi - \gamma H(\phi - \theta),
\end{aligned} \qquad (6.6)$$

these equations holding in $\mathbb{R}^2 \times \{z \in (0,1)\} \times \{t > 0\}$, where $w = u_3$. The boundary conditions are

$$u_i = 0, \quad \theta = 0, \quad \phi = 0, \qquad \text{on} \quad z = 0, 1, \qquad (6.7)$$

together with the variables u_i, π, θ and ϕ satisfying a plane tiling periodicity in the x, y directions.

One may develop a linearized instability analysis for the problem (6.6) and (6.7) and solve the resulting eigenvalue problem numerically. Likewise one may develop a nonlinear energy stability analysis for (6.6) and (6.7) although care must be taken to remember to handle the nonlinear term θ^2 in (6.6)$_1$. If one is interested in a global nonlinear energy stability analysis then a method such as a one employing a weighted energy would need to be used, cf. [414, p. 342]. I am not aware of detailed numerical calculations of stability and/or instability for (6.6) and (6.7).

6.2 Internal Heat Sources

6.2.1 Uniform Heat Generation

[326] consider the problem of thermal convection in an LTNE porous medium of Darcy type. Their equations are

$$
v_i = -\frac{K}{\mu} p_{,i} + \frac{\rho_0 g \beta K k_i}{\mu}(T_f - T_0),
$$
$$
v_{i,i} = 0,
$$
$$
\varepsilon(\rho c)_f T_{,t}^f + (\rho c)_f v_i T_{,i}^f = \varepsilon k_f \Delta T_f + h(T_s - T_f) + \varepsilon q_f''' \cos\zeta,
$$
$$
(1-\varepsilon)(\rho c)_s T_{,t}^s = (1-\varepsilon)k_s \Delta T_s - h(T_s - T_f) + (1-\varepsilon)q_s''' \sin\zeta,
$$

(6.8)

where the notation is as in chapter 2 except for the q_f''' and q_s''' terms which represent constant heat sources. The parameter ζ varies from $\zeta = 0$, which represents heat generation in only the fluid phase, to $\zeta = \pi/2$, which represents heat generation in only the solid phase. The equations of [326] hold in the horizontal layer $\mathbb{R}^2 \times \{z \in (0,L)\} \times \{t > 0\}$ with prescribed values of T_f and T_s on $z = 0, L$ and $v_3 = 0$ at $z = 0, L$.

The basic solution is, in general, a complicated one involving exponentials of z for $T_f(z)$ and $T_s(z)$. In the particular cases of $\zeta = 0$ and $\zeta = \pi/2$ the basic solution is given by [326] as

$$
\bar{T}_f = \frac{\gamma\lambda^2 + 8}{8\gamma\lambda^2} - \frac{\cosh\lambda z}{\gamma\lambda^2\cosh(\lambda/2)} - \frac{z^2}{2}
$$
$$
\bar{T}_s = \frac{\lambda^2 - 8}{8\lambda^2} + \frac{\cosh\lambda z}{\lambda^2\cosh(\lambda/2)} - \frac{z^2}{2}
$$

for $\zeta = 0$, whereas

$$
\bar{T}_f = \frac{\lambda^2 - 8}{8\lambda^2} + \frac{\cosh\lambda z}{\lambda^2\cosh(\lambda/2)} - \frac{z^2}{2}
$$
$$
\bar{T}_s = \frac{\lambda^2 + \gamma}{8\lambda^2} - \frac{\gamma\cosh\lambda z}{\lambda^2\cosh(\lambda/2)} - \frac{z^2}{2}
$$

when $\zeta = \pi/2$. Here γ is the non-dimensional parameter $\gamma = \varepsilon k_f/(1-\varepsilon)k_s$ and $\lambda = L\sqrt{h(\gamma+1)/\varepsilon k_f}$.

[326] derive equations for the linear instability of the basic state of their problem. They show that exchange of stabilities holds since the linear operator attached to the linear instability problem is symmetric. They solve the eigenvalue problem which arises numerically by a fourth order compact difference method and they provide extensive numerical results. In addition, [326] provide several asymptotic results for the case where h is small.

The problem of thermal convection in an LTNE porous material of Brinkman type with a constant internal heat source acting only on the solid skeleton is studied by [40]. If the porous material were composed of a metallic foam so that the skeleton in the porous material conducted heat easily, this could be a good application for the work of [40].

6.2.2 Internal Heat Source Penetrative Convection

We now consider an LTNE model which gives rise to penetrative convection but when the density in the buoyancy force is linear in temperature, but the penetrative effect is achieved by suitable use of constant heat sinks.

The basic equations are those of (2.9)–(2.12) with the density in the buoyancy force given by

$$\rho = \rho_f \big(1 - \alpha[T_f - T_0]\big)$$

but heat sources are added to the fluid and solid temperature equations. Thus, our basic starting point is to begin with the equations

$$v_i - \hat{\lambda}\Delta v_i = -\frac{K}{\mu}p_{,i} + \frac{\rho_f g \alpha K k_i}{\mu}T_f,$$

$$v_{i,i} = 0, \tag{6.9}$$

$$\varepsilon(\rho c)_f T_{,t}^f + (\rho c)_f v_i T_{,i}^f = \varepsilon k_f \Delta T_f + \hat{h}(T_s - T_f) + Q_f,$$

$$(1 - \varepsilon)(\rho c)_s T_{,t}^s = (1 - \varepsilon)k_s \Delta T_s - \hat{h}(T_s - T_f) + Q_s,$$

where Q_f and Q_s are heat sinks, and we have used $\hat{h}$ rather than h as in section 6.1 to denote the interaction coefficient. In this case we employ h to be the depth of the layer with d the depth of that part of the layer which is potentially destabilizing.

We seek a steady state with $\bar{v}_i \equiv 0$, $\bar{T}_f(z) = \bar{T}_s(z)$. From equations (6.9) this results in requiring

$$\bar{T}_f'' = -\frac{Q_f}{\varepsilon k_f} \qquad \text{and} \qquad \bar{T}_s'' = -\frac{Q_s}{(1 - \varepsilon)k_s} \tag{6.10}$$

and so the heat sinks must be restricted by the relation

$$\frac{Q_f}{\varepsilon k_f} = \frac{Q_s}{(1 - \varepsilon)k_s}. \tag{6.11}$$

The porous material occupies the plane horizontal layer between $z = 0$ and $z = h$, with boundary conditions

$$\begin{aligned}
v_i &= 0 & &\text{on} & z &= 0, h, \\
T_s &= T_f = T_U & &\text{on} & z &= h, \\
T_s &= T_f = T_L & &\text{on} & z &= 0,
\end{aligned} \tag{6.12}$$

where T_U and T_L are constants with $T_L > T_U$. Suppose $Q_f < 0$ and then from (6.10) we find

$$\bar{T}_f = -\frac{Q_f}{2\varepsilon k_f} z^2 + \left[\frac{Q_f h}{2\varepsilon k_f} - \frac{\Delta T}{h} \right] z + T_L, \tag{6.13}$$

where $\Delta T = T_L - T_U > 0$. We now require the temperature to have a minimum in the steady state at some point in the interval $(0, h)$, say at $z = d$, so that $\bar{T}_f'(d) = 0$. Then, the situation is potentially unstable in the sub-layer $(0, d)$. The condition $\bar{T}_f'(d) = 0$ yields

$$d = \frac{h}{2} + \frac{\beta}{A_f}$$

where $\beta = \Delta T / h$ and $A_f = -Q_f / \varepsilon k_f$. Then $T_L > \bar{T}_f(d)$ yields the restriction $\Delta T < h^2 A_f / 2$. In this way we have a situation in which the layer is potentially unstable in the part $(0, d)$ and when instability occurs it may potentially penetrate into the upper region (d, h).

To study stability of the steady state (6.13) we introduce perturbations u_i, π, θ and ϕ such that $v_i = \bar{v}_i + u_i$, $p = \bar{p} + \pi$, $T_f = \bar{T}_f + \theta$, $T_s = \bar{T}_s + \phi$. Then from (6.9) one derives equations for the perturbations. Let the length scale be h, velocity scale U, temperature scale $T^\sharp$, pressure scale P, and the time scale $\mathscr{T}$, and then introduce the non-dimensional variables

$$A = \frac{\rho_s c_s k_f}{k_s \rho_f c_f} \qquad \gamma = \frac{\varepsilon k_f}{(1 - \varepsilon) k_s} \qquad H = \frac{h^2 \hat{h}}{\varepsilon k_f}$$

$$\mathscr{T} = \frac{(\rho c)_f h^2}{\varepsilon k_f} \qquad U = \frac{\varepsilon k_f}{(\rho c)_f h} \qquad \xi = \frac{\Delta T}{A_f h^2} + \frac{1}{2}$$

$$P = \frac{h \mu U}{K} \qquad \lambda_B = \frac{\hat{\lambda}}{h^2} \qquad T^\sharp = U \sqrt{\frac{c_f h^3 A_f \mu}{2 \varepsilon k_f g \alpha K}}.$$

Introduce the Rayleigh number, Ra, as

$$Ra = R^2 = \frac{h^3 A_f g \alpha K (\rho c)_f \rho_f}{2 k_f \mu_f \varepsilon}.$$

The non-dimensional perturbation equations may then be written as

$$\begin{aligned}
&u_i - \lambda_B \Delta u_i = -\pi_{,i} + R \theta k_i, \\
&u_{i,i} = 0, \\
&\theta_{,t} + u_i \theta_{,i} = 2R(\xi - z) w + \Delta \theta + H(\phi - \theta), \\
&A \varepsilon \phi_{,t} = \Delta \phi - \gamma H(\theta - \phi),
\end{aligned} \tag{6.14}$$

where $w = u_3$, with these equations holding in $\mathbb{R}^2 \times \{z \in (0, 1)\} \times \{t > 0\}$.

The boundary conditions are

$$u_i = 0, \quad \theta = 0, \quad \phi = 0, \quad \text{on} \quad z = 0, 1, \tag{6.15}$$

with u_i, π, θ and ϕ satisfying a plane tiling periodicity in the x, y plane.

The equations (6.14) and (6.15) may be employed to develop a linear instability analysis, or to develop a fully nonlinear energy stability analysis. I am unaware as to where this has been done, but the results may well be revealing.

It is worth pointing out that the linear operator attached to (6.14) and (6.15) is the adjoint of the one for equations (6.6) and (6.7) developed in section 6.1. To see this one has to identify ξ with λ.

Chapter 7
LTNE and Multi-layers

7.1 Local Thermal Non-equilibrium Equations

Very interesting recent work of [332] and [367] has derived linear instability thresholds for the problem of thermal convection in a horizontal layer heated below where the layer is composed of identical finite thickness horizontal layers of a Darcy porous material saturated by an incompressible fluid interleaved with identical horizontal heat conducting solid layers. The work of [367] employed identical infinitesimal solid layers whereas [332] used identical layers of finite thickness. The work of [367] and [332] is in some ways continuing the fundamental work of [283, 284], and of [285–287] on thermal convection in layered porous media. Undoubtedly the work of [332] will have much application in modern heat transfer systems especially since such devices are now prevalent in the microfluidic industry. Interestingly, [123] provides a brief study of such a convection problem and he draws attention to applications in the thermal insulation of tanks of cryogenic liquids. Such tanks are widely used for fuel storage involving space research and travel, see e.g. [195].

In this section we describe work of [432] who considers the equivalent thermal convection problem to that studied in [332], but he allows the porous layers to be of local thermal non-equilibrium type. Thus we consider a horizontal layer contained between the planes $z = 0$ and $z = d$ which is comprised of alternating layers of a rigid solid and a local thermal non-equilibrium fluid saturated Darcy porous material. The solid layers each have the same thickness as do each of the layers of porous material, although the two sets of materials may have different thicknesses. The gravity field is acting in the downward (negative z) direction. For simplicity we restrict attention to a two layer configuration with the upper layer contained in $z \in (h_s, d)$ being the local thermal non-equilibrium porous material, whereas the lower layer $z \in (0, h_s)$ is composed of a rigid solid. To describe the heated below thermal convection problem for the above scenario we begin with the basic equations for thermal convection with local thermal non-equilibrium effects. For a Darcy porous medium they are,

© Springer International Publishing Switzerland 2015
B. Straughan, *Convection with Local Thermal Non-Equilibrium and Microfluidic Effects*,
Advances in Mechanics and Mathematics 32, DOI 10.1007/978-3-319-13530-4_7

see section (2.2),

$$v_i = -\frac{K}{\mu}\frac{\partial p}{\partial x_i} + \frac{\rho_f g \alpha K}{\mu} T_f k_i,$$

$$\frac{\partial v_i}{\partial x_i} = 0,$$

$$\varepsilon(\rho c)_f \frac{\partial T^f}{\partial t} + (\rho c)_f v_i \frac{\partial T^f}{\partial x_i} = \varepsilon k_f \Delta T_f + h(T_s - T_f), \tag{7.1}$$

$$(1-\varepsilon)(\rho c)_s \frac{\partial T^s}{\partial t} = (1-\varepsilon)k_s \Delta T_s - h(T_s - T_f),$$

these holding in $\mathbb{R}^2 \times \{z \in (h_s,d)\} \times \{t > 0\}$. In these equations v_i, p, T^f, T^s are velocity, pressure, fluid temperature, solid skeleton temperature, a sub or superscript f denotes the fluid component whereas sub or superscript s denotes the component of the solid skeleton. The variables $K, \mu, \rho, g, \alpha, \varepsilon, c, k$ and h are permeability, dynamic viscosity of the fluid, density, gravity, thermal expansion coefficient of the fluid, porosity, specific heat at constant pressure, thermal conductivity, and an interaction coefficient, respectively.

Let us denote by T, ρ, c and k the temperature, density, specific heat, and the thermal conductivity of the rigid solid occupying the horizontal region between the planes $z = 0$ and $z = h_s$, and then the relevant equation for this region is

$$\rho c \frac{\partial T}{\partial t} = k \Delta T, \tag{7.2}$$

this equation holding in the domain $\mathbb{R}^2 \times \{z \in (0,h_s)\} \times \{t > 0\}$.

Set $\mathbf{v} = (u,v,w)$ and then the appropriate boundary conditions are

$$\begin{aligned}
w &= 0, \quad z = h_s, d; & T &= T_L, \quad z = 0; \\
T_f &= T_U, \quad T_s = T_U, \quad z = d; \\
T_s &= T, \quad T_f = T, \quad \text{on } z = h_s; \\
\varepsilon k_f \frac{\partial T^f}{\partial z} &+ (1-\varepsilon)k_s \frac{\partial T^s}{\partial z} = k \frac{\partial T}{\partial z} \quad \text{on} \quad z = h_s.
\end{aligned} \tag{7.3}$$

Here T_L and T_U are constants with $T_L > T_U$. Since the lower region has only one temperature field whereas the upper one has two the interface, $z = h_s$, has to be treated with care, cf. [310, 313]. We suppose that both T_f and T_s are continuous with T, the temperature of the solid, at the interface. On the microscopic level it is certainly true that $T_s = T$ and $T_f = T$ at the interface, and we here assume it is also true in the continuum model. While we require continuity of the separate temperature fields T_f and T_s with T at the interface $z = h_s$, one may be tempted to pose only an average condition on the temperature fields, such as $\varepsilon T_f + (1-\varepsilon)T_s = T$ at the interface. However, one condition like this is not sufficient to make the problem determinate. The interface condition $(7.3)_4$ represents continuity of the heat flux across $z = h_s$. The configuration for convection is shown in figure 7.1.

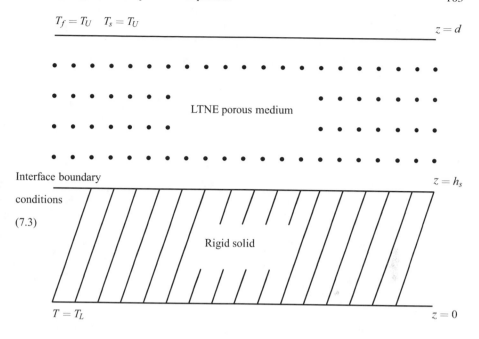

$T_f = T_U$ $\quad T_s = T_U$ $\qquad\qquad\qquad\qquad\qquad\qquad\qquad\qquad z = d$

LTNE porous medium

Interface boundary $\qquad\qquad\qquad\qquad\qquad\qquad\qquad\qquad z = h_s$

conditions

(7.3)

Rigid solid

$T = T_L$ $\qquad\qquad\qquad\qquad\qquad\qquad\qquad\qquad\qquad z = 0$

Fig. 7.1 Local thermal non-equilibrium convection in a horizontal porous layer with a rigid solid layer below, as governed by equations (7.1) and (7.2)

We require to find a steady solution of form $\bar{v}_i \equiv 0$, $\bar{T}^f = \bar{T}^f(z)$, $\bar{T}^s = \bar{T}^s(z)$ and $\bar{T} = \bar{T}(z)$. This is found to be

$$\begin{aligned}
\bar{T}^f = \bar{T}^s &= \beta + \alpha z, \qquad z \in (h_s, d), \\
\bar{T} &= \alpha_1 z + T_L, \qquad z \in (0, h_s),
\end{aligned} \tag{7.4}$$

where the coefficients α, α_1 and β are given by

$$\alpha = -\Delta T \frac{k}{kh + \hat{k}h_s}, \qquad \alpha_1 = -\Delta T \frac{\hat{k}}{kh + \hat{k}h_s},$$

$$\beta = T_U + \frac{kd\Delta T}{kh + \hat{k}h_s},$$

with $\hat{k}$ and ΔT defined by

$$\hat{k} = \varepsilon k_f + (1 - \varepsilon)k_s, \qquad \Delta T = T_L - T_U > 0.$$

To study the stability of solution (7.4) put $v_i = \bar{v}_i + u_i$, $p = \bar{p} + \pi$, $T^f = \bar{T}^f + \theta$, $T^s = \bar{T}^s + \phi$ and $T = \bar{T} + \psi$ and then from equations (7.1)–(7.4) one derives the equations and boundary conditions governing the perturbation variables. [432] employs non-dimensional variables in order to derive a non-dimensional form of the perturbation equations. The new non-dimensional variables utilized in [432] are defined in the expressions

$$u_i = U u_i^*, \qquad x_i = x_i^* d, \qquad \pi = P \pi^*, \qquad \theta = \theta^* T^\sharp,$$

$$\phi = \phi^* T^\sharp, \qquad \psi = \psi^* T^\sharp, \qquad t = t^* \mathscr{T}, \qquad P = \frac{\mu d U}{K},$$

$$U = \frac{\varepsilon k_f}{d(\rho c)_f}, \qquad \mathscr{T} = \frac{(\rho c)_f}{k_f} d^2, \qquad \gamma = \frac{\varepsilon k_f}{(1-\varepsilon)k_s},$$

$$H = \frac{h d^2}{\varepsilon k_f}, \qquad b = \frac{k}{\varepsilon k_f},$$

$$A = \frac{\rho_s c_s k_f}{k_s \rho_f c_f}, \qquad B = \frac{\rho c}{\varepsilon \rho_f c_f},$$

$$T^\sharp = U d \sqrt{\frac{c_f k \Delta T \mu}{\varepsilon k_f (kh + \hat{k} h_s) g \alpha K}}.$$

The Rayleigh number, Ra, employed by [432] has form, where R is also explicitly included since it appears directly in the equations,

$$Ra = R^2 = d^2 \rho_f^2 \frac{c_f k g \alpha K \Delta T}{\varepsilon \mu k_f (kh + \hat{k} h_s)}.$$

Drop the stars and in this case the non-dimensional perturbation equations may be shown to have the form

$$u_i = -\frac{\partial \pi}{\partial x_i} + R k_i \theta,$$

$$\frac{\partial u_i}{\partial x_i} = 0,$$

$$\frac{\partial \theta}{\partial t} + u_i \frac{\partial \theta}{\partial x_i} = R w + \Delta \theta + H(\phi - \theta), \tag{7.5}$$

$$\frac{A}{\gamma} \frac{\partial \phi}{\partial t} = \frac{1}{\gamma} \Delta \phi - H(\phi - \theta),$$

in the domain $\mathbb{R}^2 \times \{z \in (h_s/d, 1)\} \times \{t > 0\}$. In addition one requires the equation governing the temperature in the rigid solid and this is,

$$B \frac{\partial \psi}{\partial t} = b \Delta \psi, \tag{7.6}$$

in the region $\mathbb{R}^2 \times \{z \in (0, h_s/d)\} \times \{t > 0\}$. The boundary conditions which one must use now have the form,

$$w = 0, \quad z = \frac{h_s}{d}, 1; \qquad \theta = 0, \quad z = 1;$$

$$\phi = 0, \quad z = 1; \qquad \psi = 0, \quad z = 0;$$

$$\theta = \psi, \quad \phi = \psi, \quad z = \frac{h_s}{d}; \tag{7.7}$$

$$\frac{\partial \theta}{\partial z} + \frac{1}{\gamma} \frac{\partial \phi}{\partial z} = b \frac{\partial \psi}{\partial z}, \quad z = \frac{h_s}{d}.$$

The linearized problem arising from (7.5) to (7.7) which governs instability of the steady state solution (7.4) may be found by linearizing equations (7.5) and writing $u_i = u_i(\mathbf{x})e^{\sigma t}$, $\theta = \theta(\mathbf{x})e^{\sigma t}$, $\phi = \phi(\mathbf{x})e^{\sigma t}$ and $\psi = \psi(\mathbf{x})e^{\sigma t}$, and then from equations (7.5)–(7.7) we find the equations governing the linear instability are

$$u_i = -\frac{\partial \pi}{\partial x_i} + Rk_i \theta,$$

$$\frac{\partial u_i}{\partial x_i} = 0,$$

$$\sigma \theta = Rw + \Delta \theta + H(\phi - \theta), \tag{7.8}$$

$$\frac{A}{\gamma} \sigma \phi = \frac{1}{\gamma} \Delta \phi - H(\phi - \theta),$$

holding on $\mathbb{R}^2 \times \{z \in (h_s/d, 1)\}$, with

$$\sigma B \psi = b \Delta \psi, \tag{7.9}$$

holding on $\mathbb{R}^2 \times \{z \in (0, h_s/d)\}$. The boundary conditions to be employed are

$$w = 0, \quad z = \frac{h_s}{d}, 1; \qquad \theta = 0, \quad z = 1;$$

$$\phi = 0, \quad z = 1; \qquad \psi = 0, \quad z = 0;$$

$$\theta = \psi, \quad \phi = \psi, \quad z = \frac{h_s}{d}; \tag{7.10}$$

$$\frac{\partial \theta}{\partial z} + \frac{1}{\gamma} \frac{\partial \phi}{\partial z} = b \frac{\partial \psi}{\partial z}, \quad z = \frac{h_s}{d}.$$

To demonstrate exchange of stabilities we let the periodic cell for the perturbation solution be V and we suppose V may be divided into two sub-cells V_1 and V_2 which correspond to the intersection of V with that part of the domain $\{z \in (0, h_s/d)\}$ which defines V_1 and the intersection of V with that part of the domain $\{z \in (h_s/d, 1)\}$ which defines V_2. The interface at $z = h_s/d$ between V_1 and V_2 we denote by Γ. Note that $V = V_1 \cup \Gamma \cup V_2$. The configuration with V_1, V_2 and Γ is shown in figure 7.2.

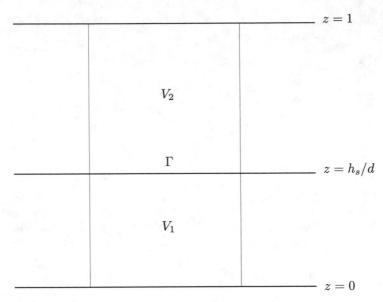

Fig. 7.2 Local thermal non-equilibrium convection in a horizontal porous layer with a rigid solid layer below. Individual "cells" V_1 and V_2 are shown with the cell interface Γ.

Now, multiply equation (7.8)$_1$ by u_i^* and integrate over V_2, where u_i^* is the complex conjugate of u_i. Likewise, multiply equation (7.8)$_3$ by θ^* and equation (7.8)$_4$ by ϕ^* and integrate each over V_2. We also multiply equation (7.9) by ψ^* and integrate over V_1. We add the results and use some integrations by parts. In particular, we need the combination

$$(\Delta\theta, \theta^*)_2 + \frac{1}{\gamma}(\Delta\phi, \phi^*)_2 + b(\Delta\psi, \psi^*)_1$$

$$= -\|\nabla\theta\|_2^2 + \oint_{\partial V_2} \frac{\partial\theta}{\partial n}\theta^* dS + \frac{1}{\gamma}\|\nabla\phi\|_2^2 \qquad (7.11)$$

$$+ \frac{1}{\gamma}\oint_{\partial V_2} \frac{\partial\phi}{\partial n}\phi^* dS - b\|\nabla\psi\|_1^2 + b\oint_{\partial V_1} \frac{\partial\psi}{\partial n}\psi^* dS.$$

In the above, $(\cdot, \cdot)_\alpha$, $\|\cdot\|_\alpha$, $\alpha = 1, 2$, denote the inner product and norm on $L^2(V_1)$ or $L^2(V_2)$. Due to boundary conditions (7.10) we have $\theta^* = \phi^* = \psi^*$ on Γ and then the boundary integrals in (7.11) may be combined as

$$-\int_\Gamma \theta^*\left(\frac{\partial\theta}{\partial z} + \frac{1}{\gamma}\frac{\partial\phi}{\partial z} - b\frac{\partial\psi}{\partial z}\right)dS = 0, \qquad (7.12)$$

where the boundary conditions (7.10) have been employed to infer the integral is zero. By employing the above procedure and adding each of the equations arising from equations $(7.8)_1$, $(7.8)_3$ and $(7.8)_4$, one may arrive at the equation,

$$\sigma \left(\|\theta\|_2^2 + \frac{A}{\gamma} \|\phi\|_2^2 + B \|\psi\|_1^2 \right) = -\|\mathbf{u}\|_2^2 - \|\nabla\theta\|_2^2$$
$$- \frac{1}{\gamma} \|\nabla\phi\|_2 - b\|\nabla\psi\|_1 + R \left[(\theta, w^*)_2 + (w, \theta^*)_2 \right] \qquad (7.13)$$
$$- H(\phi - \theta, \phi^* - \theta^*)_2 .$$

Put $\sigma = \sigma_r + i\sigma_1$ and observe by inspection that the right hand side of equation (7.13) is real. Thus, upon taking the imaginary part of equation (7.13) we find

$$\sigma_1 \left(\|\theta\|_2^2 + \frac{A}{\gamma} \|\phi\|_2^2 + B \|\psi\|_1^2 \right) = 0.$$

Hence, since we require a non-zero solution, $\sigma_1 = 0$, and the strong form of the principle of exchange of stabilities holds, cf. [150].

In addition, one may show the linear operator arising from equations (7.8)–(7.10) is symmetric and then theorem 2 of [150] applies to allow us to deduce that the linear instability boundary also yields the global nonlinear stability boundary, cf. also section 2.1.2 of this book. To establish symmetry of the linear operator L in equations (7.8)–(7.10) we write the solution vector as $(u_1, u_2, u_3, \theta, \phi, \psi)^T$ and then in the notation of section 2.1 the linear operator L may be written

$$L = \begin{pmatrix} -1 & 0 & 0 & 0 & 0 & 0 \\ 0 & -1 & 0 & 0 & 0 & 0 \\ 0 & 0 & -1 & R & 0 & 0 \\ 0 & 0 & R & \Delta - H & H & 0 \\ 0 & 0 & 0 & H & \frac{1}{\gamma}\Delta - H & 0 \\ 0 & 0 & 0 & 0 & 0 & b\Delta \end{pmatrix}$$

If we consider two solutions $\mathbf{r} = (u_1^1, u_2^1, u_3^1, \theta^1, \phi^1, \psi^1)^T$ and $\mathbf{s} = (u_1^2, u_2^2, u_3^2, \theta^2, \phi^2, \psi^2)^T$ then a calculation similar to that involved in showing exchange of stabilities allows us to establish

$$(\mathbf{r}, L\mathbf{s}) = (L\mathbf{r}, \mathbf{s})$$

and hence symmetry of L. The actual inner product involves integrals over V_1 and V_2 and requires use of the boundary conditions (7.10).

Based on the symmetry calculations the natural energy to use is

$$\frac{1}{2}\|\theta\|_2^2 + \frac{A}{2\gamma}\|\phi\|_2^2 + \frac{B}{2}\|\psi\|_1^2 .$$

If the Rayleigh number Ra is less than the linear instability threshold then from use of theorem 2 of [150] one may deduce this energy decays to zero at least exponentially. It is then an easy matter to show from equations $(7.5)_{1,2}$ that $\|\mathbf{u}\|_2$ likewise decays in a similar manner. We have described a result of [432] which demonstrates that for the class of layered convection problems involving LTNE porous materials and rigid solids one has the powerful result that the linear instability threshold completely captures the physics of the onset of convection. We have only described details of the two layer case, but a similar procedure works when there are more porous layers and alternate solid layers.

To produce quantitative stability results one still has to solve the linear instability problem. For the problem of the current section this involves a second order differential equation for w, and similar equations for θ, ϕ and ψ, yielding an 8th order eigenvalue problem for R. Clearly, we require 8 boundary conditions and this we have. We point out that a very good numerical method to solve this eigenvalue problem is the Chebyshev tau technique, cf. [115].

We give brief details of implementation of the D^2 Chebyshev tau method. Firstly take curl curl of equation $(7.8)_1$ to remove the pressure π. Then, we are faced with solving the eigenvalue problem

$$
\begin{aligned}
0 &= \Delta w - R\Delta^* \theta, \\
\sigma \theta &= Rw + \Delta \theta + H(\phi - \theta), \\
\frac{A}{\gamma} \sigma \phi &= \frac{1}{\gamma} \Delta \phi - H(\phi - \theta),
\end{aligned}
\tag{7.14}
$$

holding on $\mathbb{R}^2 \times \{z \in (h_s/d, 1)\}$, and

$$
\sigma B\psi = b\Delta \psi,
\tag{7.15}
$$

holding on $\mathbb{R}^2 \times \{z \in (0, h_s/d)\}$, together with the boundary conditions

$$
\begin{aligned}
w &= 0, \quad z = \frac{h_s}{d}, 1; \qquad \theta = 0, \quad z = 1; \\
\phi &= 0, \quad z = 1; \qquad \psi = 0, \quad z = 0; \\
\theta &= \psi, \quad \phi = \psi, \quad z = \frac{h_s}{d}; \\
\frac{\partial \theta}{\partial z} &+ \frac{1}{\gamma} \frac{\partial \phi}{\partial z} = b \frac{\partial \psi}{\partial z}, \quad z = \frac{h_s}{d}.
\end{aligned}
\tag{7.16}
$$

Let now f be a plane tiling planform with $\Delta^* f = -a^2 f$, and write $w = W(z)f(x, y)$, $\theta = \Theta(z)f(x, y)$, $\phi = \Phi(z)f(x, y)$ and $\psi = \Psi(z)f(x, y)$. Denote by D the operator d/dz, and then equations (7.14) and (7.15) become

$$
\begin{aligned}
0 &= (D^2 - a^2)W + Ra^2 \Theta, \\
\sigma \Theta &= RW + (D^2 - a^2)\Theta + H(\Phi - \Theta), \\
\frac{A}{\gamma} \sigma \Phi &= \frac{1}{\gamma}(D^2 - a^2)\Phi - H(\Phi - \Theta),
\end{aligned}
\tag{7.17}
$$

with $\{z \in (h_s/d, 1)\}$, and

$$
\sigma B\Psi = b(D^2 - a^2)\Psi,
\tag{7.18}
$$

where $\{z \in (0, h_s/d)\}$. The boundary conditions (7.16) become

$$W = 0 \quad \text{at} \quad z = \frac{h_s}{d} \quad \text{and} \quad z = 1,$$

$$\Theta = 0 \quad \text{at} \quad z = 1, \quad \Phi = 0 \quad \text{at} \quad z = 1,$$

$$\Psi = 0 \quad \text{at} \quad z = 0,$$

$$\Theta - \Psi = 0 \quad \text{and} \quad \Phi - \Psi = 0 \quad \text{at} \quad z = \frac{h_s}{d},$$

$$D\Theta + \frac{1}{\gamma}D\Phi - bD\Psi = 0, \quad \text{at} \quad z = \frac{h_s}{d}.$$

$$(7.19)$$

To find the eigenvalues σ_n and corresponding eigenfunctions one maps the domain $z \in (h_s/d)$ to $(-1, 1)$ and likewise maps the domain $z \in (0, h_s/d)$ to $(-1, 1)$ ensuring the interface $z = h_s/d$ becomes either $z = -1$ or $z = 1$. One then writes W, Θ, Φ and Ψ as a series of Chebyshev polynomials and reduces (7.17)–(7.19) to a generalized matrix eigenvalue problem of form

$$A_1 \mathbf{x} = \sigma B \mathbf{x}$$

where $\mathbf{x} = (W_0, \ldots, W_N, \Theta_0, \ldots, \Theta_N, \Phi_0, \ldots, \Phi_N, \Psi_0, \ldots, \Psi_N)^T$ with W_i, Θ_i, Φ_i and Ψ_i being the Fourier coefficients in the Chebyshev expansion for W, Θ, Φ and Ψ. This matrix generalized eigenvalue problem is then solved by the QZ technique. The boundary conditions (7.19) are incorporated into the matrix A by overwriting rows of this matrix, cf. [115], where a two layer fluid system is solved. The $4N \times 4N$ matrices A_1 and B have form

$$A_1 = \begin{pmatrix} D^2 - a^2 I & Ra^2 I & 0 & 0 \\ RI & D^2 - a^2 I - HI & HI & 0 \\ 0 & HI & \frac{1}{\gamma}(D^2 - a^2 I) - H & 0 \\ 0 & 0 & 0 & b(D^2 - a^2 I) \end{pmatrix}$$

and

$$B = \begin{pmatrix} 0 & 0 & 0 & 0 \\ 0 & I & 0 & 0 \\ 0 & 0 & \frac{1}{\gamma}AI & 0 \\ 0 & 0 & 0 & bI \end{pmatrix}$$

where, in addition each $N \times N$ block will have the last two rows of each block overwritten by boundary condition terms.

We have just sketched very briefly how the D^2 Chebyshev tau method works for system (7.17)–(7.19), in the general case. However, due to symmetry, it would be sufficient to set $\sigma = 0$ and use the D^2 Chebyshev tau method for the eigenvalues R_n.

Chapter 8
Other Convection and Microfluidic Scenarios

In this section we include a brief exposition of some recent work on microfluidic flows and/or LTNE flows in a variety of situations. This selection is in no way complete and simply indicates some of the very interesting areas which LTNE and microfluidics is currently encompassing. There will no doubt be many future novel applications of these theories and new problems solved in what is an exciting area.

8.1 Natural Convection Polymerase Chain Reaction

A very interesting use of thermal convection is suggested by e.g. [6, 7, 61, 89, 177, 218, 219, 477] and by [11]. This is to the production of copies of a DNA sequence, commencing with a few template molecules. As [11] describe this process is known as the polymerase chain reaction (PCR) and it involves taking a solution containing the template double-stranded DNA (dsDNA) together with nucleotydes which help generate new strands and a DNA polymerase enzyme. They add a few short single-stranded DNA primers and heat the solution to approximately 95°C which converts the dsDNA into two single-stranded DNA (ssDNA). The temperature of this solution is then reduced to around 50–60 °C which allows the primers to anneal to the DNA to create partially double-stranded DNA in an annealed form (aDNA). This is then reheated to 68–72 °C to create more dsDNA. The PCR reaction involves repeated cycles. What the above cited papers advocate is using a microfluidic cell and subjecting this to a Bénard convection process to effectively achieve the temperature cycle.

[11] model the microfluidic convection cell by a two-dimensional box height $2H$, width $2\gamma H$, with gravity in the downward direction. The lower wall temperature is kept fixed at a temperature T_L whilst the upper one is likewise kept fixed at a temperature of $T_U°$C with $T_L > T_U$. No-slip boundary conditions apply to the fluid velocity and the vertical walls have zero heat flux, i.e. $\partial T/\partial n = 0$ there. [11] use the Bénard convection equations, cf. equations (1.42), and solve these numerically assuming the DNA solution is in the microfluidic cell. The evolution of the DNA

© Springer International Publishing Switzerland 2015

B. Straughan, *Convection with Local Thermal Non-Equilibrium and Microfluidic Effects*, Advances in Mechanics and Mathematics 32, DOI 10.1007/978-3-319-13530-4_8

types may be expressed in terms of concentrations c_{ds} (for dsDNA), c_a (for aDNA), and c_{ss} (for ssDNA) and these are governed by reaction diffusion equations of form

$$\frac{\partial c_{ds}}{\partial t} + v_i \frac{\partial c_{ds}}{\partial x_i} = \frac{1}{Pe} \Delta c_{ds} + Da^C (\kappa_e f_e c_a - \kappa_d f_d c_{ds}),$$

$$\frac{\partial c_a}{\partial t} + v_i \frac{\partial c_a}{\partial x_i} = \frac{1}{Pe} \Delta c_a + Da^C (\kappa_a f_a c_{ss} - \kappa_e f_e c_a), \qquad (8.1)$$

$$\frac{\partial c_{ss}}{\partial t} + v_i \frac{\partial c_{ss}}{\partial x_i} = \frac{1}{Pe} \Delta c_{ss} + Da^C (2\kappa_d f_d c_{ds} - \kappa_a f_a c_{ss}),$$

where Pe and Da^C are the Péclet and Damköhler numbers, κ_e, κ_a and κ_d are coefficients and f_e, f_a and f_d are possibly functions of $\mathbf{x}$ and t given in [11].

[11] solve equations (8.1) numerically and discuss in detail the values of γ, T_L and T_U required to produce a single convection cell to aid the PCR process and optimize the double-strand DNA production.

[104] is another interesting contribution to biochemistry of DNA which involves flow in a microfluidic cell. [104] have additionally an imposed flow in their channel and also solve a system of reaction diffusion equations for the DNA. In particular, [104] compare their results with experimental work involving accumulation of DNA on biochips, and achieve favourable agreement.

Apart from modelling DNA reproduction there are many applications of microfluidics to biomaterials and, in particular, there are very interesting applications involving local thermal non-equilibrium theory, e.g. [44, 255, 339, 491, 492] and [208].

8.2 Heat Flux Boundary Conditions

When conditions of prescribed heat flux are imposed on the boundary of a porous layer rather than prescribed temperature field, interesting new effects may arise, cf. [35, 82, 129]. The purpose of this section is to include a brief account of recent analyses in a local thermal non-equilibrium porous situation where the heat flux is prescribed on the boundary(ies), according to the interesting work of [35] and [82].

8.2.1 Heat Flux Both Boundaries

[35] studies thermal convection with local thermal non-equilibrium effects in a Darcy porous layer contained between the planes $z = 0$ and $z = L$ with gravity acting in the negative $z-$direction. The equations of [35] are effectively the ones employed by [27], see equations (2.9)–(2.12), although [35] remove the pressure first so their equations are

$$\frac{\mu}{K}\operatorname{curl}\mathbf{v} = \rho_f g\alpha\operatorname{curl}(T^f\mathbf{k}),$$

$$\frac{\partial v_i}{\partial x_i} = 0,$$

$$\varepsilon(\rho c)_f \frac{\partial T^f}{\partial t} + (\rho c)_f v_i \frac{\partial T^f}{\partial x_i} = \varepsilon k_f \Delta T^f + h(T^s - T^f),$$

$$(1-\varepsilon)(\rho c)_s \frac{\partial T^s}{\partial t} = (1-\varepsilon)k_s\Delta T^s - h(T^s - T^f),$$

(8.2)

where the notation is as in section 2.2.

[35] study the stability of the steady state where $\bar{\mathbf{v}} \equiv 0$ and $\bar{T}^s, \bar{T}^f$ are linear in z and they work directly with boundary conditions on v_3 and the heat flux weighted by porosity. Thus, if q_w represents the applied heat flux the boundary conditions of [35] are

$$v_3 = 0, \qquad T^f = T^s,$$

$$-\varepsilon k_f \frac{\partial T^f}{\partial z} - (1-\varepsilon)k_s\frac{\partial T^s}{\partial z} = q_w,$$

(8.3)

these holding on the plane $z = 0$ and on $z = L$.

[35] study the instability of their basic solution and introduce perturbations u_i, θ, ϕ and they work in two space dimensions x, z, so they are able to introduce a stream function ψ such that $u = \varepsilon\,\partial\psi/\partial z$, $w = -\varepsilon\partial\psi/\partial x$. [35] are able to prove exchange of stabilities and so they effectively work with the linearized instability equations

$$\Delta\psi + \left(\frac{1+\gamma}{\gamma}\right)Ra\frac{\partial\theta}{\partial x} = 0,$$

$$\frac{\partial\psi}{\partial x} = \Delta\theta + H(\phi - \theta),$$

$$0 = \Delta\phi - \gamma H(\phi - \theta),$$

(8.4)

where $\gamma = \varepsilon k_f/(1-\varepsilon)k_s$, $H = hL^2/\varepsilon k_f$, and Ra is the Rayleigh number. Equations (8.4) are to be solved subject to the boundary conditions

$$\psi = 0, \quad \theta = \phi, \quad \gamma\frac{\partial\theta}{\partial z} + \frac{\partial\phi}{\partial z} = 0, \qquad \text{on } z = 0, 1.$$

(8.5)

[35] solve this system numerically and present many numerical findings. However, since heat flux only boundary conditions are considered they are able to employ an asymptotic analysis in the square of the wavespeed, a^2, since for zero heat flux it is known there is a range of parameters where the critical wavenumber is zero. In my opinion, an extremely interesting discovery of [35] is that there is a threshold value depending on H and γ such that the convection instability threshold switches from one of zero critical wavenumber to being one where a is non-zero. This is a behaviour which is also found in a rotating Darcy layer with prescribed heat flux when an average temperature is employed, see [129]. For the rotating problem the threshold depends on the rotation rate, i.e. the Taylor number $\mathcal{T}^2$. For the rotating

problem with a single temperature [129] obtain the asymptotic expression

$$Ra = 12(1 + \mathcal{T}^2) + a^2 \frac{2}{35}(20 - \mathcal{T}^2) + O(a^4),$$

whereas for the Darcy LTNE problem [35] derive the asymptotic expression

$$Ra = 12 + a^2 \left\{ \frac{8}{7} - \frac{72}{\gamma} \left[\frac{1}{5H(\gamma+1)} - \frac{2}{H^2(\gamma+1)^2} + \frac{24}{H^3(\gamma+1)^3} \right. \right.$$
$$\left. \left. - \frac{48}{H^{7/2}(\gamma+1)^{7/2}} \tanh\left(\frac{\sqrt{H(\gamma+1)}}{2}\right) \right] \right\} + O(a^4).$$

For the rotating heat flux problem the threshold where a switches is easily seen to be where $\mathcal{T}^2 = 20$ whereas for the LTNE heat flux problem the switching point is not so obvious but clearly exists. [35] do find this point numerically for a range of γ, H values and they are able to derive asymptotic values. Clearly, this behaviour is very interesting.

8.2.2 Heat Flux One Boundary

[82] analyse the analogous problem to that of section 8.2.1 but they apply a heat flux to the lower boundary $z = 0$ whereas they impose a constant temperature condition on the upper boundary $z = L$. Thus, instead of boundary conditions (8.3) they employ

$$v_3 = 0, \quad T^f = T^s, \qquad \text{on } z = 0,$$
$$-\varepsilon k_f \frac{\partial T^f}{\partial z} - (1 - \varepsilon)k_s \frac{\partial T^s}{\partial z} = q_w, \qquad \text{on } z = 0,$$

and

$$v_3 = 0, \quad T^f = T^s = T_U, \qquad \text{on } z = L,$$

for known values q_w and T_U.

They have a basic state in which $\bar{\mathbf{v}} \equiv 0$ and $\bar{T}^f, \bar{T}^s$ are linear in z, and they investigate linear instability in detail of this solution. They are able to prove exchange of stabilities holds, although this also follows because the linear operator attached to the problem is symmetric, a fact which is also true for the problem of section 8.2.1. Thus, due to symmetry, the results of [35] and [82] are valid also for the global nonlinear stability problem, rather than just for linear instability.

In addition to a numerical solution [82] are able to derive an implicit analytical solution to the linear instability problem. [82] find that the coefficient γ plays an important role in the critical Rayleigh and wavenumber results. These writers provide many detailed numerical results and we refer the interested reader to this publication for further details.

8.3 LTNE Convection and Horizontal Flow

In this section we briefly describe some recent work on the LTNE thermal convection problem but when the basic steady state is one where the velocity field is non-zero as opposed to the classical zero velocity solution discussed in section 2.2.

8.3.1 Horizontal Pressure Gradient

[347] studies the problem of instability of a layer of Darcy porous material heated from below allowing for LTNE effects, but he imposes a constant horizontal pressure gradient on the basic solution. He works in two space dimensions (x, y) with gravity in the $-y$ direction.

If $\bar{\mathbf{v}} - (\bar{u}, \bar{v})$ and $\bar{T}^f, \bar{T}^s$ denote the basic solution then in non-dimensional form [347] has

$$\bar{u} = \Pi, \quad \bar{v} = 0, \quad \bar{T}^f = 1 - y, \quad \bar{T}^s = 1 - y,$$

where Π is the constant pressure gradient in the x−direction. If we denote the non-dimensional variables by u_i, π, θ and ϕ then the linearized perturbation equations of [347] may be written (we have used a transformation in order to employ $R = \sqrt{Ra}$ rather than Ra itself, but the equations are equivalent to those of [347]),

$$u_i = -\frac{\partial \pi}{\partial x_i} + R\theta k_i,$$

$$\frac{\partial u_i}{\partial x_i} = 0,$$

$$\frac{\partial \theta}{\partial t} = -\Pi \frac{\partial \theta}{\partial x} + Rv + \Delta\theta + H(\phi - \theta),$$ (8.6)

$$\frac{\alpha}{\gamma} \frac{\partial \phi}{\partial t} = \frac{1}{\gamma} \Delta\phi - H(\phi - \theta),$$

where $\mathbf{k} = (0, 1)$, the terms α, γ and H are as in section 2.2, and $Ra = R^2$ is the Rayleigh number.

[347] writes that exchange of stabilities does not hold in this problem and he continues to show oscillatory convection will always ensue. One may appreciate these facts if one writes out the linear operator L associated to equations (8.6). Here we have

$$L = \begin{pmatrix} -1 & 0 & 0 & 0 \\ 0 & -1 & R & 0 \\ 0 & R & \Delta - H - \Pi\dfrac{\partial}{\partial x} & H \\ 0 & 0 & H & \dfrac{1}{\gamma}\Delta - H \end{pmatrix}$$

and we might expect a problem with the $\partial/\partial x$ term. [347] solves the linearized eigenvalue problem arising from equations (8.6) by a Galerkin numerical method. He presents a variety of numerical results analysing the effect of the diffusivity ratio α, the interphase heat transfer coefficient H, and the parameter γ. He also concentrates on the effect the horizontal flow has on the instability boundary. He finds that increasing the pressure gradient generally increases the critical Rayleigh number, Ra_c, and for $H = 1, \gamma = 10$, and $H = 10, \gamma = 0.01$, this increase is noticeable when Π varies from 0 to 25, whereas when $H = 100$ and $\gamma = 10$, the value of Ra_c increases over the range $\Pi \in [0, 100]$. The imaginary part of the growth rate is seen to decrease strongly as Π increases.

8.3.2 Uniform Horizontal Flow

[34] analyse thermal convection in a Darcy porous layer with LTNE effects and they prescribe zero heat flux on the lower boundary but allow for a viscous dissipation term in the energy balance law for the fluid component. In non-dimensional form the governing equations of [34] have form

$$\operatorname{curl} \mathbf{v} = \operatorname{curl} \left(T^f \mathbf{k} \right),$$

$$\frac{\partial v_i}{\partial x_i} = 0,$$

$$\frac{\partial T^f}{\partial t} + \frac{1}{\varepsilon} v_i \frac{\partial T^f}{\partial x_i} = \Delta T^f + \frac{Ge}{\varepsilon} v_i v_i + H(T^s - T^f), \qquad (8.7)$$

$$\Lambda \frac{\partial T^s}{\partial t} = \Delta T^s - \Omega H(T^s - T^f),$$

where Ge is the Gebhart number, Λ is the ratio of thermal coefficients, and $\Omega = \varepsilon k_f/(1 - \varepsilon)k_s$, k_f and k_s being the fluid and solid thermal conductivities.

[34] work with the horizontal layer between $y = 0$ and $y = 1$ with gravity in the $-y$ direction. Their boundary conditions are, with $\mathbf{v} = (u, v, w)$,

$$v = 0, \quad \frac{\partial T^f}{\partial y} = 0, \quad \frac{\partial T^s}{\partial y} = 0, \quad \text{on } y = 0,$$

$$v = 0, \quad T^f = 0, \quad T^s = 0, \quad \text{on } y = 1. \qquad (8.8)$$

The basic solution they analyse has form

$$\mathbf{v} = (\varepsilon Pe \cos \chi, 0, \varepsilon Pe \sin \chi),$$

with Pe the Péclet number and where χ represents an angle of deviation from the x−direction. The basic temperature profiles $\bar{T}^f$ and $\bar{T}^s$ are found by solving the coupled ordinary differential equations

$$\frac{d^2\bar{T}^f}{dy^2} = H(\bar{T}_f - \bar{T}_s) - \varepsilon\, Ge\, Pe^2,$$

$$\frac{d^2\bar{T}^s}{dy^2} = -H\Omega(\bar{T}_f - \bar{T}_s),$$

subject to boundary conditions (8.8). Both $\bar{T}^f$ and $\bar{T}^s$ are quite complicated expressions involving exponentials and may be found in equations (16) and (17) of [34].

[34] derive an approximate basic solution valid for $k_s \gg k_f$ and $\alpha_s \gg \alpha_f$. Even then the temperature $\bar{T}^f$ is not linear in y. [34] perform a linearized instability analysis of their basic solution and the linearized equations are interesting, but involved. These writers solve the resulting eigenvalue problem numerically and present various critical Rayleigh number and wavenumber results together with streamlines of the solution at criticality.

8.4 Viscoelastic LTNE Convection

8.4.1 Oldroyd-B Fluid, Darcy Material

[262] investigated the Bénard problem for a porous layer saturated with an Oldroyd-B fluid allowing for LTNE effects. They assumed the porous material is contained between the boundaries $z = 0$ and $z = d$ with $(x, y) \in \mathbb{R}^2$. For the momentum equation for a Darcy porous material saturated with an Oldroyd-B viscoelastic fluid coupled with the continuity equation they employ the model

$$\left(1 + \lambda_1 \frac{\partial}{\partial t}\right)\left(\frac{\rho_0}{\varepsilon}\frac{\partial v_i}{\partial t} + \frac{\partial p}{\partial x_i} + \rho_0 g k_i [1 - \beta(T^f - T_L)]\right)$$

$$= -\frac{\mu}{K}\left(1 + \lambda_2 \frac{\partial}{\partial t}\right) v_i, \tag{8.9}$$

$$\frac{\partial v_i}{\partial x_i} = 0,$$

where β is the coefficient of thermal expansion of the fluid and λ_1, λ_2 are positive coefficients representing the viscoelastic effects, other notation being as in section 2.2. In addition the fluid and solid skeleton temperature fields T^f and T^s satisfy the usual LTNE equations

$$\varepsilon(\rho_0 c)_f \frac{\partial T^f}{\partial t} + (\rho_0 c)_f v_i \frac{\partial T^f}{\partial x_i} = \varepsilon k_f \Delta T^f + h(T^s - T^f),$$

$$(1 - \varepsilon)(\rho_0 c)_s \frac{\partial T^s}{\partial t} = (1 - \varepsilon)k_s \Delta T^s - h(T^s - T^f). \tag{8.10}$$

With $\mathbf{v} = (u, v, w)$ the boundary conditions are

$$w = 0 \quad \text{on} \quad z = 0, d; \qquad T^f = T^s = T_U, \quad \text{on} \quad z = d;$$
$$T^f = T^s = T_L, \quad \text{on} \quad z = 0,$$

(8.11)

where T_L and T_U are constants with $T_L > T_U$.

The basic state whose stability is under consideration is one where

$$\bar{\mathbf{v}} \equiv 0, \qquad \bar{T}^f = \bar{T}^s = -\frac{\Delta T}{d} z + T_L,$$

with $\Delta T = T_L - T_U > 0$. To investigate the stability of the steady state [262] introduce perturbations and non-dimensionalize. After removing the perturbation variables u, v and the pressure perturbation they have to solve the system

$$\left(1 + \Gamma \frac{\partial}{\partial t}\right) \left[\frac{1}{Pr} \frac{\partial}{\partial t} \Delta w - Ra \Delta^* \theta\right] + \left(1 + \Gamma \Lambda \frac{\partial}{\partial t}\right) \Delta w = 0,$$
$$\left(\frac{\partial}{\partial t} - \Delta\right) \theta = w + H(\phi - \theta),$$
$$\left(\alpha \frac{\partial}{\partial t} - \Delta\right) \phi = -H(\phi - \theta),$$

(8.12)

where $\Gamma, \Gamma \Lambda$ are non-dimensional forms of λ_1, λ_2 and Ra is the Rayleigh number. The perturbation boundary conditions are

$$w = 0, \quad \theta = 0, \quad \phi = 0, \quad \text{on} \quad z = 0, 1.$$

(8.13)

[262] introduce normal modes and solve the resulting system which arises from (8.12), (8.13). They find that the stationary convection boundary is the same as the one for the analogous problem when the Oldroyd-B fluid is replaced by a Newtonian one, cf. the results of [27] described in section 2.2. However, the viscoelastic case also allows for oscillatory convection and [262] show that this is, in fact, the dominant form of convection. Many numerical results are presented in [262].

8.4.2 Oldroyd-B Fluid, Double Diffusive Convection

The analogous problem to that of section 8.4.1 but allowing also for a salt dissolved in the fluid was analysed by [266]. These writers studied the problem where the layer is heated from below and also salted from below. The basic modifications to the model of section 8.4.1 are the addition of an equation for the salt field of form

$$\varepsilon \frac{\partial S}{\partial t} + v_i \frac{\partial S}{\partial x_i} = \kappa_s \Delta S,$$

and a modification of the density in the buoyancy force term to have form

$$\rho = \rho_0 \left[1 - \beta_T (T^f - T_L) + \beta_s (S - S_L)\right].$$

[266] perform a linear instability analysis of the basic state and also a weakly nonlinear analysis. Since overstability is present in the viscoelastic problem of section 8.4.1 and also in the basic LTNE double diffusive problem of section 4.1 we might expect it here also and, indeed, [266] do find oscillatory convection for some parameter ranges. Interestingly, their weakly nonlinear results do indicate that subcritical bifurcation should occur, i.e. the critical Rayleigh number of the weakly nonlinear theory lies below that of oscillatory and stationary convection, for some parameter values.

8.4.3 Model Justification

Justification of a theoretical model for viscoelastic flow in a porous medium, as well as justification for employing an LTNE theory, really needs experimental results for completeness. In this regard, the work of [151] is very welcome. They model a porous medium as a periodic array of microchannels in a solid. The micro-channels are of two types, a geometrically symmetric one and a geometrically anti-symmetric one. They perform a series of flow experiments to test how the micro-channel model may resemble flow in a porous medium. The fluid they employ is a dilute solution of polyacrylamide. Comparison was made with flow through a type of sand. Pressure losses were measured along with flow visualization measurements. Agreements between the micro-channel results and those for flow through sand are generally good although [151] remark that microfluidic devices with special geometric configurations may be necessary to represent flow behaviour for a range of flow rates.

With regard to the LTNE behaviour the work of [470] is very useful. They give several cases where one may expect to find LTNE behaviour, and conclude in various instances that a local thermal equilibrium is unlikely to be found. In connection with LTNE behaviour we stress that the work of [364, 365] on how one may estimate the thermal interaction coefficient h is extremely useful, see also the comments in section 2.2.1. When one is faced with complicated flows like that of a viscoelastic fluid in a porous medium, assessing the validity of the model and being able to measure coefficients becomes absolutely necessary.

8.5 LTNE Thermovibrational Convection

The topic of thermal convection in a fluid saturated porous medium where the basic state is a time-periodic one has attracted some attention, see e.g. [279–282] and chapter 10 of the book by [414]. In particular, [279, 280] investigates the geophysical application to patterned ground formation due to a time-periodic surface flux generated by the Sun.

When the gravity field is itself time-periodic [389] refer to thermal convection as thermovibrational convection, or in a porous medium, thermovibrational filtration convection. [389] review some recent work in this area and we briefly describe their work on LTNE thermal convection, when the gravity field is varying periodically with time.

[389] consider a layer of porous material between the planes $z = 0$ and $z = d$, with the lower temperature fixed at the constant value T_L while the upper temperature is fixed at the constant value T_U. They allow the possibility that $T_L > T_U$ or $T_L < T_U$. Their gravity field is one of form $g(t) = g_0 + (A\Omega^2/\varepsilon)f''(t)$, where A is the amplitude of the vibration, ε is the porosity, Ω is the frequency of the vibration, and $f = \cos \omega t$, with $\omega = \Omega d^2/v$, v being the dynamic viscosity of the saturating fluid. The equations for LTNE theory they employ are the Brinkman model with the addition of a convective term, so that they are

$$\frac{1}{\varepsilon}\frac{\partial v_i}{\partial t} + \frac{1}{\varepsilon^2}v_j\frac{\partial v_i}{\partial x_j} = -\frac{1}{\rho}\frac{\partial p}{\partial x_i} - \frac{v}{K}v_i + v\Delta v_i + \alpha T^f g(t)k_i,$$

$$\frac{\partial v_i}{\partial x_i} = 0,$$

$$\varepsilon(\rho c_p)_f\frac{\partial T^f}{\partial t} + (\rho c_p)_f v_i\frac{\partial T^f}{\partial x_i} = \varepsilon k_f\Delta T^f + h(T^s - T^f),$$

$$(1-\varepsilon)(\rho c_p)_s\frac{\partial T^s}{\partial t} = (1-\varepsilon)k_s\Delta T^s - h(T^s - T^f),$$

$$(8.14)$$

where α is the coefficient of thermal expansion of the fluid. Their basic solution is one for which $\bar{v}_i \equiv 0$, $\bar{T}^f = \bar{T}^s = T_L - \Delta T z/d$, $\Delta T = T_L - T_U$, and $\bar{p}(t)$ is found from $(8.14)_1$.

[389] derive linearized non-dimensional perturbation equations for the perturbation variables (u_i, π, θ, ϕ) as

$$\frac{Da}{\varepsilon}\frac{\partial u_i}{\partial t} = -\frac{\partial \pi}{\partial x_i} - u_i + Da\Delta u_i + Gr\left[1 + \eta f''(t)\right]\theta k_i,$$

$$\frac{\partial u_i}{\partial x_i} = 0,$$

$$\frac{\partial \theta}{\partial t} = w + \frac{1}{Pr}\Delta\theta + \frac{H}{Pr}(\phi - \theta),$$

$$\alpha_1\frac{\partial \phi}{\partial t} = \frac{1}{Pr}\Delta\phi - \frac{H\gamma}{Pr}(\phi - \theta),$$

$$(8.15)$$

where Da is the Darcy number and Gr is the Grashof number. [389] employ stress free boundary conditions and so

$$w = \frac{\partial^2 w}{\partial z^2} = \theta = \phi = 0 \qquad \text{at} \quad z = 0, 1. \qquad (8.16)$$

[389] seek a solution to (8.15) and (8.16) of form

$$(w, \theta, \phi) = (W(t), \Theta(t), \Phi(t)) \sin \pi z\, e^{i(a_1 x + a_2 y)}.$$

They transform the time variable and in this manner they obtain the following set of ordinary differential equations governing the evolutionary behaviour of the solutions W, Θ and Φ,

$$
\begin{aligned}
c_1 \frac{dW}{dt} &= -(Da m^2 + 1)W + \frac{a^2}{m^2} Gr\left[1 + \eta f''(t)\right]\Theta, \\
c_2 \frac{d\Theta}{dt} &= W - \frac{m^2}{Pr}\Theta + \frac{H}{Pr}(\Phi - \Theta), \\
c_3 \frac{d\Phi}{dt} &= -\frac{m^2}{Pr}\Phi - \frac{H\gamma}{Pr}(\Phi - \Theta),
\end{aligned}
\tag{8.17}
$$

where c_1, c_2, c_3 are constants, $a^2 = a_1^2 + a_2^2$ and $m^2 = a^2 + \pi^2$.

[389] seek a Floquet solution, to the ordinary differential equations (8.17), which possesses the form

$$
(W(t), \Theta(t), \Phi(t)) = e^{\sigma t} \sum_{n=-\infty}^{\infty} (W_n, \Theta_n, \Phi_n)e^{in\omega t}.
$$

This leads to an infinite system of linear algebraic equations for the coefficients W_n, Θ_n and Φ_n. The solution employed by [389] is very interesting and employs a continued fraction method together with numerical evaluation.

A suitably defined critical Rayleigh number is calculated and results are displayed for the cases of heated below and of heated above. The effects of varying Pr and H are examined and the theoretical results are compared with some experimental data. [389] do find that local thermal non-equilibrium can have a significant effect when the gravity is varying in a time periodic manner.

8.6 Hot Fluid Injection

Recently, more articles have been concerned with showing that local thermal non-equilibrium effects really are important, and asking if and when does the LTNE model really make a change from the classical single temperature case, which is usually referred to as local thermal equilibrium, LTE, i.e. where the fluid and solid components of the porous medium have the same temperature. A particularly revealing article in this respect is that of [368] who analyse the effect of a hot fluid being injected into a cold porous matrix by using a local thermal non-equilibrium model.

To concentrate on the temperature effects [368] effectively assume the velocity field is given and begin with the standard equations for T^f and T^s with $\mathbf{v} = (U, 0, 0)$, U being a constant, cf. equations (2.9)–(2.12). They non-dimensionalize the equations for T^f and T^s and reduce to one space dimensional propagation, so they neglect y and z and work only with x. With θ and ϕ denoting the non-dimensional forms for T^f and T^s the equations of [368] are

$$
\left(\frac{\gamma+1}{\gamma+\alpha}\right)\frac{\partial\theta}{\partial t}+\left(\frac{\gamma+1}{\gamma}\right)\frac{\partial\theta}{\partial x}=\frac{\partial^2\theta}{\partial x^2}+H(\phi-\theta),
$$

$$
\alpha\left(\frac{\gamma+1}{\gamma+\alpha}\right)\frac{\partial\phi}{\partial t}=\frac{\partial^2\phi}{\partial x^2}+H\gamma(\theta-\phi).
$$

(8.18)

In these equations $H=hL^2/\varepsilon k_f$, $\gamma=\varepsilon k_f/(1-\varepsilon)k_s$, $\alpha=k_f(\rho c)_s/k_s(\rho c)_f$, these coefficients being defined in section 2.2. The velocity of the fluid effect is still present through the $((\gamma+1)/\gamma)(\partial\theta/\partial x)$ term which represents $v_i\partial T^f/\partial x_i$. The ratio $(\gamma+1)/\gamma$ is given by

$$
\frac{\gamma+1}{\gamma}=\frac{\varepsilon k_f+(1-\varepsilon)k_s}{\varepsilon k_f}
$$

and $U=(\varepsilon k_f+(1-\varepsilon)k_s)/L(\rho c)_f$ is the velocity scale for a typical length scale L.

[368] seek a similarity variable type solution and put $\eta=x/2\sqrt{t}$, $\tau=\sqrt{t}$ to transform equations (8.18) to

$$
2\tau\left(\frac{\gamma+1}{\gamma+\alpha}\right)\frac{\partial\theta}{\partial\tau}=\frac{\partial^2\theta}{\partial\eta^2}+2(\gamma+1)\left(\frac{\eta}{\gamma+\alpha}-\frac{\tau}{\gamma}\right)\frac{\partial\theta}{\partial\eta}+4\tau^2 H(\phi-\theta),
$$

$$
2\tau\alpha\left(\frac{\gamma+1}{\gamma+\alpha}\right)\frac{\partial\phi}{\partial\tau}=\frac{\partial^2\phi}{\partial\eta^2}+2\alpha\eta\left(\frac{\gamma+1}{\gamma+\alpha}\right)\frac{\partial\phi}{\partial\eta}+4\tau^2 H\gamma(\theta-\phi).
$$

(8.19)

They solve this system subject to the initial conditions

$$
\theta=\phi=0 \qquad \text{at} \quad t=0
$$

and the boundary conditions

$$
\theta=1, \quad \frac{\partial\phi}{\partial\eta}=0 \qquad \text{at} \quad \eta=0,
$$

$$
\theta,\phi\to 0 \qquad \text{as} \quad \eta\to\infty.
$$

They consider the condition $\partial\phi/\partial\eta=0$ at $\eta=0$ to amplify the LTNE effect.

[368] provide various asymptotic approaches to system (8.19) but they solve this together with the initial and boundary conditions by a numerical finite difference technique. The numerical results and explanations they give are very interesting. They keep $H=\gamma=\alpha=1$ and vary each in turn to see the effect each parameter has. For example, figure 1 of their paper shows a temperature front evolving for different values $H=1,0.1,0.01$ and 10^{-3}. The thermal front behaviour is strong and different from what would be seen with an LTE model. Other variations are discovered as γ and α vary in turn, but all have a novel effect due to LTNE. [368] also consider the small H case when the system of equations essentially becomes hyperbolic. Indeed, they do find that a temperature shock can form. [368] conclude that the state of LTE

is always approached as $t \rightarrow \infty$. However, even when the LTE state is reached the evolving front is not the same as the pure LTE one if one employs LTNE theory from the start. The LTNE system can strongly influence the temperature profile during the evolution of the thermal front and the width of this front depends strongly on the parameters H, γ and α.

Chapter 9
Convection with Slip Boundary Conditions

9.1 Thermal Convection, Slip Boundary Conditions

There is increasing interest in micro-electro-mechanical-systems (MEMS), and flow in microfluidic channels due to their applications in the electronics and related industries. In particular, at nanoscales there is increasing evidence that boundary conditions of slip type are needed rather than those of no-slip, cf. [24, 83, 118, 119, 234, 295, 350, 358, 382, 383, 402, 408, 489, 494, 496]. In this section we describe work of [473, 474] who analysed linearized instability and global nonlinear stability for the problem of Bénard convection, but he allowed the velocity at the boundary of the bounding planes to slip in the tangential directions.

Prior to including an exposition of the Bénard problem in a fluid with slip boundary conditions, we include a brief description of the Navier slip boundary condition. If $\mathbf{v}_\parallel$ denotes the tangential part of the velocity field at a solid boundary then the general Navier condition with slip is presented by [234] as

$$\mathbf{v}_\parallel = \lambda \mathbf{n} \cdot \left(\nabla \mathbf{v} + (\nabla \mathbf{v})^T \right) \cdot (\mathbf{I} - \mathbf{nn}) \tag{9.1}$$

where $\mathbf{n}$ is the unit outward normal at the boundary pointing *into the fluid*. In equation (9.1) λ is a coefficient called the slip length, see [234]. In indicial notation we may write equation (9.1) as

$$v_\parallel^i = \lambda n_a (v_{a,b} + v_{b,a})(\delta_{bi} - n_b n_i). \tag{9.2}$$

When the solid boundary is a plane, say $z = 0$, this condition simplifies. Suppose the fluid is in the region above the surface $z = 0$, i.e. in $z > 0$, and the boundary is fixed, so that $\mathbf{v} \cdot \mathbf{n} = 0$ which means $w = 0$ if $\mathbf{v} = (u, v, w)$. Then since $\mathbf{n} = (0, 0, 1)$, equation (9.2) reduces to

$$v_\alpha = \lambda (w_{,j} + v_{j,z}) \delta_{j\alpha}, \qquad \alpha = 1, 2,$$

© Springer International Publishing Switzerland 2015
B. Straughan, *Convection with Local Thermal Non-Equilibrium and Microfluidic Effects*,
Advances in Mechanics and Mathematics 32, DOI 10.1007/978-3-319-13530-4_9

and since $w = 0$ on $z = 0$, this becomes

$$v_\alpha = \lambda v_{\alpha,z}, \qquad \text{on } z = 0.$$

In other words,

$$u = \lambda \frac{\partial u}{\partial z}, \qquad \text{and} \qquad v = \lambda \frac{\partial v}{\partial z}, \qquad \text{on } z = 0. \tag{9.3}$$

Consider a linearly viscous, incompressible fluid contained in the horizontal layer between the planes $z = 0$ and $z = d$ with gravity in the negative z–direction. The lower plane is maintained at a constant temperature T_L while the upper plane is maintained at a constant temperature T_U, with $T_L > T_U$. The basic equations governing this problem are those of (1.42). The motionless conduction state in whose stability we are interested is the classical one of Bénard convection, namely

$$\bar{v}_i \equiv 0, \qquad \bar{T} = -\beta z + T_L, \qquad \beta = \frac{T_L - T_U}{d}, \tag{9.4}$$

with the steady pressure $\bar{p}$ being a quadratic function of z. The non-dimensional perturbation equations for perturbations (u_i, π, θ) to the steady state solution $(\bar{v}_i, \bar{p}, \bar{T})$ are, cf. [473, 474],

$$\frac{\partial u_i}{\partial t} + u_j \frac{\partial u_i}{\partial x_j} = -\frac{\partial \pi}{\partial x_i} + \Delta u_i + R\theta k_i,$$

$$\frac{\partial u_i}{\partial x_i} = 0, \tag{9.5}$$

$$Pr\left(\frac{\partial \theta}{\partial t} + u_i \frac{\partial \theta}{\partial x_i}\right) = Rw + \Delta\theta,$$

where Pr is the Prandtl number and $Ra = R^2$ is the Rayleigh number.

The boundary conditions are where one notices the change from the classical Bénard problem and it is here where the slip effect is manifest. In terms of $\mathbf{u} = (u, v, w)$, the non-dimensional boundary conditions may be written as, see [473, 474], and equation (9.3),

$$u = \lambda_0 \frac{\partial u}{\partial z}, \qquad v = \lambda_0 \frac{\partial v}{\partial z}, \qquad \text{on } z = 0,$$

$$u = -\lambda_1 \frac{\partial u}{\partial z}, \qquad v = -\lambda_1 \frac{\partial v}{\partial z}, \qquad \text{on } z = 1, \tag{9.6}$$

$$w = 0, \qquad \theta = 0, \qquad \text{on } z = 0, 1,$$

with u, v, w, θ and π satisfying a plane tiling periodicity in x and y. (To derive equations (9.6) one should recall the general slip boundary condition (9.2) and the fact that $\mathbf{n}$ points into the fluid.)

We firstly appeal to the operator theory for nonlinear stability given by [150], see section 2.1, and we write equations (9.5) in the form (2.5). It is easily verified that

$(u, N(u)) = 0$ where $u = (u_1, u_2, u_3, \theta)^T$, and the linear operators A and L have the forms

$$A = \begin{pmatrix} 1 & 0 & 0 & 0 \\ 0 & 1 & 0 & 0 \\ 0 & 0 & 1 & 0 \\ 0 & 0 & 0 & Pr \end{pmatrix}$$

and

$$L = \begin{pmatrix} \Delta & 0 & 0 & 0 \\ 0 & \Delta & 0 & 0 \\ 0 & 0 & \Delta & R \\ 0 & 0 & R & \Delta \end{pmatrix} \tag{9.7}$$

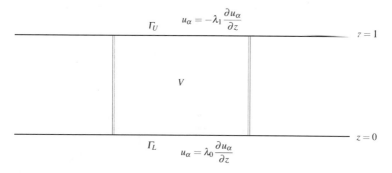

Fig. 9.1 Thermal convection in a horizontal porous layer with slip boundary conditions on the velocity field. The double vertical lines show the vertical convection cell walls and indicate periodic boundary conditions

We wish to show that L is a symmetric operator and exchange of stabilities holds and to this end we let V be a period cell for the solution u, and let the intersection of the closure of V with $z = 0$ be Γ_L while that with $z = 1$ is Γ_U. The geometric configuration for the proof of exchange of stabilities is shown in figure 9.1, where the boundary conditions are also indicated.

We shall be confronted with the calculation of $(\mathbf{v}, \Delta \mathbf{u})$ where $(\cdot, \cdot)$ is the inner product on $L^2(V)$ and $\mathbf{u}$ and $\mathbf{v}$ are subject to boundary conditions (9.6). With $\| \cdot \|$ being the norm on $L^2(V)$ we integrate by parts to see that

$$(\mathbf{v}, \Delta \mathbf{u}) = -(\nabla \mathbf{v}, \nabla \mathbf{u}) + \oint_{\partial V} v_i u_{i,j} n_j \, dS \tag{9.8}$$

where ∂V is the total boundary of V. We use the periodicity on the lateral walls of V and then equation (9.8) may be written

$$(\mathbf{v}, \Delta \mathbf{u}) = -(\nabla \mathbf{v}, \nabla \mathbf{u}) + \int_{\Gamma_U} v_\alpha \frac{\partial u_\alpha}{\partial z} \, dA - \int_{\Gamma_L} v_\alpha \frac{\partial u_\alpha}{\partial z} \, dA$$

where the repeated α denotes summation over 1 and 2 and dA denotes surface integration on Γ_L and Γ_U. The next step is to employ boundary conditions (9.6) to establish that

$$(\mathbf{v}, \Delta\mathbf{u}) = -(\nabla\mathbf{v}, \nabla\mathbf{u}) - \lambda_1 \int_{\Gamma_U} v_\alpha u_\alpha \, dA - \lambda_0 \int_{\Gamma_L} v_\alpha u_\alpha \, dA. \tag{9.9}$$

To show L given by (9.7) with boundary conditions (9.6) is a symmetric operator let $u = (u^1, v^1, w^1, \theta^1)$ and $v = (u^2, v^2, w^2, \theta^2)$ be two functions which satisfy equations (9.5) and the boundary conditions (9.6) and then compute the inner product $< u, Lv >$, the inner product being on $(L^2(V))^4$. Thus, employing equation (9.9), we may show

$$
\begin{aligned}
< u, Lv > &= (u^1, \Delta u^2) + (v^1, \Delta v^2) + (w^1, \Delta w^2) \\
&\quad + (\theta^1, \Delta\theta^2) + R(\theta^1, w^2) + R(w^1, \theta^2) \\
&= -(\nabla u^1, \nabla u^2) - (\nabla v^1, \nabla v^2) - (\nabla w^1, \nabla w^2) \\
&\quad - \lambda_1 \left(\int_{\Gamma_U} u^1 u^2 \, dA + \int_{\Gamma_U} v^1 v^2 \, dA \right) \\
&\quad - \lambda_0 \left(\int_{\Gamma_L} u^1 u^2 \, dA + \int_{\Gamma_L} v^1 v^2 \, dA \right) \\
&\quad - (\nabla\theta^1, \nabla\theta^2) + R[(\theta^1, w^2) + (w^1, \theta^2)] \\
&= (u^2, \Delta u^1) + (v^2, \Delta v^1) + (w^2, \Delta w^1) \\
&\quad + (\theta^2, \Delta\theta^1) + R(\theta^1, w^2) + R(w^1, \theta^2) \\
&= < v, Lu >, \tag{9.10}
\end{aligned}
$$

thereby establishing symmetry.

Now, linearize (9.5) and then put in the form (2.5) and the resulting linear equation has form $\sigma Au = Lu$. Multiply this equation by the complex conjugate u^* and put $\sigma = \sigma_r + i\sigma_1$. Then we find

$$\sigma < Au, u^* > = < u^*, Lu > \in \mathbb{R}, \tag{9.11}$$

where the fact that $< u^*, Lu >$ is real follows from a calculation similar to that leading to (9.10). Finally, evaluate $< Au, u^* >$ to see that

$$< Au, u^* > = \|\mathbf{u}\|^2 + Pr\|\theta\|^2$$

where $\| \cdot \|$ denotes the norm on the complex Hilbert space $L^2(V)$ and upon taking the imaginary part of (9.11) one finds

$$\sigma_1 (\|\mathbf{u}\|^2 + Pr\|\theta\|^2) = 0.$$

Thus $\sigma_1 = 0$, $\sigma \in \mathbb{R}$, and the strong form of the principle of exchange of stabilities holds.

We know L is symmetric, $<u, N(u)> = 0$, and so one may appeal to theorem 2 of [150], see section 2.1, to deduce that the linear instability boundary is the same as the global nonlinear stability one. [473, 474] shows that the energy $E(t) = \|\mathbf{u}\|^2/2 + Pr\|\theta\|^2/2$ decays at least exponentially if the Rayleigh number is below the linear instability threshold.

It remains to calcuate the instability boundary. Since exchange of stabilties is proven the linear instability boundary is governed by the equations

$$\Delta u_i + Rk_i\theta = \pi_{,i},$$
$$u_{i,i} = 0,$$
$$\Delta\theta + Rw = 0,$$

with the boundary conditions (9.6). One introduces W and Θ by $w = W(z)f(x,y)$, $\theta = \Theta(z)f(x,y)$ for f a plane tiling planform function satisfying $\Delta^* f = -a^2 f$, a being a wavenumber. [473, 474] uses the continuity equation

$$u_{,x} + v_{,y} = -w_{,z} \tag{9.12}$$

to reduce the u and v boundary conditions in (9.6) by differentiating (9.12) with respect to z, so that

$$u_{,xz} + v_{,yz} = -w_{,zz}$$

and then one employs (9.6)$_{1,2}$ to find

$$w_{,z} = \lambda_0 w_{,zz} \quad \text{at} \quad z = 0; \qquad w_{,z} = -\lambda_1 w_{,zz} \quad \text{at} \quad z = 1.$$

Hence, with $D = d/dz$, [473, 474] shows the eigenvalue problem governing the linear instability of Bénard convection in a fluid with slip boundary conditions may be written as

$$(D^2 - a^2)^2 W - Ra^2\Theta = 0,$$
$$(D^2 - a^2)\Theta + RW = 0, \tag{9.13}$$

where $z \in (0, 1)$, together with the boundary conditions

$$W = 0, \qquad \Theta = 0, \qquad z = 0, 1,$$
$$DW = \lambda_0 D^2 W, \qquad z = 0, \tag{9.14}$$
$$DW = -\lambda_1 D^2 W, \qquad z = 1.$$

[473, 474] solves system (9.13) and (9.14) numerically by a D^2 Chebyshev tau method but also by a novel compound matrix method which uses the boundary conditions on W to define one of the compound matrix variables. He establishes many numerical results among which the general trend shows that the critical Rayleigh number Ra_{crit} decreases strongly as λ_0, λ_1 increase, with λ_0, λ_1 values ranging from 10^{-5} to 10^5, and also the critical wavenumber a_{crit} likewise decreases as λ_0, λ_1 increase, over the same range of values. We refer the reader to [473, 474] for details of the numerical results.

9.2 Poiseuille Flow, Slip Boundary Conditions

While the main concern of this book is with problems of thermal convection, within the field of microfluidics the problem of flow of a clear fluid in a channel is of great importance. In particular, the instability problem of flow in a channel occupied by a linearly viscous fluid, and subject to slip boundary conditions, has an interesting history. [97] and [96] report that increasing slip length in the slip boundary condition has the effect of decreasing the critical Reynolds number for the commencement of instability. However, [233] and [407] report precisely the opposite. Namely, increasing the slip length has a stabilizing effect, thereby increasing the critical Reynolds number at which instability begins, according to linear theory. [474] employed a highly accurate Chebyshev tau method to analyse this instability problem and his findings are in agreement with those of [233] and [407]. In this section we report on the linear instability and nonlinear stability work of [474] and [476] on flow of a linearly viscous fluid in a channel.

The topic of instability of parallel shear flows in a channel and the associated numerical methods to accurately determine the instability thresholds is one of immense interest in the applied mathematics and engineering literature, cf. [22, 23, 26, 37, 109, 110, 115, 116, 154, 155, 164, 179, 206, 213, 222, 223, 267, 275, 385, 487, 488, 498]. This class of problem is very difficult to handle mathematically due to the fact that the mathematical operators which arise in the instability analysis are non-symmetric and the resulting eigenfunctions are close to being linearly dependent. This makes finding an accurate numerical solution a challenging problem, and one for which a fully accurate solution would require computations in arbitrarily large precision arithmetic, cf. [410], chapter 8.

[474] and [476] deal with the problem of Poiseuille flow of a linear viscous, incompressible fluid in an infinite channel bounded by the planes $y = \pm d$ with the flow driven in the x-direction by a constant pressure gradient $\partial p / \partial x = -G < 0$. The fluid equations are the Navier-Stokes equations which are non-dimensionalized to have form

$$\frac{\partial v_i}{\partial t} + v_j \frac{\partial v_i}{\partial x_j} = -\frac{\partial p}{\partial x_i} + \frac{1}{R} \Delta v_i,$$

$$\frac{\partial v_i}{\partial x_i} = 0,$$
(9.15)

where now the spatial region is $\{x \in \mathbb{R}\} \times \{y \in (-1, 1)\} \times \{z \in \mathbb{R}\}$, and $R = \rho d v_{max} / \mu$ is the Reynolds number.

The basic solution whose stability is under investigation is obtained employing slip boundary conditions and is thus not the classical solution $\bar{\mathbf{v}} = (1 - y^2, 0, 0)$. Instead, the basic solution is, if $\bar{\mathbf{v}} = (\bar{u}, \bar{v}, \bar{w})$, with $\bar{v} = \bar{w} = 0$,

$$\bar{u}(y) = \frac{(2+\sigma)^2(1+\sigma-y^2) - \delta(2+\delta)(\delta-2y)}{(1+\sigma)[(2+\sigma)^2 - \delta^2]},$$
(9.16)

where σ and δ are constants which depend on the slip lengths on the boundaries. If μ is the dynamic viscosity and the slip lengths on $y = \pm d$ are λ_d and λ_{-d}, then the non-dimensional slip lengths are defined by $\lambda_{-1} = \lambda_{-d}\mu/d$ and $\lambda_1 = \lambda_d\mu/d$. The parameters σ and δ are given by

$$\sigma = \lambda_1 + \lambda_{-1} \qquad \text{and} \qquad \delta = \lambda_1 - \lambda_{-1}.$$

In terms of non-dimensional perturbation variables $u_i \equiv (u, v, w)$ and π [476] derive the perturbation equations

$$u_{i,t} + \bar{u}(y)u_{i,x} + v\bar{u}_{,y}\delta_{i1} + u_ju_{i,j} = -\pi_{,i} + \frac{1}{R}\Delta u_i,$$

$$u_{i,i} = 0.$$

(9.17)

[476] show the perturbation boundary conditions are

$$u = \lambda_{-1}\frac{\partial u}{\partial y}, \qquad w = \lambda_{-1}\frac{\partial w}{\partial y}, \qquad \text{on } y = -1,$$

$$v = 0 \qquad \text{on } y = \pm 1,$$

(9.18)

$$u = -\lambda_1\frac{\partial u}{\partial y}, \qquad w = -\lambda_1\frac{\partial w}{\partial y}, \qquad \text{on } y = 1.$$

For the linear instability problem [474, 476] look for a solution of form

$$\mathbf{u} = (u(y), v(y), w(y))e^{i(\alpha x + \beta z - \alpha ct)}$$

$$\pi = \pi(y)e^{i(\alpha x + \beta z - \alpha ct)}$$

and show that a Squire's theorem holds so one may reduce the problem to one involving u, v and π only. They put $a^2 = \alpha^2 + \beta^2$, and transform u to $(\alpha u + \beta w)/a$, R to $R\alpha/a$, and π to $\pi\alpha/a$. Then, they show that the linear instability problem may be reduced to solving the equations

$$ia(\bar{u} - c)u + \bar{u}'v = -ia\pi + \frac{1}{R}(D^2 - a^2)u,$$

$$ia(\bar{u} - c)v = -D\pi + \frac{1}{R}(D^2 - a^2)v,$$

(9.19)

$$iau + Dv = 0,$$

where $D = d/dy$. The boundary conditions are derived from (9.18).

[476] further show one may eliminate u and π and derive the Orr-Sommerfeld equation

$$iaR[(\bar{u} - c)(D^2 - a^2) - \bar{u}'']v = (D^2 - a^2)^2v,$$

(9.20)

with $y \in (-1, 1)$, and subject to the boundary conditions

$$v(\pm 1) = 0, \qquad \lambda_1 v''(1) + v'(1) = 0, \qquad \lambda_{-1}v''(-1) - v'(-1) = 0. \qquad (9.21)$$

[476] and [474], chapter 3, solve system (9.20) and (9.21) for the leading eigenvalues and eigenfunctions. They show that with $\lambda_{-1} = \lambda_1 = \lambda$, the critical Rayleigh number increases strongly as λ increases, from under 6000 when $\lambda = 0$ to over 10,000 when $\lambda = 0.02$, [476], figure 1.

The problem of obtaining a global nonlinear stability threshold is also addressed by [474] and [476]. They use an energy method. We omit details but we observe that the Euler-Lagrange equatiuons which arise in their energy maximum must be solved as a system in (u, v, w) and π. One cannot invoke a Squire's theorem to reduce the problem to an Orr-Sommerfeld equation like (9.20). Thus, the nonlinear stability energy maximum problem is a truly three-dimensional eigenvalue problem. [474] and [476] look for solutions which depend on x and y, and then on y and z. While this only covers part of the full three-dimensional situation they do obtain the stabilizing effect of the slip boundary conditions in a global nonlinear stability analysis.

9.3 Poiseuille Flow in a Porous Medium

In this section we consider the problem of instability of flow in a channel occupied by a sparse porous medium when the boundary conditions are those of slip type.

The problem of instability of channel flow in a porous medium of Brinkman type with no-slip boundary conditions is one of relatively recent origin. [309] initiated this study and this work is described in [417, pp. 234–236]. [185] perform an accurate numerical analysis of instability of flow in a fluid saturated channel of porous medium. Their results largely confirm the findings of Nield [309]. It should be stressed that these papers consider only no-slip boundary conditions. The conditions of slip at the boundary, which are believed to be highly relevant in microfluidic situations, are (we believe) analysed for the first time in the work of [435].

The equations for Poiseuille flow in a Brinkman porous material are derived by [309] and by [417, p. 234]. Let v_i denote the velocity field, p the pressure, R the Reynolds number, and with M^2 being a non-dimensional Darcy coefficient the governing equations are

$$R\left(\frac{\partial v_i}{\partial t} + v_j \frac{\partial v_i}{\partial x_j}\right) = -\frac{\partial p}{\partial x_i} + \Delta v_i - M^2 v_i,$$

$$\frac{\partial v_i}{\partial x_i} = 0,$$

(9.22)

where equations (9.22) hold in the domain $\{(x, y) \in \mathbb{R}^2\} \times \{z \in (-1, 1)\} \times \{t > 0\}$.

[435] consider equations (9.22) together with boundary conditions which allow the fluid to slip on the horizontal boundaries $z = -1$ and $z = 1$. The specific boundary conditions which [435] place on solutions to equations (9.22) are, cf. [474],

sections 3.1, 3.2,

$$
\begin{aligned}
N_0 \frac{\partial v_1}{\partial y} &= v_1, & N_0 \frac{\partial v_2}{\partial y} &= v_2, & v_3 &= 0, & \text{on } z = -1, \\
N_0 \frac{\partial v_1}{\partial y} &= -v_1, & N_0 \frac{\partial v_2}{\partial y} &= -v_2, & v_3 &= 0, & \text{on } z = +1,
\end{aligned}
\tag{9.23}
$$

where N_0 is a dimensionless parameter which measures the slip length. We follow [435] and take the slip on the lower boundary to be the same as that on the upper boundary. This is purely for mathematical simplicity, since we are interested in the effect of the terms N_0 and M on the critical instability value of the Reynolds number R. As [435] remark, it is straightforward to account for different slip coefficients on the upper and lower boundaries.

The basic solution whose stability is analysed by [435] is one where the fluid is driven along the channel in the x-direction by a constant pressure gradient of form

$$
-\frac{\partial \bar{p}}{\partial x} = K^2 > 0,
$$

where an overbar denotes the basic solution. The basic velocity field corresponding to this pressure gradient has form $\bar{\mathbf{v}} = (U(z), 0, 0)$. Then one finds U must satisfy the boundary value problem

$$
-K^2 = U'' - M^2 U, \qquad -1 < z < 1,
\tag{9.24}
$$

subject to boundary conditions

$$
\begin{aligned}
N_0 U' + U &= 0, & z &= 1, \\
N_0 U' - U &= 0, & z &= -1.
\end{aligned}
\tag{9.25}
$$

[435] employ a non-dimensionalization with $K^2 = M^2 \cosh M / (\cosh M - 1)$ so that U is found to be

$$
U(z) = \frac{\cosh M}{\cosh M - 1} \left(1 - \frac{\cosh Mz}{N_0 M \sinh M + \cosh M} \right).
\tag{9.26}
$$

[435] investigate the stability of solution (9.26) and so let $\mathbf{u} = (u, v, w)$ be a perturbation to $\bar{\mathbf{v}}$ with corresponding pressure perturbation π. The perturbation equations are derived in detail in [417, p. 235], who shows that after linearization and assuming spatial and time dependence like $\exp(i\alpha x + i\beta y - ict)$ then one may show that the function $w(z)$ satisfies the Orr-Sommerfeld like equation

$$
(D^2 - a^2)^2 w - M^2 (D^2 - a^2) w = iaR(U - c)(D^2 - a^2) w - iaRU'' w,
\tag{9.27}
$$

where $D = d/dz$, $a^2 = \alpha^2 + \beta^2$, and $z \in (-1, 1)$. The boundary conditions which w must satisfy are derived by [435] from the conditions $N_0 u_{,z} = \pm u$, on the planes $z = \mp 1$, $N_0 v_{,z} = \pm v$, on the planes $z = \mp 1$, and the incompressibility condition

$u_{,x} + v_{,y} + w_{,z} = 0$. They show that the boundary conditions to employ in conjunction with equation (9.27) are

$$w = 0; \qquad N_0 \frac{\partial^2 w}{\partial z^2} = \frac{\partial w}{\partial z}; \qquad \text{on } z = -1,$$

$$w = 0; \qquad N_0 \frac{\partial^2 w}{\partial z^2} = -\frac{\partial w}{\partial z}; \qquad \text{on } z = 1. \tag{9.28}$$

In terms of the operator D the above boundary conditions become

$$w = 0, \qquad N_0 D^2 w = Dw, \qquad \text{on } z = -1;$$

$$w = 0, \qquad N_0 D^2 w = -Dw, \quad \text{on } z = 1. \tag{9.29}$$

The numerical results of [435] are based on the leading eigenvalue of the system (9.27) and (9.29). In other words, one employs the time representation in w of form e^{-ict} with $c = c_r + ic_i$ then this results in w involving terms of form $\exp(-ic_r t) . \exp(c_i t)$. The eigenvalues are found such that the largest value of c_i is $c_i = 0$ and then the result is minimized over the wavenumber a. The resulting R value is then the critical Reynolds number with corresponding wavenumber. The value $c_i = 0$ is chosen because this is the threshold at which the solution becomes unstable according to linearized theory. For, if $c_i > 0$ then w grows rapidly like $\exp(c_i t)$ and the solution is unstable.

The spectrum found by [435] is similar to that found in Poiseuille flow in a porous medium with no-slip boundary conditions by [185], the eigenvalues displaying a Y shape in the (c_r, c_i) diagram. With $M = 1$ as N_0 is increased the eigenvalues at the intersection of the three lines in the Y become more numerically unstable, and as M increases this effect is very pronounced. This numerical instability is already well known from the Orr-Sommerfeld equation in a clear viscous fluid, cf. [115], and it persists for Poiseuille flow in a Brinkman porous medium with slip boundary conditions.

Numerical results are reported in detail in [435] and they find that the critical Reynolds number R increases strongly as N_0 increases. At the same time the critical wavenumber decreases. Thus, increasing slip length is stabilizing. The increase in R is very noticeable. For example, with $M = 1$, there is a two-fold increase as N_0 goes from 0 to 0.175.

9.4 Thermal Slip

We briefly revisit the problem of thermal convection with slip boundary conditions described in section 9.1. However, we here wish to also consider the addition of "temperature-slip" boundary conditions. In microfluidics thermal-slip boundary conditions are believed very important, cf. [13, 383]. In gas dynamic flows velocity slip and thermal slip boundary conditions have been utilized for some time, see e.g. [121, pp. 467–486]. Indeed, [121] give specific values for the thermal accommodation coefficient and the momentum accommodation coefficient which appear in the boundary conditions, for several boundary materials and various gases. In microfluidics such boundary conditions are being advanced as necessary in this situation also, cf. [13, 383]. According to [383] the slip conditions at a solid boundary may be written

$$u = \left(\frac{2 - F_m}{F_m}\right) \lambda \frac{\partial u}{\partial n} \tag{9.30}$$

and

$$T - T_w = \left(\frac{2 - F_t}{F_t}\right) \left(\frac{2\gamma}{1 + \gamma}\right) \frac{\lambda}{Pr} \frac{\partial T}{\partial n}, \tag{9.31}$$

where u is a velocity component, T is the fluid temperature at the wall and T_w is the temperature of the solid wall. In addition, F_m and F_t are the momentum accommodation coefficient and the thermal accommodation coefficient, respectively, and these contain the specific information regarding the fluid-wall reaction. It is important to note that in equations (9.30) and (9.31) the normal derivative is in the direction *into the fluid*, see [234, 383]. We here employ these boundary conditions but we simply absorb the extra factors into two coefficients λ and λ_T. Note that equation (9.31) allows one to have constant wall boundary temperatures as in the classical Bénard thermal convection problem where T_w is the solid wall temperature and will be replaced by the constant upper and lower temperatures T_U and T_L.

Now consider a layer of linear viscous, incompressible fluid contained in the horizontal layer bounded by the planes $z = 0$ and $z = d$ with gravity in the negative z-direction. The equations of motion are given in section 1.2.1, equations (1.42), and are

$$\frac{\partial v_i}{\partial t} + v_j \frac{\partial v_i}{\partial x_j} = -\frac{1}{\rho} \frac{\partial p}{\partial x_i} + \nu \Delta v_i + g\alpha k_i T,$$

$$\frac{\partial v_i}{\partial x_i} = 0, \tag{9.32}$$

$$\frac{\partial T}{\partial t} + v_i \frac{\partial T}{\partial x_i} = \kappa \Delta T,$$

where the notation is as in section 1.2.1. The boundaries $z = 0$ and $z = d$ are kept at fixed constant temperatures T_L and T_U, respectively, with $T_L > T_U$. The boundary conditions may be derived as in sections 1.9 and 9.1, although one must additionally employ (9.31), and then one obtains

$$v_\alpha = \tilde{\lambda} \frac{\partial v_\alpha}{\partial z}, \qquad \alpha = 1, 2, \qquad \text{on } z = 0,$$

$$v_\alpha = -\tilde{\lambda} \frac{\partial v_\alpha}{\partial z}, \qquad \alpha = 1, 2, \qquad \text{on } z = d, \tag{9.33}$$

together with

$$T - T_L = \tilde{\lambda}_T \frac{\partial T}{\partial z}, \qquad \text{on } z = 0,$$

$$T - T_U = -\tilde{\lambda}_T \frac{\partial T}{\partial z}, \qquad \text{on } z = d,$$

(9.34)

where $\tilde{\lambda}$ and $\tilde{\lambda}_T$ reflect the extra (constant) factors in equations (9.30) and (9.31).

The steady state solution to equations (9.32)–(9.34) in whose stability we are interested may be written as

$$\bar{v}_i \equiv 0, \qquad \bar{T} = \alpha_1 - \beta z, \qquad (9.35)$$

where

$$\alpha_1 = \frac{\tilde{\lambda}_T (T_L + T_U) + d T_L}{2\tilde{\lambda}_T + d}, \qquad \beta = \frac{\Delta T}{2\tilde{\lambda}_T + d},$$

with $\Delta T = T_L - T_U > 0$. The steady pressure field $\bar{p}$ may then be found from equation (9.32)$_1$.

Perturbations (u_i, π, θ) to $(\bar{v}_i, \bar{p}, \bar{T})$ are introduced and non-dimensionalized with the scales

$$x_i = x_i^* d, \qquad t = t^* \mathcal{T}, \qquad \mathcal{T} = \frac{d^2}{v},$$

$$U = \frac{v}{d}, \qquad P = \frac{U v \rho}{d}, \qquad Pr = \frac{v}{\kappa},$$

and with the Rayleigh number given by

$$Ra = R^2 = \frac{g \alpha \beta d^4}{\kappa v}.$$

In this way one finds the non-dimensional perturbation equations which arise from (9.32) have form

$$\frac{\partial u_i}{\partial t} + u_j \frac{\partial u_i}{\partial x_j} = -\frac{\partial \pi}{\partial x_i} + \Delta u_i + R \theta k_i,$$

$$\frac{\partial u_i}{\partial x_i} = 0,$$

(9.36)

$$Pr \left(\frac{\partial \theta}{\partial t} + u_i \frac{\partial \theta}{\partial x_i} \right) = R w + \Delta \theta,$$

the spatial domain being $\mathbb{R} \times \{z \in (0,1)\}$. Non-dimensional boundary coefficients λ and λ_T are introduced by $\lambda = \tilde{\lambda}/d$ and $\lambda_T = \tilde{\lambda}_T/d$ and then the boundary conditions (9.33) and (9.34) applied to the total velocity and temperature fields yield the equations

$$u_\alpha = \lambda \frac{\partial u_\alpha}{\partial z}, \qquad \alpha = 1, 2, \qquad \text{on } z = 0,$$

$$u_\alpha = -\lambda \frac{\partial u_\alpha}{\partial z}, \qquad \alpha = 1, 2, \qquad \text{on } z = 1,$$

(9.37)

and

$$\theta = \lambda_T \frac{\partial \theta}{\partial z}, \qquad \text{on } z = 0,$$

$$\theta = -\lambda_T \frac{\partial \theta}{\partial z}, \qquad \text{on } z = 1.$$

(9.38)

We shall show the linear operator L attached to the system (9.36)–(9.38) is symmetric. Then, writing (9.36)–(9.38) in the form of equation (2.5) with $u = (u, v, w, \theta)^T$, one shows $< u, N(u) >= 0$ where $< \cdot, \cdot >$ denotes the inner product on $(L^2(V))^4$, V being a period cell for the solution to equations (9.36)–(9.38). The linear operator L may be written as

$$L = \begin{pmatrix} \Delta & 0 & 0 & 0 \\ 0 & \Delta & 0 & 0 \\ 0 & 0 & \Delta & R \\ 0 & 0 & R & \Delta \end{pmatrix}$$

Let u^1 and u^2 be solutions to (9.36)–(9.38) then

$$< u^1, Lu^2 >= (u_i^1, \Delta u_i^2) + R(\theta^1, w^2) + R(w^1, \theta^2) + (\theta^1, \Delta \theta^2) \qquad (9.39)$$

where $(\cdot, \cdot)$ denotes the inner product on $L^2(V)$. By computation employing boundary conditions (9.37) we find

$$(u_i^1, \Delta u_i^2) = -(\nabla u_i^1, \nabla u_i^2) + \int_{\Gamma_L \cup \Gamma_U} n_j u_i^1 u_{i,j}^2 dA$$

where Γ_L is that part of the boundary of V, ∂V, interesting with the plane $z = 0$, while Γ_U is that part intersecting with $z = 1$. Thus,

$$(u_i^1, \Delta u_i^2) = -(\nabla u_i^1, \nabla u_i^2) + \int_{\Gamma_U} u_\alpha^1 u_{\alpha,z}^2 dA - \int_{\Gamma_L} u_\alpha^1 u_{\alpha,z}^2 dA$$

$$= -(\nabla u_i^1, \nabla u_i^2) - \frac{1}{\lambda} \int_{\Gamma_U} u_\alpha^1 u_\alpha^2 dA - \frac{1}{\lambda} \int_{\Gamma_L} u_\alpha^1 u_\alpha^2 dA$$

$$= -(\nabla u_i^1, \nabla u_i^2) + \int_{\Gamma_U} u_{\alpha,z}^1 u_\alpha^2 dA - \int_{\Gamma_L} u_{\alpha,z}^1 u_\alpha^2 dA$$

$$= -(\nabla u_i^1, \nabla u_i^2) + \int_{\Gamma_L \cup \Gamma_U} n_j u_i^2 u_{i,j}^1 dA$$

$$= (u_i^2, \Delta u_i^1).$$

Similar integration by parts and use of boundary conditions (9.38) allows one to show

$$(\theta^1, \Delta \theta^2) = (\theta^2, \Delta \theta^1).$$

Thus, from (9.39) we may now deduce

$$< u^1, Lu^2 >=< u^2, Lu^1 >$$

and thus L is a symmetric operator.

Given that L is symmetric and $< u, N(u) > \geq 0$ we may apply theorem 2 of [150], see section 2.1.2, to deduce that the linear instability boundary is the same as the global nonlinear stability one.

Thus, to solve the global nonlinear stability boundary we have only to solve for the linear instability one. Since exchange of stabilities holds we may reduce equations (9.36) to the coupled system

$$0 = \Delta^2 w + R\Delta^* \theta,$$
$$0 = \Delta \theta + Rw. \tag{9.40}$$

The boundary conditions for (9.40) may be found as in section 9.1 using the continuity equation $u_{,x} + v_{,y} = -w_{,z}$ and then the boundary conditions are

$$\frac{\partial w}{\partial z} = \lambda \frac{\partial^2 w}{\partial z^2}, \qquad \theta = \lambda_T \frac{\partial \theta}{\partial z}, \qquad \text{on } z = 0,$$
$$\frac{\partial w}{\partial z} = -\lambda \frac{\partial^2 w}{\partial z^2}, \qquad \theta = -\lambda_T \frac{\partial \theta}{\partial z}, \qquad \text{on } z = 1. \tag{9.41}$$

If we introduce normal modes $w = W(z)f(x,y)$ and $\theta = \Theta(z)f(x,y)$ where f satisfies $\Delta^* f = -a^2 f$, a being a wavenumber, then (9.41) become

$$DW = \lambda D^2 W, \qquad \Theta = \lambda_T D\Theta, \qquad \text{on } z = 0,$$
$$DW = -\lambda D^2 W, \qquad \Theta = -\lambda_T D\Theta, \qquad \text{on } z = 1. \tag{9.42}$$

A convenient way to solve equations (9.40) is via the D^2 Chebyshev tau method, cf. [115], and then one would solve for the eigenvalues R_n and wavenumbers a_n, the system

$$(D^2 - a^2)W - A = 0,$$
$$(D^2 - a^2)A - Ra^2\Theta = 0, \tag{9.43}$$
$$(D^2 - a^2)\Theta + RW = 0,$$

subject to the boundary conditions

$$W = 0, \qquad DW = \lambda A, \qquad \Theta = \lambda_T D\Theta, \qquad \text{on } z = 0,$$
$$W = 0, \qquad DW = -\lambda A, \qquad \Theta = -\lambda_T D\Theta, \qquad \text{on } z = 1. \tag{9.44}$$

Numerical output for system (9.43) and (9.44) is given elsewhere.

9.5 Other Slip Problems

The thesis of [474] is an excellent one. In chapters 2 and 3 he describes material already examined here in sections 9.1 and 9.2. However, chapters 4–6 of [474] also contain material of relevance to microfluidics.

Chapter 4 of [474] studies the problem of instability of Poiseuille flow of a viscous incompressible fluid in a channel in a non-isothermal situation when the dynamic viscosity is a function of the temperature, i.e. $\mu = \mu(T)$. [474] employs slip boundary conditions and he derives a thermal Orr-Sommerfeld equation and presents extensive numerical results.

In chapter 5 [474] studies instability of thread annular flow in a pipe. This is like Poiseuille flow in a circular pipe, but there is a solid core which makes the fluid region an annular one. The solid core may move or rotate, and slip boundary conditions are employed. [474] provides a detailed numerical analysis for this problem and, very interestingly, he finds that instability is not always caused by the solution corresponding to the first mode. For some parameter regions it is the second or even third mode which is responsible for the onset of instability. This part of the work of [474] is also largely available in [475].

Chapter 6 of [474] is also of interest from a microfluidic viewpoint since he analyses instability of plane parallel Poiseuille flow in a channel for a [168] fluid under isothermal conditions, and also for a dipolar fluid, as derived by [54] and by [167]. The [168] model employed is the isothermal version of the one adapted to nanofluid theory in chapter 15. Both the [168] fluid and the dipolar one have fourth order spatial derivatives in the momentum equation which results in the basic flow profile being flattened rather than the quadratic one of Newtonian theory. This basic profile is what is discovered in dealing with flow of a suspension, cf. [452], and hence the detailed instability results of [474] are of value in suspension flows such as that of a nanofluid. Both chapters 4 and 6 of [474] are only available in the thesis. This thesis is, however, available on the internet at the address given in the citation.

Chapter 10
Convection in a Porous Layer with Solid Partitions

10.1 Multilayer Convection

As described at the beginning of chapter 7, [332] have produced a really interesting piece of work which derives in great detail linear instability thresholds for the problem of thermal convection in a horizontal layer heated below, where the layer is composed of identical finite thickness horizontal layers of a Darcy porous material saturated by an incompressible fluid interleaved with identical finite thickness horizontal solid layers which are themselves heat conducting. The work of [332] concentrates on determining Rayleigh number thresholds for the linear instability problem, whereas the object of [432] is to demonstrate that the linear instability problem yields exactly the same results for the global nonlinear stability one. This is, of course, a very powerful result since it shows that the linear theory has completely captured the physics of the onset of thermal convection, and no large amplitude subcritical instabilities can arise. To achieve his results [432] shows the linear operator associated to the multi-layer convection problem is symmetric, and the nonlinear terms are compatible with the theory of [150], see also section 2.1.

[432] establishes the coincidence of the linear instability and nonlinear stability boundaries firstly when the porous material is one of Darcy type but may possess an anisotropic permeability. He observes that [58] and [123] point out that in practical applications the porous layer will often be anisotropic. He also proves the coincidence of the linear instability and nonlinear stability boundaries for the analogous problem when the porous medium is of Brinkman type.

10.1.1 Darcy Porous Layers

[332] study thermal convection in a system which consists of N identical horizontal porous layers of the same thickness interleaved with $N-1$ identical layers of a rigid heat conducting solid each of the same thickness, where N is an integer greater than

© Springer International Publishing Switzerland 2015

B. Straughan, *Convection with Local Thermal Non-Equilibrium and Microfluidic Effects*,
Advances in Mechanics and Mathematics 32, DOI 10.1007/978-3-319-13530-4_10

or equal to 2. The temperatures of the upper surface of the top layer, at $z = H$, and lower surface of the bottom layer, at $z = 0$, are fixed at constant values T_U and T_L, respectively, with $T_L > T_U$. The goal of [332] is to obtain the instability threshold where convective motion begins by finding linear instability critical Rayleigh and wave numbers for various types of porous and solid properties and for several depths of the solid and porous layers. They also analyse in depth stream function patterns. The key parameters [332] find governing the behaviour of the solution to their instability problem are the ratio of the depth of rigid solid to the depth of porous layer, δ, and the ratio of the thermal conductivity of the solid to the thermal conductivity of the porous material, d, and the number of porous layers, N.

[432] studies the nonlinear stability of the above problem and argues it is in some sense sufficient to restrict attention to a two layer case of a porous layer above a rigid solid layer. Once the mathematical problem for this configuration is analysed, he argues that the extension of the equivalence between the linear instability problem and the global nonlinear stability problem is easily extended to the [332] configuration. We here follow [432] and firstly consider the problem of an infinite horizontal layer occupying the region $\{(x, y) \in \mathbb{R}^2\} \times \{z \in (0, H)\}$ with gravity acting in the negative z-direction (downward). The layer is divided into two horizontal layers namely those involving $z \in (0, h_s)$ and $z \in (h_s, H)$ where $0 < h_s < H$. A porous medium occupies the upper part of the layer whereas the lower part, $z \in (0, h_s)$, is occupied by a rigid heat conducting solid. The porous material is assumed to be of Darcy type with a permeability which is allowed to be anisotropic.

The governing equations for thermal convection in the porous medium are, cf. sections 1.4 and 1.8.1,

$$\frac{\mu}{C} M_{ij} v_j = -\frac{\partial p}{\partial x_i} - k_i g \rho_0 \left(1 - \tilde{\alpha}[T - T_0]\right),$$

$$\frac{\partial v_i}{\partial x_i} = 0, \tag{10.1}$$

$$(\rho c)_m \frac{\partial T}{\partial t} + (\rho c)_f v_i \frac{\partial T}{\partial x_i} = k_m \Delta T,$$

where $\mathbf{v}, p, T$ are the pore averaged velocity, pressure, and temperature in the porous medium, and $(\rho c)_m, k_m$ are given in section 1.8.1. The quantities $\mu, C, g, \rho_0, \tilde{\alpha}$, and T_0 are dynamic viscosity of the fluid saturating the porous medium, an appropriate constant, gravity, a reference density, the expansion coefficient of the fluid, and a reference temperature. The anisotropy tensor M_{ij} is assumed to be symmetric, dimensionless and positive-definite, i.e.

$$M_{ij} = M_{ji} \qquad \text{and} \qquad M_{ij} \xi_i \xi_j \geq \mu_0 \xi_i \xi_i, \tag{10.2}$$

for some $\mu_0 > 0$ and for all three-dimensional vectors ξ_i. A specific example of such an anisotropic tensor which occurs frequently in practice, is given in section 1.4. The example given in section 1.4 is appropriate to a tensor M_{ij} which describes a horizontally isotropic permeability. Equations (10.1) hold in the domain $\mathbb{R}^2 \times \{z \in (h_s, H)\} \times \{t > 0\}$.

For the solid domain $\mathbb{R}^2 \times \{z \in (0, h_s)\} \times \{t > 0\}$ the temperature, $T^s \equiv T_s$, of the solid satisfies the classical diffusion equation

$$(\rho c)_{\mathscr{S}} \frac{\partial T^s}{\partial t} = k_{\mathscr{S}} \Delta T^s, \tag{10.3}$$

where $k_{\mathscr{S}}$ denotes the thermal conductivity of the rigid solid, $(\rho c)_{\mathscr{S}}$ being the product of the density and specific heat of the solid.

The boundary conditions which must be satisfied are those of [332] and require no flow out of the boundaries $z = h_s, H$, fixed constant temperatures $T = T_U$ at $z = H$, $T = T_L$ at $z = 0$, where $T_L > T_U$. Additionally it is assumed that there is continuity of the temperature and of the heat flux across the interface $z = h_s$. The heat flux in the porous medium is given by $q_i = -k_m T_{,i}$ whereas in the solid one has $q_i^{\mathscr{S}} = -k_{\mathscr{S}} T_{,i}^s$, where k_m and $k_{\mathscr{S}}$ are the thermal conductivities of the porous medium and of the solid, respectively. The quantity k_m is defined after equation (1.86) and is given by $k_m = \varepsilon k_f + (1 - \varepsilon) k_s$ where k_f and k_s are the thermal conductivities of the fluid and the solid skeleton in the porous material. Therefore, denoting $\mathbf{v} = (u, v, w)$ the boundary conditions are

$$\begin{aligned}
w &= 0, \quad z = h_s, H; \\
T &= T_U, \quad z = H; \qquad T = T_L, \quad z = 0; \\
T &= T^s, \quad z = h_s; \\
k_m \frac{\partial T}{\partial z} &= k_{\mathscr{S}} \frac{\partial T^s}{\partial z}, \quad z = h_s,
\end{aligned} \tag{10.4}$$

where the condition involving $\partial T / \partial z$ represents continuity of the heat flux across the interface $z = h_s$.

The steady state solution to equations (10.1)–(10.4), namely $\bar{v}_i \equiv 0$, $\bar{T} = \bar{T}(z)$, $\bar{T}^s = \bar{T}^s(z)$, is found by [432] to be

$$\bar{T} = \alpha z + \beta, \qquad \bar{T}^s = \alpha_s z + T_L, \tag{10.5}$$

where the coefficients α, α_s and β are given by

$$\alpha_s = -\frac{k_{\mathscr{S}}}{\hat{k}} \Delta T, \qquad \alpha = -\frac{k_m}{\hat{k}} \Delta T,$$

$$\beta = T_U + \Delta T \frac{k_{\mathscr{S}}}{\hat{k}} (h + h_s).$$

In these equations ΔT and $\hat{k}$ are defined by

$$\Delta T = T_L - T_U > 0, \qquad \hat{k} = k_{\mathscr{S}} h + k_m h_s,$$

where h is the depth of the porous layer, i.e. $h + h_s = H$.

To study stability of this steady state solution [432] introduces perturbations $u_i, \theta, \theta^s, \pi$ to $\bar{v}_i, \bar{T}, \bar{T}^s, \bar{p}$ as

$$v_i = \bar{v}_i + u_i, \qquad T = \bar{T} + \theta, \qquad T^s = \bar{T}^s + \theta^s, \qquad p = \bar{p} + \pi.$$

He derives equations for u_i, θ, θ^s and non-dimensionalizes these with the variables

$$x_i = x_i^* d, \qquad t = \mathscr{T} t^*, \qquad u_i = U u_i^*, \qquad U = \frac{\kappa}{d},$$

$$\mathscr{T} = \frac{H^2}{\kappa}, \qquad \pi = \pi^* P, \qquad P = \frac{\mu U d}{C},$$

$$\theta = T^\sharp \theta^*, \qquad \theta^s = T^\sharp \theta^{s*},$$

$$K = \frac{k_{\mathscr{S}}}{k_m} \frac{(\rho c)_m}{(\rho c)_{\mathscr{S}}}, \qquad T^\sharp = U d \sqrt{\frac{(\rho c)_f \Delta T \mu}{\hat{k} g \rho_0 \tilde{\alpha} C}}.$$

He introduces the Rayleigh number $R^2 = Ra$ as

$$R = d \sqrt{\frac{\tilde{\alpha} g \rho_0 C (\rho c)_f \Delta T}{\mu \hat{k}}},$$

where $\kappa = k_m/(\rho c)_f$.

Dropping stars, the nonlinear non-dimensional perturbation equations may be written as,

$$M_{ij} u_j = -\frac{\partial \pi}{\partial x_i} + R k_i \theta,$$

$$\frac{\partial u_i}{\partial x_i} = 0, \qquad (10.6)$$

$$\frac{\partial \theta}{\partial t} + u_i \frac{\partial \theta}{\partial x_i} = R w + \Delta \theta,$$

in the domain $\mathbb{R}^2 \times \{z \in (h_s/H, 1)\} \times \{t > 0\}$. For θ^s one has the equation

$$\frac{\partial \theta^s}{\partial t} = K \Delta \theta^s, \qquad (10.7)$$

in the domain $\mathbb{R}^2 \times \{z \in (0, h_s/H)\} \times \{t > 0\}$. The non-dimensional boundary conditions take the form

$$w = 0, \quad \text{on} \quad z = \frac{h_s}{H}, 1;$$

$$\theta = 0, \quad \text{on} \quad z = 1;$$

$$\theta^s = 0, \quad \text{on} \quad z = 0; \qquad (10.8)$$

$$\theta = \theta^s, \quad \frac{\partial \theta}{\partial z} = K C_1 \frac{\partial \theta^s}{\partial z} \quad \text{on} \quad z = \frac{h_s}{H},$$

where $C_1 = (\rho c)_{\mathscr{S}}/(\rho c)_m$. The solution to (10.6)–(10.8) is also supposed periodic in the $(x, y)-$ plane and satisfies a plane tiling shape.

As in chapter 7, let V be the three-dimensional domain given by the plane tiling shape together with the region $z \in (0, 1)$. Further let V be divided into two sub-regions V_1 and V_2 which occupy the domain below the interface $z = h_s/H$ and that above, respectively. Let Γ denote the interface in V at $z = h_s/H$.

[332] note that the problem analogous to that defined by equations (10.6)–(10.8), but with $2N - 1$ combinations of layers, (N identical porous layers, $N - 1$ indentical solid layers), is such that the occurrence of linear instability must be by the process of stationary convection. We may see directly this by proceeding as follows. One firstly linearizes equations (10.6). Next one writes the solution as $u_i = e^{\sigma t} u_i(\mathbf{x})$, $\pi = e^{\sigma t} \pi(\mathbf{x})$, $\theta = e^{\sigma t} \theta(\mathbf{x})$ and $\theta^s = e^{\sigma t} \theta^s(\mathbf{x})$. The linear instability problem reduces to solving the following linear system

$$M_{ij} u_j = -\frac{\partial \pi}{\partial x_i} + R k_i \theta,$$

$$\frac{\partial u_i}{\partial x_i} = 0, \tag{10.9}$$

$$\sigma \theta = R w + \Delta \theta,$$

in the region $\mathbb{R}^2 \times \{z \in (h_s/H, 1)\}$, together with the second order linear partial differential equation

$$\sigma \theta^s = K \Delta \theta^s, \tag{10.10}$$

in the domain $\mathbb{R}^2 \times \{z \in (0, h_s/H)\}$. The relevant boundary conditions for coupling with (10.9) and (10.10) are

$$w = 0, \quad \text{on} \quad z = \frac{h_s}{H}, 1;$$

$$\theta = 0, \quad \text{on} \quad z = 1;$$

$$\theta^s = 0, \quad \text{on} \quad z = 0; \tag{10.11}$$

$$\theta = \theta^s, \quad \frac{\partial \theta}{\partial z} = K C_1 \frac{\partial \theta^s}{\partial z} \quad \text{on} \quad z = \frac{h_s}{H}.$$

For $u_i, \theta, \pi, \theta^s$ complex [432] multiplies (10.9)$_1$ by u_j^* and integrates over V_2. He likewise multiplies (10.9)$_3$ by θ^* and integrates over V_2, and adds. He further multiplies (10.10) by $C_1 \theta_s^*$, where $*$ here denotes complex conjugate, and integrates the result over V_1. By adding the results together and using the boundary conditions (10.11) one takes the imaginary part of the result to find $\sigma \in \mathbb{R}$. Thus, the strong form of the principle of exchange of stabilities holds.

We adopt the notation of [150] and write system (10.6)–(10.8) in the form, cf. section 2.1,

$$A\mathbf{u}_t = L\mathbf{u} + N(\mathbf{u})$$

where the operator variable $\mathbf{u}$ is defined by $\mathbf{u} = (u, v, w, \theta, \theta^s)^T$. One sees that the linear operator L may be inferred from the equation

$$Lu = \begin{pmatrix} -M_{11} & -M_{12} & -M_{13} & 0 & 0 \\ -M_{12} & -M_{22} & -M_{23} & 0 & 0 \\ -M_{13} & -M_{23} & -M_{33} & R & 0 \\ 0 & 0 & R & \Delta & 0 \\ 0 & 0 & 0 & 0 & KC_1\Delta \end{pmatrix} \begin{pmatrix} u \\ v \\ w \\ \theta \\ \theta^s \end{pmatrix} \qquad (10.12)$$

where in writing the operator L in full we have employed the fact that M_{ij} is symmetric. Using boundary conditions (10.8) it is not difficult to show L is a symmetric, sectorial operator.

One may demonstrate that $(\mathbf{u}, N(\mathbf{u})) = 0$ in the inner product on $L^2(V_2)$. Thus, conditions of theorem 2 of [150] hold, see also section 2.1. One may conclude that the linear instability boundary coincides with the global nonlinear stability one. [432] argues that if the Rayleigh number is in the stability region below the linear instability threshold then one may show that the thermal energy

$$\frac{1}{2} \int_{V_1} \theta_s^2 dx + \frac{C_1}{2} \int_{V_2} \theta^2 dx$$

decays exponentially. To establish also decay of u_i in a suitable measure we next multiply equation $(10.6)_1$ by u_i and integrate over V_2 to find

$$\int_{V_2} M_{ij} u_i u_j dx = R \int_{V_2} \theta w\, dx.$$

Thus, using the bound on M_{ij}, (10.2), and the arithmetic-geometric mean inequality one finds

$$\mu_0 \int_{V_2} u_i u_i dx \le \frac{R}{2\alpha} \int_{V_2} \theta^2 dx + \frac{R\alpha}{2} \int_{V_2} w^2 dx,$$

where $\alpha > 0$ is to be selected. Pick $\alpha = 2\mu_0/R$ and use the fact that $w^2 \le u_i u_i$ to now see that

$$\int_{V_2} u_i u_i dx \le \frac{R^2}{2\mu_0^2} \int_{V_2} \theta^2 dx.$$

Thus, for a Rayleigh number below the linear instability threshold one also has exponential decay of $\int_{V_2} u_i u_i dx$.

We have only dealt with one Darcy porous region and one solid region in the above example. However, [432] points out that the main result that the linear instability boundary is the same as the nonlinear one is also true for the thermal convection problem involving an arbitrary number of identical porous layers alternating with identical rigid heat conducting layers, in which the porous layers are of the same depths, are composed of the same fluid and same solid skeleton material, and the rigid partitions are all of the same material and same depth which is, in general, different to that of the porous layers. The proof given above carries over to this more general situation, *mutatis mutandis*. Since the porous layers are identical and the solid layers are likewise identical we can take the growth rates for each porous layer to be the same and this gives rise to the same cell shape in each layer, and to the same critical wavenumber.

[332] performed an extensive analysis of the linear instability boundary for the case of N identical porous layers interleaved with $N - 1$ identical heat conducting rigid solid layers. Their analysis shows that the multilayer scenario has the potential to yield a much lower critical Rayleigh number than the single layer case. This is thus very important from a heat transfer point of view and shows that it is potentially preferable to employ a multilayer system in a heat transfer device. For example, with $\delta = 10^{-4}$ and $d = 1$, [332] show that the critical Rayleigh number obtained with the first mode in a four layer porous system is 18.73965 as opposed to $4\pi^2 \approx 39.47842$ for a single layer. The critical wavenumber is likewise smaller having value 1.61159 as opposed to $\pi \approx 3.14159$ for a single layer. [332] display several streamline profiles and one sees the effect of decreasing wavenumber in that the cell widths increase and they also discover that the convective motion is more intense in the cells in the porous layers nearer the centre of the "sandwich". This is seen because the streamlines are much closer together in the cells nearer the centre. [332] also contains an extensive analysis of various combinations of the layer ratio δ, and the thermal conductivity ratio d, for N having values 2, 3, 4 and 5. I believe the paper of [332] is a very important contribution to convection in porous media and to heat transfer theory.

10.1.2 Brinkman Layers

[432] also considers the analogous problem to that of [332] but where the identical porous layers are of Brinkman type. Again, he argues that it is sufficient to sketch the proof of equivalence of the nonlinear stability boundary and the linear instability boundary for a one porous layer—one solid layer case, and the more general result for a multilayer system follows in a very similar manner, *mutatis mutandis*. The analogous system of non-dimensional perturbation equations to those of (10.6) and (10.7) one derives if the porous layer is one of Brinkman type are,

$$M_{ij}u_j = -\frac{\partial \pi}{\partial x_i} + Rk_i\theta + \lambda \Delta u_i,$$

$$\frac{\partial u_i}{\partial x_i} = 0, \qquad (10.13)$$

$$\frac{\partial \theta}{\partial t} + u_i \frac{\partial \theta}{\partial x_i} = Rw + \Delta \theta,$$

in the region $\mathbb{R}^2 \times \{z \in (h_s/H, 1)\} \times \{t > 0\}$, and

$$\frac{\partial \theta^s}{\partial t} = K\Delta\theta^s, \qquad (10.14)$$

in the domain $\mathbb{R}^2 \times \{z \in (0, h_s/H)\} \times \{t > 0\}$. The boundary conditions must be different to (10.8) due to the presence of higher derivatives in the Brinkman term.

For no slip at the boundary of the porous layer the boundary conditions are

$$u_i = 0, \quad \text{on} \quad z = \frac{h_s}{H}, 1;$$

$$\theta = 0, \quad \text{on} \quad z = 1;$$

$$\theta^s = 0, \quad \text{on} \quad z = 0; \tag{10.15}$$

$$\theta = \theta^s, \quad \frac{\partial \theta}{\partial z} = KC_1 \frac{\partial \theta^s}{\partial z} \quad \text{on} \quad z = \frac{h_s}{H}.$$

Note that we still allow anisotropic permeability.

One may write system (10.13)–(10.15) in the form of equation (2.5). To do this we put $u = (u_1, u_2, u_3, \theta, \theta^s)^T$ in equation (2.5) and then the linear operator L has form

$$L = \begin{pmatrix} -M_{11} + \lambda\Delta & -M_{12} & -M_{13} & 0 & 0 \\ -M_{12} & -M_{22} + \lambda\Delta & -M_{23} & 0 & 0 \\ -M_{13} & -M_{23} & -M_{33} + \lambda\Delta & R & 0 \\ 0 & 0 & R & \Delta & 0 \\ 0 & 0 & 0 & 0 & KC_1\Delta \end{pmatrix} \tag{10.16}$$

where we employ the fact that M_{ij} is symmetric. The operator A is

$$A = \begin{pmatrix} 0 & 0 & 0 & 0 & 0 \\ 0 & 0 & 0 & 0 & 0 \\ 0 & 0 & 0 & 0 & 0 \\ 0 & 0 & 0 & 1 & 0 \\ 0 & 0 & 0 & 0 & C_1 \end{pmatrix} \tag{10.17}$$

Let $u = (u_1^1, u_2^1, u_3^1, \theta^1, \theta_1^s)^T$ and $v = (u_1^2, u_2^2, u_3^2, \theta^2, \theta_2^s)^T$ be potential solutions to equations (10.13)–(10.15). We may use a procedure similar to that in section 7.1 to show that

$$(u, Lv) = (Lu, v)$$

where the inner product is on $L^2(V_2)$ for the u_i and θ terms in u whereas the inner product is on $L^2(V_1)$ for the θ^s terms in u. The main difference with the proof in section 7.1 is that one encounters a term of form

$$(u_i^1, \Delta u_i^2)_2 = -(\nabla u_i^1, \nabla u_i^2)_2 = (\Delta u_i^1, u_i^2)_2$$

where the subscript 2 indicates the inner product on $L^2(V_2)$. Thus, L is a symmetric linear operator.

The linearized version of equations (10.13)–(10.15) is written as

$$\sigma Au = Lu \tag{10.18}$$

Let u^* be the complex conjugate of u, i.e. if

$$u = (u_1^r + iu_1^i, u_2^r + iu_2^i, u_3^r + iu_3^i, \theta^r + i\theta^i, \theta_r^s + i\theta_i^s)^T$$

then u^* is its complex conjugate. Next, multiply equation (10.18) by u^* and form the inner product to find

$$\sigma(u^*, Au) = (u^*, Lu) \qquad (10.19)$$

and since L is symmetric one shows $(u^*, Lu) \in \mathbb{R}$. Thus, $\sigma(u^*, Au) \in \mathbb{R}$. Write $\sigma = \sigma_r + i\sigma_1$ and then use the form for A given in (10.17) and thus if we take the imaginary part of (10.19) we find

$$\sigma_1(\|\theta\|_2^2 + C_1\|\theta^s\|_1^2) = 0$$

where $\|\cdot\|_2$ and $\|\cdot\|_1$ denote the norms on $L^2(V_2)$ and $L^2(V_1)$, respectively. This shows that $\sigma \in \mathbb{R}$ and so exchange of stabilities holds. One may now apply theorem 2 of [150], see also section 2.1, to deduce that the linear instability boundary is the same as the global nonlinear energy stability one (since here $(u, N(u)) = 0$). Thus, even for a Brinkman layer next to a heat conducting rigid solid layer one has the important result that a linear instability analysis captures correctly the physics of the onset of convection.

We reiterate that although we have treated only a two layer problem we can deduce the same results for a combination of $N-1$ identical solid layers, between N identical Brinkman layers, the porous and solid layers having, in general, different depths, providing these layers are governed by the equations presented here. [432] points out that one could also derive the same sort of result if the porous layers were of LTNE type with a Brinkman term present.

We stress that we deduce complete information on the nonlinear stability problem by solving the linear instability one. One still has, of course, to solve the linear instability problem and we point out that an ideal method to do this is via a Chebyshev tau method coupled to a matrix solver like the QZ-algorithm, cf. [115]. To solve a many layer problem one simply maps each sub-layer alternately to $(-1, 1)$ then $(1, -1)$, expands the appropriate functions in Chebyshev series in each layer and truncates employing the boundary and interface equations to complete the matrix eigenvalue problem. For example, we might consider a 5-layer problem of porous depths h, solid depths h_s, with layers 2 and 4 composed of a rigid solid, and layers 1,3,5 composed of a porous material with Brinkman terms. This is analogous to the situation sketched in [332, p. 236], excepting they consider a Darcy porous material. The situation is pictured in figure 10.1

To solve the linear instability problem appropriate to figure 10.1 we use equations (10.13)–(10.15) and linearize after putting $u_i = e^{\sigma t} u_i(\mathbf{x})$, $\pi = e^{\sigma t}\pi(\mathbf{x})$, $\theta = e^{\sigma t}\theta(\mathbf{x})$, $\theta^s = e^{\sigma t}\theta^s(\mathbf{x})$. Suppose M_{ij} corresponds to a horizontally isotropic permeability, cf. section 1.4, and then we may arrange the non-dimensional form of M_{ij} so that $M_{ij} = \text{diag}(1, 1, \xi)$, for some positive number ξ in general different from 1. Then we take the curl twice of equation (10.13)$_1$ and in this way we reduce the instability threshold to solving the eigenvalue problem consisting of the equations

$$\frac{\partial^2 w}{\partial z^2} + \xi \Delta^* w = R\Delta^*\theta + \lambda \Delta^2 w,$$

$$\sigma\theta = Rw + \Delta\theta, \qquad (10.20)$$

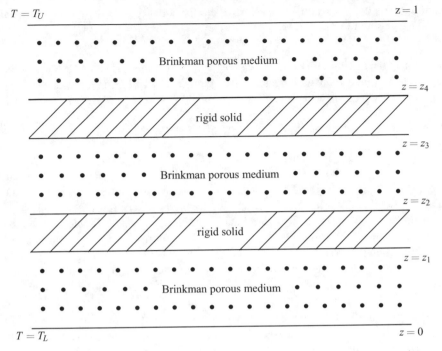

$T = T_U$ $z = 1$

Fig. 10.1 Layered Brinkman configuration for thermal convection. In this figure, $z_1 = h/H$, $z_2 = (h + h_s)/H$, $z_3 = (2h + h_s)/H$, $z_4 = 2(h + h_s)/H$. The layers marked "Brinkman porous medium" contain a porous medium saturated with fluid.

in each of the porous layers 1, 3 and 5, and the equation

$$\sigma \theta^s = K \Delta \theta^s, \tag{10.21}$$

in each of the rigid layers 2 and 4, where Δ^* is the horizontal Laplacian.

Now write $w = W(z)f(x,y)$, $\theta = \Theta(z)f(x,y)$, $\theta^s = \Theta^s(z)f(x,y)$, where f is a plane tiling planform with $\Delta^* f = -a^2 f$. Denote by $D = d/dz$ and then equations (10.20), (10.21) reduce to

$$(D^2 - \xi a^2)W = -Ra^2\Theta + \lambda(D^2 - a^2)^2 W,$$
$$(D^2 - a^2)\Theta + RW = \sigma\Theta, \tag{10.22}$$

holding in each of the layers $z \in (0, h/H)$, $z \in ((h + h_s)/H, (2h + h_s)/H)$, and $z \in (2(h + h_s)/H, 1)$ together with

$$K(D^2 - a^2)\Theta^s = \sigma\Theta^s, \tag{10.23}$$

with this holding in each of the layers $z \in (h/H, (h + h_s)/H)$ and $z \in ((2h + h_s)/H, 2(h + h_s)/H)$. From (10.15) the above differential equations are to be solved in conjunction with the boundary conditions

$$\Theta^s = 0, \qquad z = 0; \qquad \Theta = 0, \qquad z = 1;$$
$$W = DW = 0,$$
$$\text{on} \quad z = 0, \frac{h}{H}, \frac{(h+h_s)}{H}, \frac{(2h+h_s)}{H}, \frac{2(h+h_s)}{H}, 1; \qquad (10.24)$$
$$\Theta = \Theta^s, \qquad D\Theta = KC_1 D\Theta^s,$$
$$\text{on} \quad z = \frac{h}{H}, \frac{(h+h_s)}{H}, \frac{(2h+h_s)}{H}, \frac{2(h+h_s)}{H}.$$

Thus, we see that the eigenvalue system corresponding to the instability problem for the configuration in figure 10.1 consists of three 6th order differential equations in layers 1, 3 and 5, together with two 2nd order differential equations in layers 2 and 4. This thus makes an eigenvalue problem of order 22. Equations (10.22) and (10.23) are to be solved with the boundary conditions (10.24) which involve one boundary condition each on Θ and on Θ^s, twelve boundary conditions on W, and eight boundary conditions coupling Θ and Θ^s. The Chebyshev tau D^2 method is ideally suited to solving such an eigenvalue problem, cf. [115]. Details of solving similar eigenvalue problems which involve porous and fluid layers may be found in [412] and [413]. (We have deliberately indicated how to solve equations (10.22)–(10.24) leaving the growth rate in. This will give a guide as to how to solve such a system in a more general situation. However, in the present case since we know $\sigma \in \mathbb{R}$ we could set it equal to zero and solve directly for R.)

In this chapter we have neglected the fluid inertia. [432] remarks that inclusion of such inertia may be important in some problems, see e.g. section 1.7, and he observes that it presents no difficulty whatsoever to establish the results of this section in the presence of fluid inertia.

Chapter 11
Convection with Protruding Baffles

In this book we are paying particular attention to effects which may raise or lower the critical Rayleigh number in a thermal convection problem, as opposed to the classical case of a clear fluid, or a clear fluid saturating a porous medium. In this chapter we consider work of [87, 88] who study thermal convection in a Darcy porous medium in a rectangular box but when the box has either a vertical partition (also called a vertical baffle), [88], or a horizontal partition, [87]. The findings of [87, 88] are very interesting since they show that the addition of a partition may strongly inhibit convective fluid motion, the critical Rayleigh number is substantially increased, and this is of much interest for thermal insulation. It could be of much use in microfluidic convection devices to suggest what sort of baffle to add, where to position the baffle, and what size it should be.

11.1 Vertical Partition

The geometric configuration of [88] which employs a vertical partition in a thermal convection problem is shown in figure 11.1.

[88] choose a rectangular, two-dimensional box of height H and horizontal width σH. They suppose a partition (or baffle) is at a distance λH from the left boundary, and has height βH, $0 < \lambda < \sigma$, $0 \leq \beta \leq 1$, as shown in figure 11.1. The temperature on the horizontal wall, $y = 0$, is fixed at a constant value T_L whereas that on the upper horizontal wall, $y = H$, is fixed at a constant value T_U, with $T_L > T_U$. Gravity acts in the negative y−direction and the baffle has zero thickness. The vertical side walls at $x = 0$ and $x = \sigma H$, along with the baffle are assumed adiabatic with $\partial T/\partial x = 0$ there.

The study of [88] is interested in finding the critical Rayleigh number for convective motion in the configuration of figure 11.1, and is thus concerned with assessing the effect the partition has on the onset of convection. [88] present non-dimensional equations for a perturbation to the conduction state and these may be written as

© Springer International Publishing Switzerland 2015

B. Straughan, *Convection with Local Thermal Non-Equilibrium and Microfluidic Effects*,
Advances in Mechanics and Mathematics 32, DOI 10.1007/978-3-319-13530-4_11

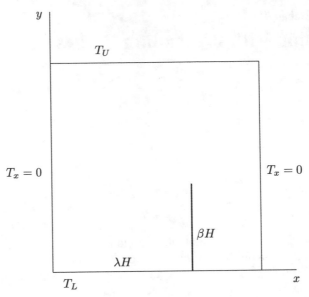

Fig. 11.1 Vertical partition configuration for thermal convection in a box. The box has vertical height H and width σH. Gravity acts in the negative $y-$direction, and the vertical sidewalls are adiabatic.

$$u_\alpha = -\frac{\partial \pi}{\partial x_\alpha} + RTk_\alpha,$$

$$\frac{\partial u_\alpha}{\partial x_\alpha} = 0, \tag{11.1}$$

$$\frac{\partial T}{\partial t} + u_\alpha \frac{\partial T}{\partial x_\alpha} = \Delta T + Rv,$$

where $\alpha = 1$ or 2, u_α are the velocity components which are also denoted by u and v, T is the temperature perturbation, R^2 is the Rayleigh number, $\Delta = \partial^2/\partial x^2 + \partial^2/\partial y^2$, where $\mathbf{k} = (0, 1)$, and a repeated subscript Greek letter denotes summation over 1 and 2.

[88] assume exchange of stabilities holds and work with the linearized form of (11.1). However, one may show exchange of stabilities holds and so we put $u_\alpha = u_\alpha(x_\beta)e^{\sigma t}$, $\pi = \pi(x_\beta)e^{\sigma t}$, $T = T(x_\beta)e^{\sigma t}$, $\beta = 1, 2$, in (11.1) and linearize to find

$$u_\alpha = -\frac{\partial \pi}{\partial x_\alpha} + RTk_\alpha,$$

$$\frac{\partial u_\alpha}{\partial x_\alpha} = 0, \tag{11.2}$$

$$\sigma T = \Delta T + Rv,$$

these equations holding on the box domain shown in figure 11.1.

To prove exchange of stabilities we assume u_α, π and T are complex and multiply equation $(11.2)_1$ by u_α^* (complex conjugate), equation $(11.2)_3$ by T^*, add the results and integrate over the domain of figure 11.1. It is convenient to split the domain in figure 11.1 into three regions, labelled 1,2 and 3, as shown in figure 11.2.

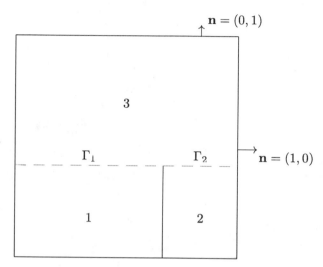

Fig. 11.2 Three regions configuration for the thermal convection problem in figure 11.1.

In figure 11.2 the surface Γ_1 is the horizontal line at height βH which stretches from $x = 0$ to $x = \lambda H$, whereas Γ_2 is the horizontal line at height βH which stretches from $x = \lambda H$ to $x = \sigma H$.

After adding the equations as indicated above if we let Ω denote the domain composed of regions 1,2 and 3 in figure 11.2, then we find

$$\int_\Omega u_\alpha u_\alpha^* dA + \sigma \int_\Omega T T^* dA = R \int_\Omega (T w^* + w T^*) dA$$
$$- \int_\Omega \frac{\partial T}{\partial x_\alpha} \frac{\partial T^*}{\partial x_\alpha} dA - \oint_\Gamma \pi n_\alpha u_\alpha^* d\ell + \oint_\Gamma n_\alpha T^* \frac{\partial T}{\partial x_\alpha} d\ell, \tag{11.3}$$

where Γ is the total boundary of all sub-domains 1,2 and 3. We show the boundary integrals are zero. For example, in region 1, we have on the boundary $y = 0, 0 \le x \le \lambda H$, $v = 0$ and $T = 0$, on the vertical boundaries we have $u = 0$ and $T_{,x} = 0$, whereas on Γ_1, u, v, π and T are continuous. Thus, in region 1 on the lower boundary $\mathbf{n} = (0, -1)$ and since $v = 0$ there the integral involving π is zero. On the vertical walls $u = 0$ and $\mathbf{n} = (\pm 1, 0)$ and so also the integral involving π is zero. On Γ_1 we have $\mathbf{n} = (0, 1)$ but due to continuity the πv term cancels with that from region 3. Hence the integral involving π is zero. Similarly, in region 1, on the $y = 0$ boundary $T^* = 0$ there, so the integral involving $\partial T / \partial x_\alpha$ is zero. On the vertical walls, $\mathbf{n} = (\pm 1, 0)$, and $\partial T / \partial x_\alpha$ reduces to $\partial T / \partial x$ which is zero, so the vertical wall contribution to the integral is zero. Finally, $\int_{\Gamma_1} u_\alpha T^* \partial T / \partial x_\alpha \, d\ell$ vanishes when

combined with the contribution from region 3. In a similar manner we may remove
the boundary integrals for regions 2 and 3. Thus, equation (11.3) reduces to

$$\sigma \int_\Omega TT^* dA = R \int_\Omega (Tw^* + wT^*) dA$$
$$- \int_\Omega u_\alpha u_\alpha^* dA - \int_\Omega \frac{\partial T}{\partial x_\alpha} \frac{\partial T^*}{\partial x_\alpha} dA . \tag{11.4}$$

Put $\sigma = \sigma_r + i\sigma_1$ and take the imaginary part of equation (11.4) to see that

$$\sigma_1 \int_\Omega |T|^2 dA = 0.$$

For non-zero T we require $\sigma_1 = 0$ and so exchange of stabilties holds.

[88] actually work with the linearized perturbation equations (11.2) but they use
the alternative non-dimensionalization which replaces T in (11.2) by T_1 and then
puts $T = T_1/R$ so that in component form, with $\sigma = 0$ because exchange of stabilities
holds, the equations of [88] are

$$u = -\frac{\partial \pi}{\partial x}$$
$$v = -\frac{\partial \pi}{\partial y} + Ra\,T$$
$$-v = \frac{\partial^2 T}{\partial x^2} + \frac{\partial^2 T}{\partial y^2} \tag{11.5}$$
$$\frac{\partial u}{\partial x} + \frac{\partial v}{\partial y} = 0,$$

where $Ra = R^2$. They define a streamfunction ψ by $u = \partial\psi/\partial y$, $v = -\partial\psi/\partial x$ and
then remove π from (11.5)$_{1,2}$ to find

$$\frac{\partial u}{\partial y} - \frac{\partial v}{\partial x} = -Ra\frac{\partial T}{\partial x} . \tag{11.6}$$

In this way they show the streamfunction satisfies the equations

$$\Delta \psi = -Ra\frac{\partial T}{\partial x} , \tag{11.7}$$
$$\frac{\partial \psi}{\partial x} = \Delta T .$$

Eliminating T they find

$$\Delta^2 \psi + Ra\frac{\partial^2 \psi}{\partial x^2} = 0. \tag{11.8}$$

The boundary conditions they derive are

$$\psi = 0, \qquad \frac{\partial^2 \psi}{\partial y^2} = 0,$$

on the top and bottom boundaries, and

$$\psi = 0, \qquad \frac{\partial^2 \psi}{\partial x^2} = 0,$$

on the side walls and also on the baffle.

[88] observe that due to the presence of the baffle they cannot employ normal modes to solve equation (11.8) for the critical Rayleigh number of linearized instability. Instead, they divide the domain in figure 11.1 into two regions separated by a vertical line extending from $y = 0$ to $y = H$ and coinciding with the baffle. They define ψ to be ψ_1 to the left of the baffle and they define ψ to be ψ_2 to the right of the baffle. They then write

$$\psi_1(x,y) = \sum_{n=1}^{\infty} \sin(n\pi y)\phi_n^1(x)$$

and

$$\psi_2(x,y) = \sum_{n=1}^{\infty} \sin(n\pi y)\phi_n^2(x)$$

and proceed to find differential equations for ϕ_n^1 and ϕ_n^2 from equation (11.8). By using the boundary conditions on ψ they are able to find an equation for Ra which requires numerical solution. In fact, they find that the critical Rayleigh number Ra_c is given by

$$Ra_c = 2\pi^2 + \left(\frac{M\pi}{\sigma}\right)^2 + \frac{\pi^4 \sigma^2}{M^2 \pi^2}$$

where M represents the number of convection cells in the horizontal direction and $\sqrt{M(M-1)} \le \sigma \le \sqrt{M(M+1)}$. Extensive interesting numerical results involving various baffle lengths and placements of the baffle are provided in [88].

11.2 Horizontal Partition

[87] analyse the analogous problem to that of section 11.1 but when the vertical partition is replaced by a horizontal one. Then they study thermal convection in a fluid saturated Darcy porous medium in a two-dimensional box of height H and width σH with an adiabatic baffle (partition) and adiabatic vertical walls and with the top temperature T_U (constant) while the lower temperature is T_L (constant), with $T_L > T_U$. The geometric configuration is as shown in figure 11.3.

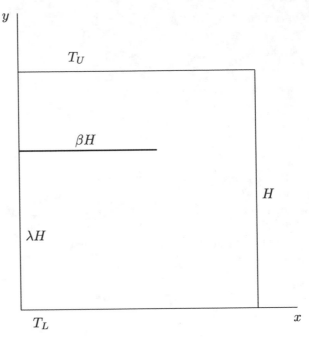

Fig. 11.3 Horizontal partition configuration for thermal convection in a box. The box has height H and width σH.

   The solution procedure for the linearized instability problem is similar to that of section 11.1 but with the roles of x and y reversed.

   Again, [87] provide many numerical results varying the length and position of the partition, and again they find greatly increased critical Rayleigh numbers, a fact which should be of interest in insulation technology. [87] present many streamline graphs which show a possibility of various convection cell structures at criticality.

Chapter 12
Anisotropic Inertia Effect

12.1 Introduction

Anisotropic porous media occur everywhere. We saw in figure 1.8 a photograph of a piece of oak where there is clearly a preferred direction along the grain. However, the anisotropy will not, in general, be as regular as that in figure 1.8 where the oak is approximately transversely isotropic. Figure 12.1 shows a piece of polished granite which displays an almost random anisotropy. One may expect properties like the permeability and the thermal diffusivity to vary strongly with changes of direction in a porous material like that of figure 12.1. Another example of irregular anisotropy is provided in figure 12.2 where a piece of polished elm is displayed. Here one may expect strong permeability variation along the grain lines as opposed to directions orthogonal to these lines. Furthermore, where the knots are in 12.2 one would expect the permeability to be much greater in a direction orthogonal to that of the rest of the wood. Figures 12.1 and 12.2 display porous materials which show strong irregular anisotropy and an understanding of fluid flow in an anisotropic porous medium is highly desirable.

In this chapter we consider the problem of double diffusive convection in a layer of Darcy porous material with a single temperature, but the emphasis is on assessing the effect of an anisotropic inertia coefficient. In section 1.7.5 we have already indicated strong evidence to show that, in general, the inertia coefficient may be very anisotropic, especially in a microfluidic setting. Indeed, thermal convection in porous media is an area which is attracting much attention when the material properties are anisotropic. Since many real porous materials display strong anisotropy this focus of attention is natural and is studied by e.g. [71–73, 172, 176, 193, 214, 226, 228, 370, 404] and [485], and the references therein. These articles concentrate mainly on anisotropy of the permeability tensor or thermal conductivity, and thus the emphasis on the inertia coefficient, as performed by [430], is warranted.

[430] develops a linear instability analysis together with a complimentary non-linear energy stability analysis for the thermosolutal convection problem in a satu-

B. Straughan, *Convection with Local Thermal Non-Equilibrium and Microfluidic Effects*, Advances in Mechanics and Mathematics 32, DOI 10.1007/978-3-319-13530-4_12

rated porous medium allowing the inertia coefficient in Darcy's equation to be an anisotropic tensor. When he performs numerical calculations he specializes to the case where the inertia term is composed of constants and is diagonal, but this still allows for the very important case of horizontal anisotropy in the inertia coefficient, cf. section 1.4.

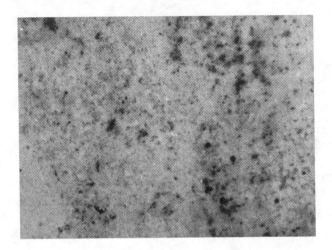

Fig. 12.1 Polished granite showing strong anisotropy

12.2 Thermosolutal Convection

The equations for a perturbation solution for thermosolutal convection in a saturated Darcy porous medium are given in (1.99). When the inertia coefficient, permeability, thermal diffusivity, and salt diffusion coefficient are isotropic the non-dimensional form for the perturbation equations for the velocity, u_i, pressure, π, temperature, θ, and salt field, ϕ, may be written in the form

$$a_0 \frac{\partial u_i}{\partial t} = -u_i + R\theta k_i - R_s \phi k_i - \frac{\partial \pi}{\partial x_i},$$

$$\frac{\partial u_i}{\partial x_i} = 0,$$

$$\frac{\partial \theta}{\partial t} + u_i \frac{\partial \theta}{\partial x_i} = Rw + \Delta \theta,$$

$$\varepsilon Le \frac{\partial \phi}{\partial t} + Le\, u_i \frac{\partial \phi}{\partial x_i} = \pm R_s w + \Delta \phi,$$

$$(12.1)$$

Fig. 12.2 A section of a piece of elm showing strong anisotropy

where $a_0 > 0$ is the inertia coefficient, $w = u_3$, R^2 is the Rayleigh number, R_s^2 is the salt Rayleigh number, Le is the Lewis number, and $\varepsilon = \varepsilon M$. The plus sign is taken in equation $(12.1)_4$ when the porous layer is heated and salted from below whereas the negative sign is adopted when the layer is heated from below and salted from above.

[430] considers the situation where the inertia coefficient a_0 is replaced by a symmetric, positive-definite tensor $a_{ij}(= a_{ji})$. He specifically rewrites equations (12.1) with the anisotropic tensor in the cases of heated and salted below, and heated below whilst simultaneously salted above. These two fundamental cases are mathematically very different, the distinction is worth making, and so we repeat the perturbation equations here.

When the porous layer is heated from below and salted from above the non-dimensional perturbation equations with an anisotropic inertia coefficient are

$$a_{ij}\frac{\partial u_j}{\partial t} = -u_i + R\theta k_i - R_s\phi k_i - \frac{\partial \pi}{\partial x_i},$$

$$\frac{\partial u_i}{\partial x_i} = 0,$$

$$\frac{\partial \theta}{\partial t} + u_i\frac{\partial \theta}{\partial x_i} = Rw + \Delta\theta,$$

$$\varepsilon_1\frac{\partial \phi}{\partial t} + Le\, u_i\frac{\partial \phi}{\partial x_i} = -R_s w + \Delta\phi,$$

(12.2)

where for convenience we have set $\varepsilon_1 = \varepsilon Le$. When the porous layer is heated from below and also salted from below the non-dimensional perturbation equations with an anisotropic inertia coefficient are

$$a_{ij}\frac{\partial u_j}{\partial t} = -u_i + R\theta k_i - R_s\phi k_i - \frac{\partial \pi}{\partial x_i},$$

$$\frac{\partial u_i}{\partial x_i} = 0,$$

$$\frac{\partial \theta}{\partial t} + u_i\frac{\partial \theta}{\partial x_i} = Rw + \Delta\theta,$$

$$\varepsilon_1\frac{\partial \phi}{\partial t} + Le\, u_i\frac{\partial \phi}{\partial x_i} = R_s w + \Delta\phi.$$

(12.3)

Equations (12.2) or (12.3) hold on the domain $\{(x,y) \in \mathbb{R}^2\} \times \{z \in (0,1)\} \times \{t > 0\}$. The boundary conditions to be satisfied are

$$w = 0, \quad \theta = 0, \quad \phi = 0 \quad \text{on} \quad z = 0,1$$

(12.4)

together with $\{u_i, \pi, \theta, \phi\}$ satisfying a plane tiling periodic shape in the (x,y) directions.

In this chapter we consider a generalization of the problem of [430] in that we consider an anisotropic inertia coefficient but we also allow the thermal diffusivity to be anisotropic. In this case one may show that the equations to be employed instead of (12.2) are

$$a_{ij}\frac{\partial u_j}{\partial t} = -u_i + R\theta k_i - R_s\phi k_i - \frac{\partial \pi}{\partial x_i},$$

$$\frac{\partial u_i}{\partial x_i} = 0,$$

$$\frac{\partial \theta}{\partial t} + u_i\frac{\partial \theta}{\partial x_i} = Rw + \mathscr{L}_{ij}\frac{\partial^2 \theta}{\partial x_i \partial x_j},$$

$$\varepsilon_1\frac{\partial \phi}{\partial t} + Le\, u_i\frac{\partial \phi}{\partial x_i} = -R_s w + \Delta\phi.$$

(12.5)

In equation $(12.5)_3$ the tensor $\mathscr{L}_{ij}$ represents the anisotropic thermal diffusivity.

When we consider the tricky heated and salted below case equations (12.3) for an anisotropic thermal diffusivity are replaced by

$$a_{ij}\frac{\partial u_j}{\partial t} = -u_i + R\theta k_i - R_s\phi k_i - \frac{\partial \pi}{\partial x_i},$$

$$\frac{\partial u_i}{\partial x_i} = 0,$$

$$\frac{\partial \theta}{\partial t} + u_i\frac{\partial \theta}{\partial x_i} = Rw + \mathscr{L}_{ij}\frac{\partial^2 \theta}{\partial x_i \partial x_j}, \qquad (12.6)$$

$$\varepsilon_1\frac{\partial \phi}{\partial t} + Le\, u_i\frac{\partial \phi}{\partial x_i} = R_s w + \Delta\phi.$$

For simplicity we restrict attention here to tensors $a_{ij}, \mathscr{L}_{ij}$ with constant coefficients. We suppose a_{ij} and $\mathscr{L}_{ij}$ are symmetric and positive-definite in the sense that

$$a_{ij} = a_{ji} \qquad \text{and} \qquad \mathscr{L}_{ij} = \mathscr{L}_{ji}, \qquad (12.7)$$

together with

$$a_{ij}\xi_i\xi_j \geq a_0\xi_i\xi_i \qquad \text{and} \qquad \mathscr{L}_{ij}\xi_i\xi_j \geq \ell_0\xi_i\xi_i \qquad (12.8)$$

for all vectors ξ_i, where $a_0 > 0$ and $\ell_0 > 0$ are constants.

12.2.1 Heated Below, Salted Above

We wish to analyse instability of the steady motionless solution and so we commence with equations (12.5) and the boundary conditions (12.4). This case is special in that the right hand side of equations (12.2) constitutes a symmetric linear operator. In this situation we know that the linear instability boundary coincides with the nonlinear stability one, cf. sections 2.1, 2.1.2. The importance of this result is that we may ascert that no subcritical instabilities are possible and the linear theory completely captures the physics of the onset of convective fluid motion, cf. the account of the general symmetric case in section 2.1.2, or see [414], chapter 4, or [417], section 4.3, or complete details may be found in [150].

It remains to show how exponential decay follows in the nonlinear case when the inertia and the thermal diffusivity are anisotropic. One then has to derive the linear instability boundary (which is equivalent to the nonlinear stability one). To establish decay of a suitable energy we let $\|\cdot\|$ and $(\cdot,\cdot)$ denote the norm and inner product on $L^2(V)$, where V is a periodic cell for the solution. Multiply equation $(12.5)_1$ by u_i and integrate over V, then multiply equation $(12.5)_3$ by θ and integrate over V, and finally multiply equation $(12.5)_4$ by ϕ, and integrate over V. After some integration by parts and use of the boundary conditions (12.4) one may arrive at the energy identities,

$$\frac{d}{dt}\frac{1}{2}(a_{ij}u_i,u_j) = -\|\mathbf{u}\|^2 + R(\theta,w) - R_s(\phi,w),$$ (12.9)

together with

$$\frac{d}{dt}\frac{1}{2}\|\theta\|^2 = R(\theta,w) - (\mathscr{L}_{ij}\theta_{,i},\theta_{,j}),$$ (12.10)

and

$$\frac{d}{dt}\frac{\varepsilon_1}{2}\|\phi\|^2 = -R_s(\phi,w) - \|\nabla\phi\|^2.$$ (12.11)

Adding equations (12.9), (12.10) and (12.11) one may arrive at

$$\frac{dE}{dt} = I - D,$$ (12.12)

where the "energy" functional E is given by

$$E = \frac{1}{2}(a_{ij}u_i,u_j) + \frac{1}{2}\|\theta\|^2 + \frac{\varepsilon_1}{2}\|\phi\|^2,$$

the quantity I is an indefinite production term,

$$I = 2R(\theta,w) - 2R_s(\phi,w),$$

whereas D is the dissipation term,

$$D = \|\mathbf{u}\|^2 + \|\nabla\phi\|^2 + (\mathscr{L}_{ij}\theta_{,i},\theta_{,j}).$$

Define R_E by

$$\frac{1}{R_E} = \max_{\mathscr{H}}\frac{I}{D}$$ (12.13)

where $\mathscr{H}$ is the space of admissible solutions. Suppose in addition to (12.7) and (12.8) a_{ij} is bounded above in the sense that

$$a_{ij}\xi_i\xi_j \leq \hat{a}\xi_i\xi_i, \qquad \forall\xi_i,$$ (12.14)

for some constant $\hat{a} > 0$. Then from (12.12) one may show that provided $R_E > 1$, with the aid of Poincaré's inequality,

$$\frac{dE}{dt} \leq -D\left(1 - \frac{1}{R_E}\right) \leq -\lambda E$$ (12.15)

for a constant $\lambda > 0$. Decay of E, and hence of θ and ϕ, follows from (12.15), in an L^2 sense. Decay of $\mathbf{u}$ in $L^2(V)$ follows by use of inequality (12.8)$_1$.

Thanks to symmetry, the solution of the variational problem (12.13) is equivalent to the solution of the linear instability problem with the time dependent terms in (12.5) omitted. This assertion also relies on the fact that exchange of stabilities holds, cf. section 2.1.2. [430] shows that when $\mathscr{L}_{ij} = \delta_{ij}$ the linear instability equations arising from (12.5) may be reduced to

$$\Delta w = R\Delta^*\theta - R_s\Delta^*\phi$$
$$\Delta\theta = -Rw$$
$$\Delta\phi = R_s w$$

where $\Delta^* = \partial^2/\partial x^2 + \partial^2/\partial y^2$. Solution of these equations incorporating the boundary conditions leads to the instability threshold

$$R^2 + R_s^2 = 4\pi^2. \tag{12.16}$$

This is a straight line in the (R^2, R_s^2) plane. For (R^2, R_s^2) below the straight line given by the threshold (12.16) one has nonlinear stability and no convective motion ensues. If (R^2, R_s^2) are above this line convective motion follows.

We emphasize that this is a strong result which shows that the linear instability boundary is exactly the same as the nonlinear one and no subcritical instabilities may arise.

12.3 Heated and Salted Below

When the fluid layer is heated from below and simultaneously salted from below then heating is potentially destabilizing whereas the salt gradient is potentially stabilizing which leads to competition between the two effects regarding convective overturning instability. This problem is mathematically very different from the one of section 12.2.1. The stability problem in this case is governed by the non-dimensional perturbation equations (12.6) and (12.4).

To develop a standard energy stability analysis one multiplies equation $(12.6)_1$ by u_i and integrates over V, then one multiplies equation $(12.6)_3$ by θ and integrates over V, and finally one multiplies equation $(12.6)_4$ by ϕ and likewise integrates over the period cell V. After some integration by parts and use of the boundary conditions (12.4) one derives the "energy identities"

$$\frac{d}{dt}\frac{1}{2}(a_{ij}u_i, u_j) = -\|\mathbf{u}\|^2 + R(\theta, w) - R_s(\phi, w),$$
$$\frac{d}{dt}\frac{1}{2}\|\theta\|^2 = R(w, \theta) - (\mathscr{L}_{ij}\theta_{,i}, \theta_{,j}), \tag{12.17}$$
$$\frac{d}{dt}\frac{\varepsilon_1}{2}\|\phi\|^2 = R_s(w, \phi) - \|\nabla\phi\|^2.$$

If we now add equations (12.17) then the R_s terms cancel out and instead of equation (12.12) one obtains the result,

$$\frac{d}{dt}\left[\frac{1}{2}(a_{ij}u_i, u_j) + \frac{1}{2}\|\theta\|^2 + \frac{\varepsilon_1}{2}\|\phi\|^2\right] = 2R(\theta, w) - \|\mathbf{u}\|^2$$
$$- (\mathscr{L}_{ij}\theta_{,i}, \theta_{,j}) - \|\nabla\phi\|^2. \tag{12.18}$$

One may again define suitable functions I and D and proceed as in the previous section. The problem is that the effect of salting below, manifest by the R_s terms, is lost in any calculation. Since in the present case the salt effect stabilizes this means that the standard energy technique cannot capture nonlinearly this stabilizing effect.

The basic mathematical problem of this section arises because the linear operator on the right of equations (12.6) is not symmetric. This is easily seen if we write the linear operator L as

$$
L = \begin{pmatrix}
-1 & 0 & 0 & 0 & 0 \\
0 & -1 & 0 & 0 & 0 \\
0 & 0 & -1 & R & -R_s \\
0 & 0 & R & \mathscr{L}_{ij}\dfrac{\partial^2}{\partial x_i \partial x_j} & 0 \\
0 & 0 & R_s & 0 & \Delta
\end{pmatrix}
$$

The R_s terms lead to a skew-symmetric component in the linear operator and this leads to loss of the R_s and ϕ terms in the energy equation (12.18). In this section the linear operator in (12.6) is no longer symmetric and so we cannot assert $\sigma \in \mathbb{R}$. In fact, exchange of stabilities does not hold.

[430] derives linear instability thresholds for the case where $\mathscr{L}_{ij} = \delta_{ij}$. In this book we present analysis and numerical results for the linear instability problem arising from equations (12.6). These are new results and are not published elsewhere.

Let us suppose now $(a_{ij}) = \text{diag}(a_\parallel, a_\parallel, a_\perp)$ with $a_\parallel, a_\perp$ positive constants. Thus, the inertia in the vertical direction may be different to that in the horizontal directions. In addition, we suppose $\mathscr{L}_{ij} = \text{diag}(\ell, \ell, 1)$ where $\ell = \ell_\parallel / \ell_\perp$. This is equivalent to starting with a thermal diffusivity of the form $L_{ij} = \text{diag}(\ell_\parallel, \ell_\parallel, \ell_\perp)$ and using $\ell_\perp$ in the non-dimensionalization. The coefficients $\ell_\perp$ and $\ell_\parallel$ are positive constants. In this way we allow for a thermal diffusivity which may be different in the vertical direction to what it is in the horizontal directions. One may show that linear instability theory arising from equations (12.6) results in the following eigenvalue problem for σ,

$$
\begin{aligned}
\sigma \mathscr{L} w + \Delta w &= R\Delta^* \theta - R_s \Delta^* \phi \\
\mathscr{L}_1 \theta &= -Rw \\
\mathscr{L}_2 \phi &= -R_s w
\end{aligned}
\tag{12.19}
$$

where the linear operators $\mathscr{L}, \mathscr{L}_1$ and $\mathscr{L}_2$ are given by

$$
\begin{aligned}
\mathscr{L} &\equiv a_\parallel \frac{\partial^2}{\partial z^2} + a_\perp \Delta^*, \\
\mathscr{L}_1 &\equiv \frac{\partial^2}{\partial z^2} + \ell \Delta^* - \sigma I, \\
\mathscr{L}_2 &\equiv \Delta - \sigma \varepsilon_1 I,
\end{aligned}
\tag{12.20}
$$

and the boundary conditions are those of (12.4).

System (12.19) may be reduced to the single equation for w,

$$(\sigma\mathcal{L} + \Delta)\mathcal{L}_1\mathcal{L}_2 w = -R^2\Delta^*\mathcal{L}_2 w + R_s^2\Delta^*\mathcal{L}_1 w. \tag{12.21}$$

The stationary convection boundary is found from (12.21) by selecting $\sigma = 0$ and putting $w = W(z)f(x,y)$ where f is a plane tiling form satisfying $\Delta^* f + a^2 f = 0$ and $W = \sin\pi z$. One may show that the stationary convection boundary reduces to the relation $R = R(a^2)$ where

$$R^2 = \frac{\Lambda\Lambda_1}{a^2} + R_s^2\frac{\Lambda_1}{\Lambda}. \tag{12.22}$$

In expression (12.22) the terms Λ and Λ_1 are given by

$$\Lambda = \pi^2 + a^2 \qquad \text{and} \qquad \Lambda_1 = \pi^2 + \ell a^2. \tag{12.23}$$

To find the actual critical value of R^2 we minimize expression (12.22) in a^2 numerically. The critical value is denoted by R_{stat}^2 and numerical values of this are presented in tables 12.1–12.8 along with corresponding values of the wavenumber.

It is worth observing that when $\ell = 1$ [430] shows that

$$R^2 = 4\pi^2 + R_s^2. \tag{12.24}$$

To calculate the oscillatory convection boundary we put $\sigma = i\sigma_1$, $\sigma_1 \in \mathbb{R}$, in equation (12.21) and then we take real and imaginary parts of the resulting equation. This yields the equations

$$R^2 a^2\Lambda = R_s^2 a^2\Lambda_1 + \Lambda^2\Lambda_1 - \sigma_1^2\big[\varepsilon_1(M\Lambda_1 + \Lambda) + M\Lambda\big] \tag{12.25}$$

and

$$R^2 a^2\varepsilon_1 = R_s^2 a^2 - \varepsilon_1 M\sigma_1^2 + \varepsilon_1\Lambda\Lambda_1 + \Lambda(M\Lambda_1 + \Lambda). \tag{12.26}$$

In equations (12.25) and (12.26) the terms Λ and Λ_1 are given by (12.23) while $M = a_\|\pi^2 + a_\perp a^2$.

From equations (12.25) and (12.26) we obtain the following expressions for R^2 and σ_1^2,

$$R^2 = R_s^2\frac{\big[M\Lambda + \varepsilon_1 M(\Lambda_1 - \Lambda) + \varepsilon_1\Lambda\big]}{\varepsilon_1^2(M\Lambda_1 + \Lambda)}$$
$$+ \frac{M\Lambda^2}{a^2\varepsilon_1^2} + \frac{\Lambda\Lambda_1}{a^2} + \frac{\Lambda(M\Lambda_1 + \Lambda)}{\varepsilon a^2} \tag{12.27}$$

and

$$\sigma_1^2 = \frac{R_s^2 a^2 - R^2 a^2\varepsilon_1 + \varepsilon_1\Lambda\Lambda_1 + \Lambda(M\Lambda_1 + \Lambda)}{\varepsilon_1 M}. \tag{12.28}$$

To find whether oscillatory convection occurs one minimizes (12.27) in a^2 and compares the result to the stationary convection boundary. If the resulting value of R^2, say $R_{osc}^2 < R_{stat}^2$, where R_{stat}^2 is the stationary convection threshold, then oscillatory convection is dominant provided $\sigma_1^2 > 0$, as may be verified or not by computing (12.28) at the value of a which minimizes R^2.

When $\Lambda_1 = \Lambda$, i.e. $\ell = 1$, then equation (12.27) reduces to

$$R^2 = \frac{R_s^2}{\varepsilon_1^2}\left(\frac{\varepsilon_1 + M}{1 + M}\right) + \frac{\Lambda^2}{a^2}\frac{(1+\varepsilon_1)(M+\varepsilon_1)}{\varepsilon_1^2}.$$

This is effectively the expression employed by [430], in equation (18) of that work. We take the opportunity to note that in equation (18) of [430] the R^2/ε_1 term should read R^2/ε_1^2. The numerical results presented in [430] are valid because the correct expression for R^2 was employed in the numerical codes.

When there is no inertia, i.e. $a_\perp = a_\parallel = 0$ then one may show that the critical value of R^2 arising from equation (12.27) is given by

$$R^2 = \frac{R_s^2}{\varepsilon_1} + \pi^2\left(\frac{1+A}{A}\right)\left(1 + \ell A + \frac{1+A}{\varepsilon_1}\right). \tag{12.29}$$

In equation (12.29) the coefficient A is given by

$$A = \sqrt{\frac{1+\varepsilon_1}{1+\varepsilon_1\ell}}.$$

Observe that this still includes anisotropy of the thermal conductivity.

One may then develop a nonlinear energy stability theory from the energy equation (12.18). To do this we define E, I and D by

$$E = \frac{1}{2}(a_{ij}u_i, u_j) + \frac{1}{2}\|\theta\|^2 + \frac{\varepsilon_1}{2}\|\phi\|^2 \tag{12.30}$$

and

$$I = 2(\theta, w) \tag{12.31}$$

together with

$$D = \|\mathbf{u}\|^2 + (\mathcal{L}_{ij}\theta_{,i}, \theta_{,j}). \tag{12.32}$$

Then equation (12.18) may be rewritten as

$$\frac{dE}{dt} = RI - D - \|\nabla\phi\|^2. \tag{12.33}$$

Define now R_E by

$$\frac{1}{R_E} = \max_H \frac{I}{D} \tag{12.34}$$

where H is the space of admissible solutions. Then from (12.33) we may obtain

$$\frac{dE}{dt} \le -D\left(1 - \frac{R}{R_E}\right) - \|\nabla\phi\|^2. \tag{12.35}$$

Suppose now $R < R_E$ and then using inequality $(12.8)_2$ together with Poincaré's inequality one may show

$$D \geq \|\mathbf{u}\|^2 + \ell_0 \pi^2 \|\theta\|^2. \qquad (12.36)$$

Since $R < R_E$ we employ (12.36) together with Poincaré's inequality again to find from (12.35) that with $1 - R/R_E = b(>0)$ we have

$$\frac{dE}{dt} \leq -b(\|\mathbf{u}\|^2 + \ell_0 \pi^2 \|\theta\|^2) - \pi^2 \|\phi\|^2. \qquad (12.37)$$

Recalling the energy function, E, is in this case given by (12.30), we use inequality (12.14) to deduce

$$\frac{dE}{dt} \leq -\frac{b}{\hat{a}}(a_{ij} u_i, u_j) - \ell_0 \pi^2 b \|\theta\|^2 - \pi^2 \|\phi\|^2. \qquad (12.38)$$

Define $k = \min\{2b/\hat{a}, 2\ell_0 \pi^2 b, 2\pi^2/\varepsilon_1\}$ and then from inequality (12.38) we obtain the energy inequality

$$\frac{dE}{dt} \leq -kE. \qquad (12.39)$$

Exponential decay of E follows from inequality (12.39) and then nonlinear stability is ensured via decay of $\mathbf{u}, \theta, \phi$ in $L^2(V)$ norm. The decay of $\|\mathbf{u}\|$ follows by use of inequality $(12.8)_1$.

The nonlinear stability threshold is determined by calculating R_E from the maximum problem in (12.34). The relevant Euler-Lagrange equations for this maximum problem are calculated and then one may show that this leads to the nonlinear stability threshold

$$R_E^2 = \pi^2(\sqrt{\ell} + 1)^2. \qquad (12.40)$$

Thus, for $R < \pi(\sqrt{\ell} + 1)$ we have global nonlinear stability.

12.3.1 Deductions

[430] presents numerical results for the case of anisotropy in the inertia tensor only, i.e. $a_\perp$ and $a_\parallel$ are allowed to be different. Thus, [430] deals only with the situation where $\ell = 1$. He reports numerical results for $\varepsilon_1 = 2, 5$ and 10. We here include some calculations for $\ell = 1$ but when $\varepsilon_1 = 6$ and 12. The results of this section are presented here for the first time and are not published elsewhere. We additionally examine the effect of $\ell \neq 1$ when $a_\parallel$ and $a_\perp$ are different. We specifically choose $\varepsilon_1 = 12$ and consider the cases of $\ell = 0.5$ and $\ell = 0.1$. Thus, we investigate the effect of variation of anisotropy in the inertia tensor a_{ij} and in the thermal conductivity $\mathscr{L}_{ij}$, upon the threshold for the onset of thermal convection.

12.3.2 Deductions When $\ell = 1$

We commence with the case studied by [430] where $\mathscr{L}_{ij} = \delta_{ij}$ and so we initially consider only the effect of anisotropic inertia.

Tables 12.1 and 12.2 display the critical values when $\varepsilon_1 = 6$ and $a_\perp = 1, a_\| = 10^{-2}$ then when $\varepsilon_1 = 6$ and $a_\perp = 10^{-2}, a_\| = 1$, respectively. Throughout tables 12.1–12.4 the critical wavenumber when stationary convection occurs is $a_{stat} = \pi$ so $a^2 \approx 9.8696$. In tables 12.1–12.4 the notation employed is that R^2 denotes the value for oscillatory convection, R^2_{osc} denotes the corresponding value of R^2 with no inertia, i.e. when $a_\perp = a_\| = 0$, a^2 is the value of the square of the wavenumber for oscillatory convection, σ_1 is the imaginary part of σ at the onset of convection, and R^2_s is the corresponding value of the salt Rayleigh number.

The first pair of values in each of tables 12.1–12.4 is where stationary and oscillatory convection occur at the same value of R^2_s. For R^2_s less than this value convection is stationary, whereas if R^2_s is larger then convection is oscillatory. For example, from table 12.1 the transition value of R^2_s is $R^2_{sT} = 58.0376$. For $R^2_s < R^2_{sT}$ instability begins by stationary convection while if $R^2_s > R^2_{sT}$ then oscillatory convection is witnessed. The corresponding transition value of R^2 in this case is given by $R^2 = 97.516$.

We see that at the transition value of R^2_s the wavenumber changes strongly. For the case of table 12.1 the wavenumber changes from $a^2 = 9.8696$ to $a^2 = 4.3$ as R^2_s passes through R^2_{sT}. This means that the convection cell becomes correspondingly wider since the width of the convection cell is proportional to $1/a$. We see the effect of $a_\perp$ and $a_\|$ in tables 12.1 and 12.2. The critical Rayleigh number transition value is smaller when $a_\perp = 1, a_\| = 10^{-2}$, then when $a_\perp = 10^{-2}, a_\| = 1$, and the critical wavenumber is smaller too. Thus a stronger inertia in the vertical direction is inducing a smaller oscillatory convection wavenumber and a much wider convection cell.

The same pattern of behaviour is found from tables 12.3 and 12.4 to what is observed from tables 12.1 and 12.2. However, the Rayleigh number values at transition are smaller and the transition takes place at a much smaller value of R^2_s. The change in cell width is not so great as in tables 12.1 and 12.2, although it is still substantial. When $\varepsilon_1 = 12$ the instability curves are displayed in figure 12.3. For $a_\perp = 1, a_\| = 10^{-2}$ the motionless solution is linearly stable if the values of (R^2, R^2_s) lie below the curve composed of S and a whereas the motionless solution is unstable if (R^2, R^2_s) lies above this curve. Likewise when $a_\perp = 10^{-2}, a_\| = 1$ the instability curve is composed of S and b. The S branch indicates stationary convection, whereas a or b denote oscillatory convection.

When $a_\perp = 0, a_\| = 0$, the instability curve in figure 12.3 is formed by a part of S which is then taken over by the curve marked "no inertia".

The instability curve when $\varepsilon_1 = 12, a_\perp = 1, a_\| = 10^{-2}, \ell = 1$ is displayed in figure 12.4. Here the instability curve is the one formed by the branches S and osc. (The curve osc corresponds to curve a in figure 12.3.) The S denotes stationary convection whereas osc indicates oscillatory convection. If (R^2, R^2_s) lies above this instability curve then the motionless solution is unstable. The global nonlinear stability curve is shown as the curve marked "Energy" in figure 12.4. If (R^2, R_s^2) lies below this curve then the motionless solution is unconditionally stable. The region marked "sub-critical" is where there is a *possibility* that finite amplitude nonlinear instabilities may arise. Figure 12.5 is a similar figure to that of figure 12.4 but involves curve b of figure 12.3, where curve b is marked *osc* in figure 12.5. Similar figures to figures 12.3 and 12.4 may be produced for the data corresponding to tables 12.1 and 12.2 where $\varepsilon_1 = 6$.

12.3.3 Deductions When $\ell \neq 1$

Data corresponding to $\varepsilon_1 \neq 1$ is given in tables 12.5–12.8 and in figures 12.6–12.9. The format is effectively the same as those of tables 12.1–12.4 and figures 12.3 and 12.4. However, we now also assess the effect of an anisotropic thermal conductivity since we take $\varepsilon_1 = 12$ but $\ell = 0.5$ and $\ell = 0.1$.

In tables 12.5 and 12.6 the value of ℓ is 0.5. Here we see that the transition value for the switch from stationary to oscillatory convection is $R_{sT}^2 = 40.58$ when $a_\perp = 1, a_\parallel = 10^{-2}$, and $R_{sT}^2 = 43.29$ when $a_\perp = 10^{-2}, a_\parallel = 1$. Thus, the pattern of tables 12.1–12.4 is again followed, but the transition values are much closer. This is also clearly displayed in figure 12.6 where the transitions are where the S-a curves meet and where the S-b curves meet. The oscillatory convection wavenumber is, however, very different when $a_\perp = 1, a_\parallel = 10^{-2}$ to when $a_\perp = 10^{-2}, a_\parallel = 1$, changing from 6.6 to 13.6. Thus, a combination of stronger vertical thermal conductivity and stronger vertical inertia is leading to a strong variation in convection cell widths. Overall, however, the pattern of convection as depicted in figures 12.6 and 12.7 corresponds to that when $\varepsilon_1 = 12$ and $\ell = 1$.

A strong change in behaviour is seen in tables 12.7 and 12.8 to that of the previous six tables. This change is also seen in figures 12.8 and 12.9 when compared to the previous figures. We see from tables 12.7 and 12.8, or from figure 12.8, that the transition to oscillatory convection branches switches position with regard to $a_\perp$ and $a_\parallel$ when $\ell = 0.1$. To explain further, in this case we observe from table 12.7 that $R_{sT}^2 = 111.04$ when $a_\perp = 1$ and $a_\parallel = 10^{-2}$ whereas it is $R_{sT}^2 = 66.35$ when $a_\perp = 10^{-2}$ and $a_\parallel = 1$. This is the reverse of what is seen in the previous tables. The anisotropic thermal conductivity is also having a pronounced effect on the critical wavenumber, or convection cell size, when $\ell = 0.1$. The oscillatory convection critical wavenumber value squared is $a^2 = 11$ when $a_\perp = 1, a_\parallel = 10^{-2}$ whereas it is $a^2 = 31.3$ when $a_\perp = 10^{-2}, a_\parallel = 1$. Thus, the much stronger vertical thermal conductivity than the horizontal one in conjunction with a much weaker vertical inertia than the horizontal one is leading to a narrower convection cell.

When $\ell = 0.1$ and $\varepsilon_1 = 12$, the oscillatory convection Rayleigh number values also display a strong decrease as R_s^2 increases, as seen in figures 12.8 and 12.9.

In conclusion, [430] observes that an anisotropic inertia in Darcy's law for the important heated and salted below thermosolutal convection problem can play a strong role in determining the onset of convective motion and on the size of the convection cells which arise. A larger inertia in the vertical direction would appear to aid convective motion and ensure the convection cells are relatively wider in the horizontal directions. In this chapter we have extended the work of [430] to include an anisotropic thermal diffusion coefficient in addition to an anisotropic inertia coefficient. The effects of these additional anisotropies upon the onset of thermosolutal convection in a saturated porous layer of Darcy type has been reported in detail, for both the case of heated from below and salted from above and heated from below and salted from below.

Table 12.1 Critical values of Rayleigh number and wavenumber, in linear instability theory. $\varepsilon_1 = 6$, $a_\perp = 1, a_\parallel = 10^{-2}$, $\ell = 1$. Showing R^2 and the analogous oscillatory curve *with no inertia*.

R^2	R^2_{stat}	R^2_{osc}	a^2	σ_1^2	R^2_s
97.516	97.516	55.731	4.3	0.843	58.0376
97.567	98.478	55.891	4.3	0.950	59
97.621	99.478	56.058	4.3	1.060	60
97.674	100.478	56.225	4.3	1.171	61
97.728	101.478	56.391	4.3	1.281	62
97.782	102.478	56.558	4.3	1.392	63
97.835	103.478	56.725	4.3	1.503	64
97.889	104.478	56.891	4.3	1.613	65
98.156	109.478	57.725	4.3	2.166	70
98.424	114.478	58.558	4.3	2.720	75

Table 12.2 Critical values of Rayleigh number and wavenumber, in linear instability theory. $\varepsilon_1 = 6$, $a_\perp = 10^{-2}, a_\parallel = 1$, $\ell = 1$. Showing R^2 and the analogous oscillatory curve *with no inertia*.

R^2	R^2_{stat}	R^2_{osc}	a^2	σ_1^2	R^2_s
126.077	126.077	60.491	9.8	0.147×10^{-3}	86.599
126.093	126.478	60.558	9.8	0.0499	87
126.133	127.478	60.725	9.8	0.174	88
126.174	128.478	60.891	9.8	0.298	89
126.214	129.478	61.058	9.8	0.422	90
126.417	134.478	61.891	9.8	1.043	95
126.619	139.478	62.725	9.8	1.663	100

Table 12.3 Critical values of Rayleigh number and wavenumber, in linear instability theory. $\varepsilon_1 = 12$, $a_\perp = 1$, $a_\| = 10^{-2}$, $\ell = 1$. Showing R^2 and the analogous oscillatory curve *with no inertia*.

R^2	R_{stat}^2	R_{osc}^2	a^2	σ_1^2	R_s^2
68.751	68.751	45.208	5.3	0.254	29.2721
68.764	69.478	45.268	5.3	0.300	30
68.783	70.478	45.352	5.3	0.363	31
68.802	71.478	45.435	5.3	0.427	32
68.821	72.478	45.518	5.3	0.490	33
68.840	73.478	45.602	5.3	0.553	34
68.859	74.478	45.685	5.3	0.617	35
68.953	79.478	46.102	5.3	0.933	40
69.047	84.478	46.518	5.3	1.249	45
69.142	89.478	46.935	5.3	1.566	50

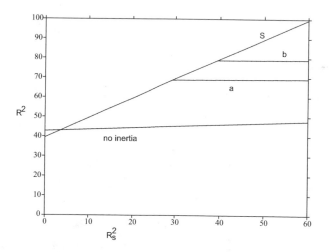

Fig. 12.3 R^2 versus R_s^2 at criticality. $\varepsilon_1 = 12$; $\ell = 1$; S is stationary convection; b is oscillatory convection with $a_\perp = 10^{-2}, a_\| = 1$; a is oscillatory convection with $a_\perp = 1, a_\| = 10^{-2}$; *no inertia* is oscillatory convection with $a_\perp = 0, a_\| = 0$.

Table 12.4 Critical values of Rayleigh number and wavenumber, in linear instability theory. $\varepsilon_1 = 12$, $a_\perp = 10^{-2}$, $a_\parallel = 1$, $\ell = 1$. Showing R^2 and the analogous oscillatory curve *with no inertia*.

R^2	R_{stat}^2	R_{osc}^2	a^2	σ_1^2	R_s^2
78.842	78.842	46.049	9.8	0.379×10^{-4}	39.3632
78.850	79.478	46.102	9.8	0.0435	40
78.864	80.478	46.185	9.8	0.112	41
78.878	81.478	46.268	9.8	0.180	42
78.892	82.478	46.352	9.8	0.248	43
78.906	83.478	46.435	9.8	0.317	44
78.920	84.478	46.518	9.8	0.385	45
78.990	89.478	46.935	9.8	0.726	50
79.129	99.478	47.768	9.8	1.409	60

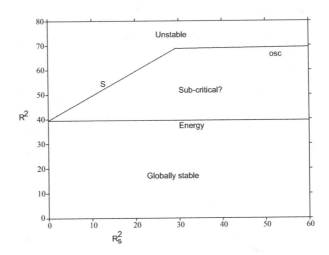

Fig. 12.4 R^2 versus R_s^2 at criticality. $\varepsilon_1 = 12$, $\ell = 1$, $a_\perp = 1$, $a_\parallel = 10^{-2}$. Curve *Energy* is the global stability curve, *osc* denotes the oscillatory convection curve, and S is stationary convection. "Sub-critical?" is the unknown region which *may* allow sub-critical instabilities.

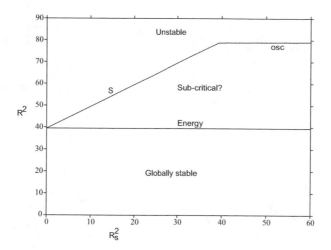

Fig. 12.5 R^2 versus R_s^2 at criticality. $\varepsilon_1 = 12, \ell = 1, a_\perp = 10^{-2}, a_\parallel = 1$. Curve *Energy* is the global stability curve, *osc* denotes the oscillatory convection curve, and S is stationary convection. "Sub-critical?" is the unknown region which *may* allow sub-critical instabilities.

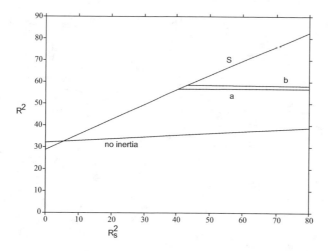

Fig. 12.6 R^2 versus R_s^2 at criticality. $\varepsilon_1 = 12$; $\ell = 0.5$; S is stationary convection; b is oscillatory convection with $a_\perp = 10^{-2}, a_\parallel = 1$; a is oscillatory convection with $a_\perp = 1, a_\parallel = 10^{-2}$; *no inertia* is oscillatory convection with $a_\perp = 0, a_\parallel = 0$.

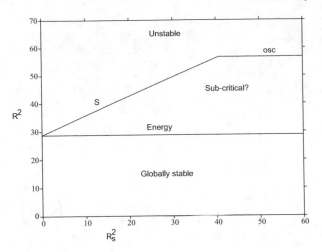

Fig. 12.7 R^2 versus R_s^2 at criticality. $\varepsilon_1 = 12, \ell = 0.5, a_\perp = 1, a_\parallel = 10^{-2}$. Curve *Energy* is the global stability curve, *osc* denotes the oscillatory convection curve, and S is stationary convection. "Sub-critical?" is the unknown region which *may* allow sub-critical instabilities.

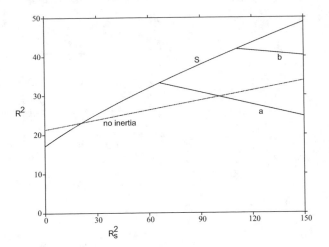

Fig. 12.8 R^2 versus R_s^2 at criticality. $\varepsilon_1 = 12$; $\ell = 0.1$; S is stationary convection; b is oscillatory convection with $a_\perp = 10^{-2}, a_\parallel = 1$; a is oscillatory convection with $a_\perp = 1, a_\parallel = 10^{-2}$; *no inertia* is oscillatory convection with $a_\perp = 0, a_\parallel = 0$.

Table 12.5 Critical values of Rayleigh number and wavenumber, in linear instability theory. $\varepsilon_1 = 12, a_\perp = 1, a_\parallel = 10^{-2}, \ell = 0.5$. Showing R^2 and the analogous oscillatory curve *with no inertia*.

R^2	R^2_{stat}	R^2_{osc}	a^2_{osc}	a^2_{stat}	σ^2_1	R^2_s
	29.468			14.1	0	1
	30.174			14.2	0	2
	30.879			14.3	0	3
	32.285			14.6	0	5
	35.781			15.2	0	10
	42.70			16.5	0	20
	49.53			17.8	0	30
	56.27			19.2	0	40
56.66	56.66	35.52	6.6	19.2	1.335	40.58
56.66	56.94	35.55	6.6	19.3	1.368	41
56.66	57.61	35.64	6.6	19.4	1.448	42
56.67	58.27	35.72	6.6	19.6	1.527	43
56.67	58.94	35.81	6.6	19.7	1.606	44
56.67	59.61	35.89	6.6	19.8	1.686	45
56.69	62.93	36.31	6.6	20.5	2.082	50
56.71	69.52	37.14	6.7	21.8	2.874	60

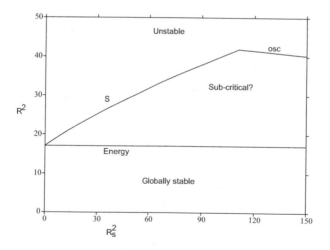

Fig. 12.9 R^2 versus R^2_s at criticality. $\varepsilon_1 = 12, \ell = 0.1, a_\perp = 1, a_\parallel = 10^{-2}$. Curve *Energy* is the global stability curve, *osc* denotes the oscillatory convection curve, and S is stationary convection. "Sub-critical?" is the unknown region which *may* allow sub-critical instabilities.

Table 12.6 Critical values of Rayleigh number and wavenumber, in linear instability theory. $\varepsilon_1 = 12$, $a_\perp = 10^{-2}, a_\parallel = 1, \ell = 0.5$. Showing R^2 and the analogous oscillatory curve *with no inertia*.

R^2	R^2_{stat}	R^2_{osc}	a^2_{osc}	a^2_{stat}	σ^2_1	R^2_s
	29.468			14.1	0	1
	30.174			14.2	0	2
	30.879			14.3	0	3
	32.285			14.6	0	5
	35.781			15.2	0	10
	42.70			16.5	0	20
	49.53			17.8	0	30
	56.27			19.2	0	40
58.468	58.468	35.75	13.6	19.6	1.723	43.29
58.46	58.94	35.81	13.6	19.7	1.814	44
58.45	59.61	35.89	13.6	19.8	1.942	45
58.39	62.93	36.31	13.6	20.5	2.583	50
58.34	66.23	36.72	13.6	21.1	3.223	55
58.28	69.52	37.14	13.7	21.8	3.896	60

Table 12.7 Critical values of Rayleigh number and wavenumber, in linear instability theory. $\varepsilon_1 = 12$, $a_\perp = 1, a_\parallel = 10^{-2}, \ell = 0.1$. Showing R^2 and the analogous oscillatory curve *with no inertia*.

R^2	R^2_{stat}	R^2_{osc}	a^2_{osc}	a^2_{stat}	σ^2_1	R^2_s
	17.41			32.0	0	1
	17.72			33.0	0	2
	18.63			35.0	0	5
	20.72			39.0	0	10
	22.75			47.0	0	20
	25.23			54.0	0	30
	27.57			60.0	0	40
	29.79			66.0	0	50
	33.98			77.0	0	70
	39.83			91.0	0	100
41.89	41.89	30.55	11.0	95.0	10.629	111.04
41.85	42.07	30.63	11.0	96.0	10.75	112
41.73	42.62	30.88	11.0	97.0	11.12	115
41.52	43.53	31.30	11.0	99.0	11.73	120

Table 12.8 Critical values of Rayleigh number and wavenumber, in linear instability theory. $\varepsilon_1 = 12$, $a_\perp = 10^{-2}$, $a_\parallel = 1$, $\ell = 0.1$. Showing R^2 and the analogous oscillatory curve *with no inertia*.

R^2	R^2_{stat}	R^2_{osc}	a^2_{osc}	a^2_{stat}	σ^2_1	R^2_s
	17.41			32.0	0	1
	17.72			33.0	0	2
	18.63			35.0	0	5
	20.72			39.0	0	10
	22.75			47.0	0	20
	25.23			54.0	0	30
33.23	33.23	26.83	31.3	74.6	25.86	66.35
33.17	33.37	26.88	31.0	75.0	25.90	67
33.07	33.57	26.97	32.0	75.0	27.59	68
32.97	33.78	27.05	32.0	76.0	28.17	69
32.86	33.98	27.13	32.0	77.0	28.75	70
32.35	34.99	27.55	33.0	79.0	32.98	75
31.72	36.18	28.05	34.0	82.0	38.09	80

Chapter 13
Bidispersive Porous Media

13.1 Double Porosity

There has been much recent interest in so called double porosity materials. This is where a solid body may have a double porosity structure. There is a macro porosity but also a micro porosity which may be linked with fissures in the solid skeleton. A very good example of a double porosity material may be seen in the pictures in [273] who show a pile of rocks which are themselves full of cracks. The macro porosity slowly degrades so that another photograph some ten years later displays a pile of finer material which is consistent with that of the micro porous structure. Such double porosity materials are important in the field of elasticity due to applications in land slides, civil engineering, or even in bone. Theoretical approaches to modelling double porosity materials have a long history and were started with the work of [29] and are described in some detail in [442, 443]. Stability and uniqueness studies for theories of double porosity materials have been given by [444] and by [429], with acceleration wave analyses by [153] and by [431].

We should mention that types of double porosity materials have been of interest in the field of chemistry for some time, see e.g. [67, 445, 464], where such materials are also known as bidispersive. In particular bidisperse porous media have been studied within the context of use as a catalyst in the production of high octane petrol, cf. [446, 447]. Bidisperse porous media and in particular the nano-scale pore structure have been found to be important in other areas of fuel production, especially in coal, cf. [199]. In the light of the many applications for bidisperse porous media, theories of fluid flow through a doubly porous body, or a bidispersive porous body, have been gaining impetus in recent years.

A theoretical study which is of particular relevance to the present section is the development of a theory for thermal convection in a bidispersive porous material, advanced by [315–319]. I believe this represents a potential to describe a class of porous materials which will have many useful applications. The theory of bidispersive porous media uses a collection of small spherical porous beads which are constructed into a larger "raspberry-like" spherical body. These raspberry-like struc-

© Springer International Publishing Switzerland 2015
B. Straughan, *Convection with Local Thermal Non-Equilibrium and Microfluidic Effects*,
Advances in Mechanics and Mathematics 32, DOI 10.1007/978-3-319-13530-4_13

tures are then arranged into a porous medium formed of porous spherical beads. A picture of the configuration is given on page 3069 of [317], and a schematic picture is given in figure 1.9 in section 1.1.1 of this book. Another example of a bidispersive porous medium is provided by figure 13.1. This configuration could be achieved by arranging small spherical beads as shown in figure 13.1. The configuration shown does not display a close packing of the small spherical beads although such a bidispersive porous medium could be created in practice, cf. figure 1.9 in section 1.1.1 of this book.

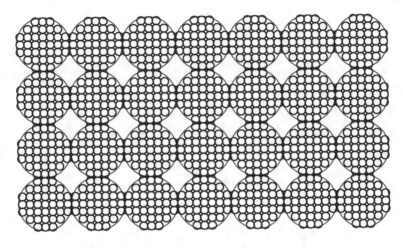

Fig. 13.1 An example of a bidispersive porous medium. The micro porosity defined by the smaller spheres is denoted by ε whereas the macro porosity defined by the larger "fictitious" spheres is denoted by ϕ.

The bidispersive porous medium theory employs a micro scale, denoted by p, and a macro scale, denoted by f. The p is associated with the pores in the raspberry-like structure while the f associates to the bulk (macro) material. [317] developed the first analysis of thermal convection in a bidispersive porous medium. Their fundamental work demonstrates the highly important fact that the critical Rayleigh number is much greater in a bidispersive porous medium than in a classical one. This gives a strong indication that a bidispersive porous medium will be significantly better to employ in thermal insulation than a regular porous medium. Since convection occurs at a higher Rayleigh number in the bidispersive porous medium the possibility of enhanced heat transfer due to convective fluid movement is therefore delayed. [418] continued the work of [317] and recalculated linear instability thresholds by a different numerical method. He additionally developed an energy stability theory for convection in a bidispersive porous medium. The analysis of [418] strongly reinforces the work of [317] which we believe is important.

13.2 Thermal Convection in Bidispersive Porous Media

The key equations governing the evolutionary behaviour of thermal convection in a bidispersive porous medium are based upon momentum equations involving the macro and micro fluid velocities U_i^f and U_i^p and upon energy balance equations for the macro and micro temperature fields T^f and T^p. The porosity at the macro level is denoted by ϕ whereas the porosity at the micro level is denoted by ε. The momentum equations advanced by [317] may be written as

$$
\begin{aligned}
&-\frac{\mu}{K_f}U_i^f + \tilde{\mu}\Delta U_i^f - \zeta(U_i^f - U_i^p) - \frac{\partial p^f}{\partial x_i} \\
&+ \rho_F g\hat{\beta}k_i\left(\frac{\phi}{\phi+\varepsilon(1-\phi)}T^f + \frac{\varepsilon(1-\phi)}{\phi+\varepsilon(1-\phi)}T^p\right) = 0,
\end{aligned}
\tag{13.1}
$$

and

$$
\begin{aligned}
&-\frac{\mu}{K_p}U_i^p + \tilde{\mu}\Delta U_i^p - \zeta(U_i^p - U_i^f) - \frac{\partial p^m}{\partial x_i} \\
&+ \rho_F g\hat{\beta}k_i\left(\frac{\phi}{\phi+\varepsilon(1-\phi)}T^f + \frac{\varepsilon(1-\phi)}{\phi+\varepsilon(1-\phi)}T^p\right) = 0.
\end{aligned}
\tag{13.2}
$$

The energy balance equations at the macro and micro levels are

$$
\phi(\rho c)_f\left(\frac{\partial T^f}{\partial t} + U_i^f\frac{\partial T^f}{\partial x_i}\right) = \phi k_f\Delta T^f + h(T^p - T^f),
\tag{13.3}
$$

and

$$
(1-\phi)(\rho c)_p\left(\frac{\partial T^p}{\partial t} + U_i^p\frac{\partial T^p}{\partial x_i}\right) = (1-\phi)k_p\Delta T^p + h(T^f - T^p).
\tag{13.4}
$$

The other coefficients appearing in equations (13.1)–(13.4) are μ, K_f, K_p, $\tilde{\mu}$, ζ, ρ_F, g, $\hat{\beta}$, $(\rho c)_f$, $(\rho c)_p$, k_f and k_p. These represent the viscosity of the fluid, permeability at the macro level, permeability at the micro (raspberry) level, the effective viscosity appropriate to the Brinkman term, an interaction coefficient, the density of the fluid at a reference temperature, gravity, the coefficient of thermal expansion of the fluid, the product of the density and the specific heat at constant pressure of the macro and micro phases, the thermal conductivity of the fluid in the macro and micro phases, respectively, with h being a thermal interaction coefficient. Note that inertia is neglected in equations (13.1) and (13.2).

In addition, the fluid is regarded as incompressible in both the macro and micro parts of the porous body. Thus the velocity fields U_i^f and U_i^p satisfy the continuity equations

$$
\frac{\partial U_i^f}{\partial x_i} = 0
\tag{13.5}
$$

and

$$\frac{\partial U_i^p}{\partial x_i} = 0. \tag{13.6}$$

In equations (13.1), (13.2) [317] and [418] adopt a single pressure p. However, since both $\mathbf{U}^f$ and $\mathbf{U}^p$ are divergence free this means that in the system composed of (13.1)–(13.6) there are 10 governing equations but only 9 variables, which would lead to an overdeterminate system. Thus, we follow the approach of elastic double porosity materials, see e.g. [442, 443] and allow for a pressure on the macro scale, p^f, together with a pressure on the micro scale, p^m. This does not alter the stability results of [317] or [418].

To study thermal convection as in [317] consider a fluid saturated bidispersive porous medium contained in the layer $\{(x,y) \in \mathbb{R}^2\} \times \{z \in (0,d)\}$. The upper plane $z = d$ is maintained at constant temperature, T_U, while the lower plane $z = 0$ is maintained at constant temperature, T_L, with $T_L > T_U$. The motionless solution in whose stability one is interested is

$$\bar{T}^f = -\frac{\Delta T}{d} z + T_L = \bar{T}^p, \qquad \bar{\mathbf{U}}^f = \bar{\mathbf{U}}^p = \mathbf{0}, \tag{13.7}$$

where $\Delta T = T_L - T_U > 0$. The steady pressure fields $\bar{p}^f$ and $\bar{p}^m$ are determined from (13.1), (13.2), with our knowledge of (13.7).

Following standard stability theory, perturbations $u_i^f, u_i^p, \theta^f, \theta^p, \pi^f, \pi^m$ are introduced in the manner

$$T^f = \bar{T}^f + \theta^f, \qquad T^p = \bar{T}^p + \theta^p, \qquad U_i^f = \bar{U}_i^f + u_i^f,$$
$$U_i^p = \bar{U}_i^p + u_i^p, \qquad p^f = \bar{p}^f + \pi^f, \qquad p^m = \bar{p}^m + \pi^m. \tag{13.8}$$

Then employing the governing evolutionary equations (13.1)–(13.6), together with the basic solution (13.7) and the definition of a perturbation (13.8), one may derive the equations satisfied by the perturbation variables.

[418] employs a different non-dimensionalization to that of [317], although it is entirely equivalent. We here follow that of [418] so put $D = \phi + \varepsilon(1 - \phi)$, and choose the time, velocity and temperature scales as

$$\mathscr{T} = \frac{(\rho c)_f d^2}{k_f}, \qquad U = \frac{k_f}{(\rho c)_f d}, \qquad T^\sharp = U\sqrt{\frac{(\rho c)_f \Delta T \, d\mu D}{k_f \rho_F g \hat{\beta} \phi K_f}}.$$

It transpires that other non-dimensional variables are required and these are defined in the following expressions

$$\sigma_f = \frac{\zeta d(\rho c)_f}{k_f}, \qquad H = \frac{hd^2}{\phi k_f}, \qquad \gamma = \frac{\phi k_f}{(1 - \phi)k_p},$$
$$\alpha = \frac{k_f}{k_p}\frac{(\rho c)_p}{(\rho c)_f}, \qquad K_r = \frac{K_p}{K_f}, \qquad Da_f = \frac{\tilde{\mu} K_f}{\mu d^2}.$$

[418] employs two Rayleigh numbers which are equivalent but one may be preferable to the other in physical interpretations. These numbers are Ra_F and Ra where

$$Ra_F = \frac{(\rho c)_f \Delta T d \rho_F g \hat{\beta} K_f}{k_f \mu \phi}, \qquad \text{and} \qquad Ra = Ra_F \left(\frac{\gamma}{\gamma + 1} \right). \tag{13.9}$$

It is convenient to use the parameter R defined in the relation $Ra_F = R^2$. It is also worth noting that Ra was suggested by [317].

Recollect that both velocity fields U_i^f and U_i^p are incompressible. With this in mind one may derive the non-dimensional perturbation equations which follow from (13.1) to (13.6). They comprise the momentum and continuity macro perturbation equations

$$-u_i^f + Da_f \Delta u_i^f - \sigma_f (u_i^f - u_i^p) - \frac{\partial \pi^f}{\partial x_i}$$
$$+ \frac{\phi}{\sqrt{D}} Rk_i \theta^f + \frac{\varepsilon (1 - \phi) R}{\sqrt{D}} k_i \theta^p = 0, \tag{13.10}$$

and

$$\frac{\partial u_i^f}{\partial x_i} = 0. \tag{13.11}$$

There are the momentum and continuity micro perturbation equations

$$-u_i^p + K_r Da_f \Delta u_i^p - K_r \sigma_f (u_i^p - u_i^f) - K_r \frac{\partial \pi^m}{\partial x_i}$$
$$+ \frac{\phi}{\sqrt{D}} RK_r k_i \theta^f + \frac{\varepsilon (1 - \phi) R}{\sqrt{D}} K_r k_i \theta^p = 0, \tag{13.12}$$

and

$$\frac{\partial u_i^p}{\partial x_i} = 0. \tag{13.13}$$

Further we have the macro energy balance perturbation equation

$$\frac{\partial \theta^f}{\partial t} + u_i^f \frac{\partial \theta^f}{\partial x_i} = \frac{\phi}{\sqrt{D}} Ru_3^f + \Delta \theta^f + H(\theta^p - \theta^f), \tag{13.14}$$

and the micro energy balance perturbation equation

$$\alpha \left(\frac{\partial \theta^p}{\partial t} + u_i^p \frac{\partial \theta^p}{\partial x_i} \right) = \frac{\phi}{\sqrt{D}} Ra u_3^p + \Delta \theta^p + \gamma H(\theta^f - \theta^p). \tag{13.15}$$

The perturbation equations (13.10)–(13.15) are required to hold on the domain $\{x, y \in \mathbb{R}^2\} \times \{z \in (0,1)\} \times \{t > 0\}$ and the solution is subject to the boundary conditions

$$u_i^f = 0, \qquad u_i^p = 0, \qquad \theta^f = 0, \qquad \theta^p = 0, \qquad \text{at } z = 0, 1. \qquad (13.16)$$

It is of interest to consider the linear operator attached to equations (13.10)–(13.15). If we write these equations in the abstract form of equation (2.5) we have a choice of ordering for the variable u. Suppose we here take

$$u = (u_1^f, u_2^f, u_3^f, u_1^p, u_2^p, u_3^p, \theta^f, \theta^p)$$

then divide equation (13.12) by K_r and divide equation (13.15) by γ. The linear operator consistent with equation (2.5) is then

$$L = \begin{pmatrix} A & 0 & 0 & \sigma_f & 0 & 0 & 0 & 0 \\ 0 & A & 0 & 0 & \sigma_f & 0 & 0 & 0 \\ 0 & 0 & A & 0 & 0 & \sigma_f & \dfrac{\phi R}{\sqrt{R}} & \dfrac{\varepsilon(1-\phi)R}{\sqrt{D}} \\ \sigma_f & 0 & 0 & B & 0 & 0 & 0 & 0 \\ 0 & \sigma_f & 0 & 0 & B & 0 & 0 & 0 \\ 0 & 0 & \sigma_f & 0 & 0 & B & \dfrac{\phi R}{\sqrt{D}} & \dfrac{\varepsilon(1-\phi)R}{\sqrt{D}} \\ 0 & 0 & \dfrac{\phi R}{\sqrt{D}} & 0 & 0 & 0 & \Delta - H & H \\ 0 & 0 & 0 & 0 & 0 & \dfrac{\phi R\alpha}{\sqrt{D}\gamma} & H & \dfrac{1}{\gamma}\Delta - H \end{pmatrix}$$

where A and B are given by

$$A = -1 + Da_f \Delta - \sigma_f$$

and

$$B = -\frac{1}{K_r} + Da_f \Delta - \sigma_f$$

The interest in analysing this operator is to see whether it is symmetric so one could infer information about nonlinear stability from the linear instability results. However, L is not symmetric.

13.2.1 Instability

[418] shows that the linear instability problem arising from (13.10) to (13.15) reduces to solving following system of equations for $u_3^f(\mathbf{x}), u_3^p(\mathbf{x}), \theta^f(\mathbf{x})$ and $\theta^p(\mathbf{x})$,

for the eigenvalue σ,

$$\Delta^2 u_3^f - a_1 \Delta u_3^f + a_3 \Delta u_3^p + a_4 R\Delta^* \theta^f + a_5 R\Delta^* \theta^p = 0,$$

$$\Delta^2 u_3^p - a_2 \Delta u_3^p + a_3 \Delta u_3^f + a_4 R\Delta^* \theta^f + a_5 R\Delta^* \theta^p = 0,$$

$$\sigma \theta^f = \frac{\phi}{\sqrt{D}} R u_3^f + \Delta \theta^f + H(\theta^p - \theta^f), \qquad (13.17)$$

$$\sigma \alpha \theta^p = \frac{\phi}{\sqrt{D}} R \alpha u_3^p + \Delta \theta^p + \gamma H(\theta^f - \theta^p).$$

In equations (13.17) the coefficients $a_1, \cdots, a_5$ are defined by

$$a_1 = \frac{1 + \sigma_f}{Da_f}, \quad a_2 = \frac{1 + \sigma_f K_r}{Da_f K_r}, \quad a_3 = \frac{\sigma_f}{Da_f},$$

$$a_4 = \frac{\phi}{Da_f \sqrt{D}}, \quad a_5 = \frac{\varepsilon(1 - \phi)}{Da_f \sqrt{D}},$$

and equations (13.17) are defined on $\{(x,y) \in \mathbb{R}^2\} \times \{z \in (0,1)\}$. One now represents $u_3^f = u_3^f(z)f(x,y)$ with similar forms for $u_3^p, \theta^f, \theta^p$, where f is a planform which tiles the plane, cf. such that $\Delta^* f = -a^2 f$ where a is a wavenumber.

It is convenient to solve the eigenvalue problem (13.17) together with the boundary conditions (13.16) by the Chebyshev-tau D^2 method. To do this we let $D = d/dz$ in (13.17) and we introduce new variables $\Omega_1(z)$ and $\Omega_2(z)$ such that we effectively replace Δu_3^f and Δu_3^p by these variables. In fact, we write $\Omega_1 = (D^2 - a^2)u_3^f$ and $\Omega_2 = (D^2 - a^2)u_3^p$ and we regard Ω_1 and Ω_2 as independent. In this manner one is faced with solving the equations

$$(D^2 - a^2)u_3^f - \Omega_1 = 0,$$

$$(D^2 - a^2)\Omega_1 - a_1\Omega_1 + a_3\Omega_2 - a_4 Ra^2 \theta^f - a_5 Ra^2 \theta^p = 0,$$

$$(D^2 - a^2)u_3^p - \Omega_2 = 0,$$

$$(D^2 - a^2)\Omega_2 - a_2\Omega_2 + a_3\Omega_1 - a_4 Ra^2 \theta^f - a_5 Ra^2 \theta^p = 0, \qquad (13.18)$$

$$(D^2 - a^2)\theta^f + H\theta^p - H\theta^f + \frac{\phi R}{\sqrt{D}} u_3^f = \sigma \theta^f,$$

$$(D^2 - a^2)\theta^p + \gamma H \theta^f - \gamma H \theta^p + \frac{\phi R}{\sqrt{D}} \alpha u_3^p = \sigma \alpha \theta^p.$$

Equations (13.18) are to be solved subject to the boundary conditions

$$u_3^f = Du_3^f = u_3^p = Du_3^p = \theta^f = \theta^p = 0, \quad z = 0, 1, \qquad (13.19)$$

for two fixed surfaces, or

$$u_3^f = \Omega_1 = u_3^p = \Omega_2 = \theta^f = \theta^p = 0, \quad z = 0, 1, \qquad (13.20)$$

when the surfaces are stress free.

[418] made an extensive search for oscillatory convection, i.e. for cases where σ is definitely complex. He tried many values of the parameters in equations (13.18). However, he always found the spectrum for σ (for all eigenvalues σ_n) to be *real*, despite the fact that the linear operator displayed earlier is not symmetric.

13.2.2 Nonlinear Stability

[418] developed a nonlinear energy stability theory for equations (13.10)–(13.15). He observes that there are various partial symmetries in equations (13.10)–(13.15) to potentially exploit. He chose to exploit symmetries between the $Rk_i\theta^f, Ru_3^f, Rk_i\theta^p$ and Ru_3^p terms in equations (13.10), (13.14), (13.12) and (13.15), respectively. To achieve this he multiplies equation (13.10) by u_i^f and integrates over the period cell V. He likewise multiples equation (13.12) by u_i^p and integrates over the period cell V. After some integrations by parts and use of equations (13.11) and (13.13) one may then arrive at the energy identities

$$
\begin{aligned}
0 = {} & -\|\mathbf{u}^f\|^2 - Da_f\|\nabla\mathbf{u}^f\|^2 - \sigma_f(u_i^f - u_i^p, u_i^f) \\
& + \frac{\phi R}{\sqrt{D}}(\theta^f, u_3^f) + \frac{\varepsilon(1-\phi)R}{\sqrt{D}}(\theta^p, u_3^f),
\end{aligned}
\tag{13.21}
$$

and

$$
\begin{aligned}
0 = {} & -\|\mathbf{u}^p\|^2 - K_r Da_f\|\nabla\mathbf{u}^p\|^2 - \sigma_f K_r(u_i^p - u_i^f, u_i^p) \\
& + \frac{\phi R}{\sqrt{D}}K_r(\theta^f, u_3^p) + \frac{\varepsilon(1-\phi)R}{\sqrt{D}}K_r(\theta^p, u_3^p).
\end{aligned}
\tag{13.22}
$$

Furthermore, one now multiplies equation (13.14) by θ^f and integrates over V and then one multiplies equation (13.15) by θ^p and integrates over V. In this way one may show that

$$
\frac{d}{dt}\frac{1}{2}\|\theta^f\|^2 = \frac{\phi R}{\sqrt{D}}(u_3^f, \theta^f) - \|\nabla\theta^f\|^2 - H(\theta^f - \theta^p, \theta^f),
\tag{13.23}
$$

and

$$
\frac{d}{dt}\frac{\alpha}{2}\|\theta^p\|^2 = \frac{\phi R\alpha}{\sqrt{D}}(u_3^p, \theta^p) - \|\nabla\theta^p\|^2 - \gamma H(\theta^p - \theta^f, \theta^p).
\tag{13.24}
$$

[418] then formed $\lambda((13.23) + \gamma^{-1}(13.24)) + (13.21) + K_r^{-1}(13.22)$, for a coupling parameter, $\lambda > 0$, to obtain the energy equation

$$
\frac{dE}{dt} = RI - \mathscr{D}.
\tag{13.25}
$$

In equation (13.25) the energy function E, the indefinite production term I, and the dissipation D, are defined by

$$E = \frac{\lambda}{2}\|\theta^f\|^2 + \frac{\lambda\alpha}{2\gamma}\|\theta^p\|^2, \tag{13.26}$$

$$I = (1+\lambda)\frac{\phi}{\sqrt{D}}(u_3^f, \theta^f) + \frac{\phi}{\sqrt{D}}(u_3^p, \theta^f)$$
$$+ \left(\frac{\lambda\phi\alpha}{\gamma\sqrt{D}} + \frac{\varepsilon(1-\phi)}{\sqrt{D}}\right)(\theta^p, u_3^p) + \frac{\varepsilon(1-\phi)}{\sqrt{D}}(\theta^p, u_3^f), \tag{13.27}$$

and

$$\mathscr{D} = \lambda\|\nabla\theta^f\|^2 + \frac{\lambda}{\gamma}\|\nabla\theta^p\|^2 + \lambda H\|\theta^f - \theta^p\|^2$$
$$+ Da_f(\|\nabla\mathbf{u}^f\|^2 + \|\nabla\mathbf{u}^p\|^2)$$
$$+ \|\mathbf{u}^f\|^2 + \frac{1}{K_r}\|\mathbf{u}^p\|^2 + \sigma_f\|\mathbf{u}^f - \mathbf{u}^p\|^2. \tag{13.28}$$

To determine a global nonlinear stability threshold [418] calculated the maximum

$$\frac{1}{R_E} = \max_H \frac{I}{\mathscr{D}}$$

where H is the space of admissible solutions and he deduced that $R < R_E$ leads to global nonlinear stability.

[418] introduces the variables $\zeta_1 = \sqrt{\lambda}\,\theta^f$, $\zeta_2 = \sqrt{\lambda}\,\theta^p$ and calculates the Euler-Lagrange equations for the maximum of $I/\mathscr{D}$ in terms of the functions $u_i^f, u_i^p, \zeta_1, \zeta_2$. The Euler-Lagrange equations are found to be

$$Da_f\Delta^2 u_3^f - (1+\sigma_f)\Delta u_3^f + \sigma_f\Delta u_3^p$$
$$= R_E\left(\frac{1+\lambda}{2\sqrt{\lambda}}\frac{\phi}{\sqrt{D}}a^2\zeta_1 + \frac{\varepsilon(1-\phi)}{2\sqrt{\lambda D}}a^2\zeta_2\right) \tag{13.29}$$

$$Da_f\Delta^2 u_3^p - \left(\frac{1}{K_r}+\sigma_f\right)\Delta u_3^p + \sigma_f\Delta u_3^f$$
$$= R_E\left[\left(\frac{\phi\alpha\sqrt{\lambda}}{2\gamma\sqrt{D}} + \frac{\varepsilon(1-\phi)}{2\sqrt{\lambda D}}\right)\zeta_2 + \frac{\phi}{2\sqrt{\lambda D}}\zeta_1\right]a^2 \tag{13.30}$$

$$\Delta\zeta_1 + H(\zeta_2 - \zeta_1) = -\frac{R_E}{2}\left[\frac{(1+\lambda)\phi}{\sqrt{\lambda D}}u_3^f + \frac{\phi}{\sqrt{\lambda D}}u_3^p\right] \tag{13.31}$$

$$\Delta\zeta_2 + H\gamma(\zeta_1 - \zeta_2) =$$
$$-R_E\frac{\gamma}{2}\left[\left(\frac{\phi\alpha\sqrt{\lambda}}{\gamma\sqrt{D}} + \frac{\varepsilon(1-\phi)}{\sqrt{\lambda D}}\right)u_3^p + \frac{\varepsilon(1-\phi)}{\sqrt{\lambda D}}u_3^f\right]. \tag{13.32}$$

[418] solves equations (13.29)–(13.32) by a D^2 Chebyshev tau numerical method, cf. [115]. However, he explains that care has to be taken since the solution may oscillate. Full details of the procedure are given in [418].

13.2.3 Numerical Results

Some linear instability and nonlinear energy stability thresholds are given by [418]. Extensive numerical computations for the linear instability problem are given by [317, figs 2–7, pp. 3072, 3073].

We here present some new calculations for the linear instability and nonlinear energy stability thresholds and these are presented in tables 13.1–13.3. In the tables Ra_F^L and Ra_F^E are the critical Rayleigh numbers of linear instability theory and of global nonlinear energy stability theory, with Ra and Ra_F^L defined as in (13.9), i.e.

$$Ra = Ra_F^L \frac{\gamma}{(1+\gamma)}.$$

The nonlinear energy Rayleigh number Ra_F^E corresponds to Ra_F as in (13.9). This means that one should compare Ra_F^E with Ra_F^L when comparing the nonlinear stability boundary with the linear instability boundary, for the results given here.

13.2.4 Variation of σ_f and Da_f

In tables 13.1–13.3 the values of α, K_r, H and γ are fixed as

$$\alpha = 1.5 \times 10^{-3}, K_r = 10^{-4}, H = 1, \qquad \text{and} \qquad \gamma = 1.$$

The values of $\phi, \varepsilon, \sigma_f$ and Da_f are as indicated in tables 13.1 and 13.3. In table 13.2 the values of $\phi, \varepsilon, \sigma_f$ and Da_f are as presented in table 13.1 and in the same order, i.e. in the first row $\phi = 0.4, \varepsilon = 0.3$, in the second $\phi = 0.3, \varepsilon = 0.4$, and so on. In table 13.2 the term λ is the critical value of the coupling parameter. We note from table 13.2 that the critical Rayleigh numbers of nonlinear stability theory Ra_F^E are somewhat close to those of linear instability theory Ra_F^L and so we may have some confidence that the linear theory is capturing the physics of the onset of convection reasonably well. One sees that for values of $\phi = 0.3, \varepsilon = 0.4$, the critical Rayleigh numbers are substantially increased compared to those for other values. Also, for fixed values of the parameters σ_f and Da_f, it is to be observed that a smaller macro porosity coupled with a larger micro porosity leads, in general, to much larger critical Rayleigh numbers.

13.2.5 Variation of σ_f and H

In tables 13.4–13.6 we include further numerical results for bidispersive porous convection. The Rayleigh and wave numbers have the same meaning as in tables 13.1–13.3. However, we now fix Da_f and allow σ_f and H to vary in order to analyse the effect of changing the interaction coefficients. The interaction coefficients are also allowed to be smaller than in tables 13.1–13.3. In tables 13.4–13.6 the constants α, K_r, Da_f and γ are fixed with values

$$\alpha = 1.5 \times 10^{-3}, K_r = 1, Da_f = 1, \gamma = 1.$$

In table 13.5 the values of $\phi, \varepsilon, \sigma_f$ and H are in the same order as those of table 13.4, i.e. the first row has $\sigma_f = 1, H = 0.1$, second row has $\sigma_f = 0.1, H = 1$, and so on.

From table 13.5 we observe that the nonlinear stability values are some way from those of linear instability theory. In addition, the effect of varying σ_f and H is to produce little variation in Ra_F^L. Indeed, we see no variation in Ra_F^L when H is fixed at $H = 0.1$ and σ_f varies from 1 to 0.1 to 0.01. The variation in H appears to decrease Ra_F^L.

Table 13.1 Critical values for two free surfaces, Brinkman bidispersive theory. Ra_F^L corresponds to definition (13.9)

Ra_F^L	Ra	a_L	ϕ	ε	σ_f	Da_f
2791.5	1395.7	2.31	0.4	0.3	1	1
4822.2	2411.1	2.29	0.3	0.4	1	1
3761.0	1880.5	2.31	0.3	0.2	1	1
3207.5	1603.8	2.31	0.3	0.1	1	1
862.9	431.4	2.62	0.3	0.3	1	10^{-1}
497.7	248.9	3.03	0.3	0.3	1	10^{-2}
2925.2	1462.6	2.95	0.3	0.3	10	10^{-1}
2535.2	1267.6	3.13	0.3	0.3	10	10^{-2}

13.3 Darcy Bidispersive Porous Media

To the best of my knowledge the results of this section are not available elsewhere.

In this section we consider the equivalent thermal convection problem to that of section 13.2 but we take the porous medium model to be one of Darcy type. Thus, the governing equations are again (13.1)–(13.6), but we set $\tilde{\mu} = 0$.

For thermal convection the motionless solution of interest is given by equations (13.7). The non-dimensionalization is the same as that of section 13.2 but now $Da_f = 0$. Hence, the non-dimensional perturbation equations governing the stability of solution (13.7) may be derived from (13.10) to (13.15). The non-dimensional perturbation equations for thermal convection in a bidispersive porous material of

Table 13.2 Critical values of the Rayleigh and wave numbers together with the critical value of the coupling parameter λ, for two free surfaces. Brinkman bidispersive theory. Ra_F^L and Ra_F^E correspond to definition (13.9)

Ra_F^L	Ra_F^E	a_L	a_E	λ
2.7915×10^3	2.6686×10^3	2.31	2.31	1.12
4.8222×10^3	4.1334×10^3	2.29	2.30	1.41
3.7610×10^3	3.5845×10^3	2.31	2.31	1.13
3.2075×10^3	3.1652×10^3	2.31	2.31	1.03
8.629×10^2	7.826×10^2	2.62	2.62	1.25
4.977×10^2	4.511×10^2	3.03	3.04	1.25
2.9252×10^3	2.6526×10^3	2.95	2.96	1.24
2.5352×10^3	2.2983×10^3	3.13	3.14	1.25

Table 13.3 Critical values of the Rayleigh numbers and of the linear wave number for two fixed surfaces, employing Brinkman bidispersive theory. The number Ra_F^L corresponds to definition (13.9)

Ra_F^L	Ra	a_L	ϕ	ε	σ_f	Da_f
6680.7	3340.4	3.15	0.4	0.3	1	1
11612.6	5806.3	3.13	0.3	0.4	1	1
9003.0	4501.5	3.15	0.3	0.2	1	1
7654.3	3827.2	3.16	0.3	0.1	1	1
1487.9	744.0	3.19	0.3	0.3	1	10^{-1}
589.4	294.7	3.26	0.3	0.3	1	10^{-2}
3707.2	1853.6	3.25	0.3	0.3	10	10^{-1}
2707.1	1353.5	3.23	0.3	0.3	10	10^{-2}

Table 13.4 Critical values for the Rayleigh numbers and of the linear wave number for two free surfaces, employing Brinkman bidispersive theory. The number Ra_F^L corresponds to definition (13.9)

Ra_F^L	Ra	a_L	ϕ	ε	σ_f	H
3979.9	1990.0	2.26	0.3	0.3	1	0.1
4043.3	2021.7	2.27	0.3	0.3	0.1	1
3979.9	1990.0	2.26	0.3	0.3	0.1	0.1
3979.9	1990.0	2.26	0.3	0.3	0.01	0.1
3972.9	1986.4	2.26	0.3	0.3	0.1	0.01
3972.9	1986.4	2.26	0.3	0.3	0.01	0.01

Table 13.5 Critical values for two free surfaces, Brinkman bidispersive theory. Ra_F^L and Ra_F^E correspond to definition (13.9)

Ra_F^L	Ra_F^E	a_L	a_E	λ
3.9799×10^3	3.1216×10^3	2.26	2.28	1.75
4.0433×10^3	3.0587×10^3	2.27	2.27	1.78
3.9799×10^3	2.9902×10^3	2.26	2.26	1.73
3.9799×10^3	2.9761×10^3	2.26	2.26	1.73
3.9729×10^3	2.9827×10^3	2.26	2.26	1.73
3.9729×10^3	2.9686×10^3	2.26	2.26	1.73

Table 13.6 Critical values for two fixed surfaces, Brinkman bidispersive theory

Ra_F^L	Ra	a_L	ϕ	ε	σ_f	H
9933.7	4966.9	3.12	0.3	0.3	1	0.1
10057.3	5028.7	3.13	0.3	0.3	0.1	1
9933.7	4966.9	3.12	0.3	0.3	0.1	0.1
9933.7	4966.9	3.12	0.3	0.3	0.01	0.1
9920.3	4960.1	3,12	0.3	0.3	0.1	0.01
9920.3	4960.1	3.12	0.3	0.3	0.01	0.01

Darcy type are, therefore,

$$-u_i^f - \sigma_f(u_i^f - u_i^p) - \frac{\partial \pi^f}{\partial x_i}$$
$$+ \frac{\phi}{\sqrt{D}} Rk_i \theta^f + \frac{\varepsilon(1-\phi)R}{\sqrt{D}} k_i \theta^p = 0,$$

$$\frac{\partial u_i^f}{\partial x_i} = 0,$$

$$-u_i^p - K_r \sigma_f(u_i^p - u_i^f) - K_r \frac{\partial \pi^m}{\partial x_i}$$
$$+ \frac{\phi}{\sqrt{D}} RK_r k_i \theta^f + \frac{\varepsilon(1-\phi)R}{\sqrt{D}} K_r k_i \theta^p = 0, \qquad (13.33)$$

$$\frac{\partial u_i^p}{\partial x_i} = 0,$$

$$\frac{\partial \theta^f}{\partial t} + u_i^f \frac{\partial \theta^f}{\partial x_i} = \frac{\phi}{\sqrt{D}} Ru_3^f + \Delta \theta^f + H(\theta^p - \theta^f),$$

$$\alpha \left(\frac{\partial \theta^p}{\partial t} + u_i^p \frac{\partial \theta^p}{\partial x_i} \right) = \frac{\phi}{\sqrt{D}} R\alpha u_3^p + \Delta \theta^p + \gamma H(\theta^f - \theta^p).$$

The boundary conditions which u_i^f, u_i^p, θ^f and θ^p must satisfy are

$$u_3^f = 0, \qquad u_3^p = 0, \qquad \theta^f = 0, \qquad \theta^p = 0, \qquad \text{at } z = 0, 1, \qquad (13.34)$$

together with $(u_i^f, u_i^p, \pi^f, \pi^m, \theta^f, \theta^p)$ satisfying a plane tiling periodicity in the x, y directions.

13.3.1 Linear Instability for Darcy Theory

To study linear instability we remove π^f and π^m from equations (13.33) and then we linearize the resulting system. One then assumes an exponential form in t such that $u_3^f(\mathbf{x}, t) = e^{\sigma t} u_3^f(\mathbf{x})$ with similar forms for $u_3^p, \pi^f, \pi^m, \theta^f, \theta^p$. One has then to solve the following eigenvalue problem for σ

$$\Delta u_3^f - \left(\frac{\sigma_f}{1 + \sigma_f}\right) \Delta u_3^p - \frac{\phi R}{(1 + \sigma_f)\sqrt{D}} \Delta^* \theta^f$$
$$- \frac{\varepsilon(1 - \phi)R}{(1 + \sigma_f)\sqrt{D}} \Delta^* \theta^p = 0,$$

$$\Delta u_3^p - \left(\frac{\sigma_f}{K_r^{-1} + \sigma_f}\right) \Delta u_3^f - \frac{\phi R}{(K_r^{-1} + \sigma_f)\sqrt{D}} \Delta^* \theta^f$$
$$- \frac{\varepsilon(1 - \phi)R}{(K_r^{-1} + \sigma_f)\sqrt{D}} \Delta^* \theta^p = 0, \qquad (13.35)$$

$$\Delta \theta^f - H\theta^f + H\theta^p + \frac{\phi R}{\sqrt{D}} u_3^f = \sigma \theta^f,$$

$$\Delta \theta^p - \gamma H \theta^p + \gamma H \theta^f + \frac{\phi R \alpha}{\sqrt{D}} u_3^p = \alpha \sigma \theta^p.$$

System (13.35) is to be solved in conjunction with the boundary conditions (13.34).

The idea is to introduce a periodic plane tiling function $f(x, y)$ such that $\Delta^* f + a^2 f = 0$ for a being a wavenumber. One then writes

$$u_3^f = W^f(z) f(x, y), \qquad u_3^p = W^p(z) f(x, y),$$
$$\theta^f = \Theta^f(z) f(x, y), \qquad \theta^p = \Theta^p(z) f(x, y), \qquad (13.36)$$

in equations (13.35). Then system (13.35) becomes a system of four second order differential equations in the variable z. Due to the boundary conditions one seeks a solution of form

$$W^f = W_1 \sin n\pi z, \qquad W^p = W_2 \sin n\pi z,$$
$$\Theta^f = \Theta_1 \sin n\pi z, \qquad \Theta^p = \Theta_2 \sin n\pi z, \qquad (13.37)$$

and from (13.35) we may derive a fourth order determinant and equate this to zero. This determinant is evaluated and one looks first for a stationary convection solution in which $\sigma = 0$. Then we put $\sigma = i\sigma_1$, $\sigma_1 \in \mathbb{R}$, and take real and imaginary parts of the resulting determinant equation. This yields the oscillatory convection solution and an expression for σ_1^2.

Define $\Lambda = n^2\pi^2 + a^2$ and define $K_1,\ldots,K_4$ by

$$K_1 = \frac{\sigma_f}{1+\sigma_f}, \qquad K_2 = \frac{\sigma_f}{K_r^{-1}+\sigma_f},$$

$$K_3 = \frac{1}{K_r}+\sigma_f, \qquad K_4 = 1+\sigma_f. \tag{13.38}$$

After some calculation one may show that the stationary convection boundary arises from the equation

$$R_{stat}^2 = \frac{\Lambda^3}{a^2}(1-K_1K_2)\frac{\left[\Lambda + H(1+\gamma)\right]}{P}, \tag{13.39}$$

where the function $P = P(a^2,n^2)$ is given by

$$\begin{aligned}
P = &\frac{1}{K_3D}\left[H\phi^2\alpha + \phi\alpha\varepsilon(1-\phi)(\Lambda+H)\right.\\
&\left.+\phi^2 K_1\Lambda(\Lambda+\gamma H)+\gamma HK_1\Lambda\varepsilon\phi(1-\phi)\right]\\
&+\frac{K_2\Lambda}{K_4D}\left[H\phi^2\alpha + \phi\varepsilon(1-\phi)\alpha(\Lambda+H)\right]\\
&+\frac{\phi\Lambda}{K_4D^{3/2}}\left[\phi\Lambda + \phi\gamma H + \gamma H\varepsilon(1-\phi)\right].
\end{aligned} \tag{13.40}$$

The oscillatory convection boundary arises from the expression

$$R_{osc}^2 = \frac{\Lambda^2}{Qa^2}(1-K_1K_2)\left[\Lambda\left(\frac{1+\alpha}{\alpha}\right)+H\left(\frac{\gamma+\alpha}{\alpha}\right)\right] \tag{13.41}$$

where the term Q is

$$Q = \frac{\varepsilon\phi(1-\phi)}{K_3D}+\Lambda\left\{\frac{\phi^2 K_1}{K_3D}+\frac{K_2\phi\varepsilon(1-\phi)}{K_4D}+\frac{\phi^2}{K_4D^{3/2}}\right\}.$$

Finally, σ_1^2 is given by the equation

$$\sigma_1^2 = \frac{\Lambda^2(1-K_1K_2)\left[\Lambda^2+\Lambda H(1+\gamma)\right]-R^2a^2P}{\alpha\Lambda^2(1-K_1K_2)\left[\Lambda^2+\Lambda H(1+\gamma)\right]} \tag{13.42}$$

where R^2 is the value obtained from equation (13.41).

To compute the stationary convection instability threshold we minimize R^2 from equation (13.39) in n and in a^2, and likewise we minimize R^2 from equation (13.41) to give the oscillatory convection threshold. For the parameters we have investigated we always find $n = 1$ yields the minimum. Also, for the parameters employed in this section we always found R^2_{stat} to be much lower than R^2_{osc} and $\sigma_1 = 0$. This means that initiation of thermal convection is always by stationary convection, at least for the parameter ranges explored here. Numerical results for the critical value of R^2_{stat} and the corresponding critical value of a are given in tables 13.7 and 13.8, and these are discussed in section 13.3.3

13.3.2 Nonlinear Stability for Darcy Theory

To establish a nonlinear stability boundary we follow the analysis of section 13.2.2, but we note that here $Da_f = 0$. The analysis of section 13.2.2 proceeds also here with the dissipation in equation (13.28) changing to

$$
\begin{aligned}
\mathscr{D} = & \lambda \|\nabla \theta^f\|^2 + \frac{\lambda}{\gamma} \|\nabla \theta^p\|^2 + \lambda H \|\theta^f - \theta^p\|^2 \\
& + \|\mathbf{u}^f\|^2 + \frac{1}{K_r} \|\mathbf{u}^p\|^2 + \sigma_f \|\mathbf{u}^f - \mathbf{u}^p\|^2.
\end{aligned}
\tag{13.43}
$$

The boundary conditions to be employed are those of equations (13.34).

The Euler-Lagrange equations which arise are those of (13.29)–(13.32) but the Da_f terms are not present. We represent the solution to (13.29)–(13.32) with $Da_f = 0$, as in (13.36) and (13.37). Then we find that the Euler-Lagrange equations reduce to the equation

$$
\mathscr{D}_1 = 0 \tag{13.44}
$$

where $\mathscr{D}_1$ is the 4×4 determinant given by

$$
\mathscr{D}_1 = \begin{vmatrix} K_4 \Lambda & -\sigma_f \Lambda & -\dfrac{R_E(1+\lambda)\phi a^2}{2\sqrt{D\lambda}} & -\dfrac{R_E \varepsilon (1-\phi) a^2}{2\sqrt{D\lambda}} \\[3mm] -\sigma_f \Lambda & K_3 \Lambda & -\dfrac{R_E \phi a^2}{2\sqrt{D\lambda}} & -R_E G a^2 \\[3mm] \dfrac{R_E(1+\lambda)\phi}{2\sqrt{D\lambda}} & \dfrac{R_E \phi}{2\sqrt{D\lambda}} & -(\Lambda + H) & H \\[3mm] \dfrac{R_E \gamma \varepsilon (1-\phi)}{2\sqrt{D\lambda}} & R_E \gamma G & H\gamma & -(\Lambda + H\gamma) \end{vmatrix}
$$

In the expression for $\mathscr{D}_1$ the term G has representation

$$
G = \frac{\phi \alpha \sqrt{\lambda}}{2\gamma \sqrt{D}} + \frac{\varepsilon (1-\phi)}{2\sqrt{\lambda D}}.
$$

After some calculation one shows that the equation (13.44) reduces to solving the equation for R_E^2,

$$AR_E^4 - BR_E^2 + C = 0. \tag{13.45}$$

In equation (13.45) the coefficients A and C are given by

$$A = \frac{\phi^2 a^4 \gamma}{16D^2} \left[\varepsilon(1-\phi) + \frac{(1+\lambda)\phi\alpha}{\gamma} \right]^2,$$

and

$$C = \Lambda^2(\Lambda^2 + H\Lambda + \gamma H\Lambda) \left[\frac{1}{K_r} + \sigma_f \left(1 + \frac{1}{K_r} \right) \right].$$

The coefficient B has form

$$C = \Lambda(\sigma_f Y_1 + Y_2),$$

where Y_1 and Y_2 are defined by

$$\begin{aligned}
Y_1 = & \frac{1}{2\sqrt{\lambda D}} HG\gamma\phi \left[(3+2\lambda)a^2 + 1 \right] + \gamma G^2(\Lambda + H) \\
& + \frac{\phi^2 a^2(\Lambda + \gamma H)}{4\lambda D} + \frac{Ga^2\gamma\varepsilon(1-\phi)(\Lambda+H)}{\sqrt{\lambda D}} \\
& + \frac{\phi a^2}{2\lambda D} \left[\phi(1+\lambda)(\Lambda + \gamma H) + H\gamma\varepsilon(1-\phi) \right],
\end{aligned}$$

and

$$\begin{aligned}
Y_2 = & \frac{H\gamma G\phi(1+a^2)}{2\sqrt{\lambda D}} + \gamma G^2(\Lambda + H) + \frac{\phi^2 a^2(\Lambda + \gamma H)}{4\lambda D} \\
& + \frac{K_3 a^2}{4\lambda D} \left\{ H\gamma\varepsilon(1-\phi)(1+\lambda)\phi + \phi^2(1+\lambda)^2(\Lambda + \gamma H) \right. \\
& \left. + \gamma\varepsilon^2(1-\phi)^2(\Lambda + H) + \gamma H(1+\lambda)\phi\varepsilon(1-\phi) \right\}.
\end{aligned}$$

We observe that $A > 0, B > 0$ and $C > 0$ and since we seek the minimum solution for R_E^2 we must have

$$R_E^2 = \frac{B - \sqrt{B^2 - 4AC}}{2A}. \tag{13.46}$$

To find the energy stability threshold we now solve the optimization problem

$$\max_{\lambda} \min_{a^2} R_E^2(a^2, \lambda), \tag{13.47}$$

where R_E^2 is evaluated using equation (13.46). Since λ is a coupling parameter at our disposal we have the freedom to select it as in (13.47).

Numerical output from the process (13.47) is included in section 13.3.3.

13.3.3 Numerical Results for Darcy Theory

Numerical results for the Darcy bidispersive convection problem are presented in tables 13.7 and 13.8. In table 13.7 the constants K_r, σ_f, γ, H and α are kept fixed with the values $K_r = 1, \sigma_f = 1, \gamma = 1, H = 1$ and $\alpha = 1.5 \times 10^{-3}$ and ϕ and ε have values as shown in the table. In table 13.8 we fix the constants $\phi, \varepsilon, K_r, \gamma$ and α with values $\phi = 0.3, \varepsilon = 0.3, K_r = 1, \gamma = 1, \alpha = 1.5 \times 10^{-3}$. In table 13.8 the coefficients H and σ_f vary with values as displayed in the table. We point out that we always found stationary convection to be dominant in the linear instability analysis. The Rayleigh number Ra_{stat} corresponds to the definition Ra_F in (13.9). The number Ra_E is the nonlinear energy critical Rayleigh number corresponding to the definition Ra_F in (13.9).

We firstly observe that we found $n = 1$ to yield the minimum values of Ra for the parameter ranges explored in tables 13.7 and 13.8. For example if we consider the first row in table 13.7 then $Ra_{stat} = 214.988$ with $a_L = 2.983$ are the values obtained with $n = 1$. If we allow $n = 2$ then instead we find $Ra_{stat}(n = 2) = 943.263$ with $a_L(n = 2) = 6.237$.

From table 13.7 we see that if we keep ε fixed then as ϕ decreases Ra_{stat} increases. This clearly has implications for thermal insulation where a higher Rayleigh number means less heat transfer. We could equally employ the alternate Rayleigh number in (13.9) which as $\gamma = 1$ here is simply half what Ra_{stat} is in tables 13.7 and 13.8. Even if we employ the alternate Rayleigh number the values reported here are much higher than the value $Ra = 4\pi^2$ of classical Darcy porous convection.

If we fix ϕ and allow ε to decrease then from table 13.7 we find that Ra_{stat} decreases. This is quantified by noting that with $\phi = 0.3$ when $\varepsilon = 0.5$ we have $Ra_{stat} = 298.373$ whereas when $\varepsilon = 0.2$ one sees $Ra_{stat} = 175.360$. The values for ε fixed and ϕ varying are with $\varepsilon = 0.3, \phi = 0.5$, then $Ra_{stat} = 114.066$, whereas with $\varepsilon = 0.3, \phi = 0.2$, then $Ra_{stat} = 375.986$.

From table 13.8 we observe that if we fix $\sigma_f = 1$ then vary $H = 1, 0.1, 0.01$ the variation in Ra_{stat} is minimal, from 214.988 to 214.252. When we fix $H = 1$ and vary $\sigma_f = 1, 0.1, 0.01$ the variation in Ra_{stat} is large, changing from 214.988 to 160.411 then to 149.261. This should be compared to table 13.4 for a Brinkman bidispersive porous material where essentially the opposite effect is seen. Namely, from table 13.4 when $\sigma_f = 0.1$ and H changes from $1, 0.1$ to $0.01, Ra_{stat}$ takes values 4043.3, 3979,9 then 3972.9. When $H = 0.1$ and $\sigma_f = 1, 0.1, 0.01, Ra_{stat}$ remains at the value 3979.9. The variation of Ra_{stat} is much greater in the Darcy bidispersive case, but the effects of σ_f and H are exactly opposite. It would appear that the lack of the higher derivative Laplacian Brinkman terms in the Darcy bidispersive theory is allowing the friction interaction term $\sigma_f(u_i^f - u_i^p)$ to have a much stronger effect on thermal convection.

It is also very noticeable that the nonlinear energy stability values obtained in the Darcy bidispersive theory are much lower than their counterparts in Brinkman theory. For example, from tables 13.5 and 13.2 we see that in the Brinkman problem the worst energy stability value is when $\sigma_f = 0.01$ and $H = 0.01$ but the Rayleigh number value is 74.7% below that of linear instability theory. The closest energy stability value for the Brinkman bidispersive problem is when $\phi = 0.3, \varepsilon = 0.1$ and then the energy stability critical Rayleigh number is 98.7% of that of linear instability. By contrast, from tables 13.7–13.8 the worst energy stability result is when $\phi = 0.3, \varepsilon = 0.5$ which has an energy critical Rayleigh number only 37.1% of that of the linear instability value. Even the best energy stability result in this case is when $H = 1, \sigma_f = 0.01$ where Ra_E is 67.7% of the linear instability critical Rayleigh number. Again, the higher order Laplacian terms in the Brinkman theory appear to aid the energy stability analysis. Their absence in Darcy bidispersive theory appears to allow the friction terms involving $\sigma_f(u_i^f - u_i^p)$ to impede an energy stability analysis.

One should observe that we have in this section only employed a nonlinear energy stability analysis with one coupling parameter λ. Perhaps an alternative strategy would improve the nonlinear energy stability thresholds. For example, we could multiply each of equations $(13.33)_1$–$(13.33)_4$ by u_i^f, u_i^p, θ^f and θ^p, respectively. Then integrate each resulting equation over V. This would yield four separate energy identities, which we denote, say, by A, B, C and J. Now, for three coupling parameters $\lambda_1, \lambda_2, \lambda_3 > 0$ at our disposal we form the single equation $A + \lambda_1 B + \lambda_2 C + \lambda_3 J$. Then one could define functions E, I and $\mathscr{D}$ by

$$E = \frac{\lambda_2}{2}\|\theta^f\|^2 + \frac{\lambda_3 \alpha}{2}\|\theta^p\|^2,$$

$$I = \sigma_f(1 + \lambda_1 K_r)(u_i^f, u_i^p) + \frac{\phi R}{\sqrt{D}}(1 + \lambda_2)(\theta^f, u_3^f)$$
$$+ \left(\frac{\phi R \lambda_3}{\sqrt{D}} + \frac{\varepsilon(1 - \phi)RK_r\lambda_1}{\sqrt{D}}\right)(\theta^p, u_3^p)$$
$$+ \frac{\varepsilon(1 - \phi)R}{\sqrt{D}}(\theta^p, u_3^f) + \frac{\phi R K_r \lambda_1}{\sqrt{D}}(\theta^f, u_3^p)$$
$$+ (\lambda_2 + \gamma\lambda_3)H(\theta^p, \theta^f)$$

and

$$\mathscr{D} = (1 + \sigma_f)\|\mathbf{u}^f\|^2 + \lambda_1(1 + K_r)\|\mathbf{u}^p\|^2$$
$$+ \lambda_2(\|\nabla\theta^f\|^2 + H\|\theta^f\|^2)$$
$$+ \lambda_3(\|\nabla\theta^p\|^2 + \gamma H\|\theta^p\|^2).$$

A nonlinear energy stability analysis may then proceed to find the nonlinear stability boundary R_E defined by

$$\frac{1}{R_E} = \max_H \frac{I}{\mathscr{D}}$$

where H is the space of admissible solutions. One then calculates the Euler-Lagrange equations for R_E and one has to solve the multi-dimensional optimization problem

$$Ra_E = \max_{\lambda_1, \lambda_2, \lambda_3} \min_{a^2} R_E^2(a^2, \lambda_1, \lambda_2, \lambda_3).$$

This is clearly a more challenging and much larger numerical problem than the energy one solved here. Nevertheless, its solution should produce improved nonlinear stability values when compared to those of tables 13.7 and 13.8.

Table 13.7 Critical values for Darcy bidispersive theory, as defined in section 13.3

Ra_{stat}	Ra_E	a_L	a_E	λ	ϕ	ε
214.988	97.846	2.983	3.16	2.05	0.3	0.3
375.986	167.124	2.933	3.13	2.51	0.2	0.3
175.360	88.687	3.000	3.18	1.89	0.3	0.2
286.456	151.740	2.983	3.15	2.13	0.2	0.2
148.899	65.964	3.000	3.18	1.88	0.4	0.3
256.098	105.072	2.950	3.15	2.25	0.3	0.4
170.546	70.361	2.983	3.17	1.98	0.4	0.4
114.066	48.496	3.017	3.19	1.81	0.5	0.3
298.373	110.718	2.933	3.13	2.48	0.3	0.5
139.079	54.075	2.983	3.17	1.91	0.5	0.5

Table 13.8 Critical values for Darcy bidispersive theory, as defined in section 13.3

Ra_{stat}	Ra_E	a_L	a_E	λ	H	σ_f
214.988	97.846	2.983	3.16	2.05	1	1
214.326	96.052	2.966	3.13	2.01	0.1	1
160.411	103.929	3.017	3.16	2.52	1	0.1
158.575	102.038	3.000	3.13	2.48	0.1	0.1
214.252	95.859	2.966	3.13	2.01	0.01	1
149.261	101.083	3.017	3.16	2.61	1	0.01
147.088	99.050	3.000	3.13	2.55	0.01	0.01

13.4 Tridispersive Media

[227] and [322] have generalized their theory of bidispersive porous media outlined in section 13.2 to derive a theory for a tridispersive porous medium. In fact, Professor Merab Svanadze has informed me that bone is really a triply porous material in which there is a macro porosity, then a micro porosity at a much smaller scale, and then a further micro (cracks) porosity at a third level. In fact, the tridisperse theory of [227] has precisely three levels of porosity which they denote by ϕ_1 (macro porosity), ϕ_2 (meso porosity) and finally ϕ_3 (micro porosity). With the increasing use of microfluidics the theory of [227] could become very important. We should point out that triple porosity models and their effects are of much importance in the oil industry, especially in modelling underground reservoirs, see e.g. [8, 327] and the references therein.

The theory of [227] has three porosity levels, but additionally is based on a velocity at each scale, v_i^1, v_i^2 and v_i^3, and a corresponding temperature at each scale, T^1, T^2 and T^3. The equations of [227] may be written as, where we employ separate pressures p^1, p^2, p^3 on the three scales, cf. section 13.2,

$$0 = -\frac{\partial p^1}{\partial x_i} - \frac{\mu}{K_1}v_i^1 - \zeta_{12}(v_i^1 - v_i^2) - gk_i\rho(T),$$

$$0 = -\frac{\partial p^2}{\partial x_i} - \frac{\mu}{K_2}v_i^2 - \zeta_{12}(v_i^2 - v_i^1) - \zeta_{23}(v_i^2 - v_i^3) - gk_i\rho(T),$$

$$0 = -\frac{\partial p^3}{\partial x_i} - \frac{\mu}{K_3}v_i^3 - \zeta_{23}(v_i^3 - v_i^2) - gk_i\rho(T),$$

$$\frac{\partial v_i^1}{\partial x_i} = 0,$$

$$\frac{\partial v_i^2}{\partial x_i} = 0,$$

$$\frac{\partial v_i^3}{\partial x_i} = 0,$$

$$\phi_1(\rho c)_1\frac{\partial T_1}{\partial t} + \phi_1(\rho c)_1 v_i^1 T_{,i}^1 = \phi_1 k_1 \Delta T_1 + h_{12}(T_2 - T_1),$$

$$(1 - \phi_1)\phi_2(\rho c)_2\frac{\partial T_2}{\partial t} + (1 - \phi_1)\phi_2(\rho c)_2 v_i^2 T_{,i}^2$$
$$= (1 - \phi_1)\phi_2 k_2 \Delta T_2 + h_{12}(T_1 - T_2) + h_{23}(T_3 - T_2),$$

$$(1 - \phi_1)(1 - \phi_2)(\rho c)_3\frac{\partial T^3}{\partial t} + (1 - \phi_1)(1 - \phi_2)(\rho c)_3 v_i^3 T_{,i}^3$$
$$= (1 - \phi_1)(1 - \phi_2)k_3 \Delta T_3 + h_{23}(T_2 - T_3),$$

$$(13.48)$$

where

$$\rho(T) = \rho_F\left(1 - \alpha[T - T_0]\right)$$

with T being the quantity

$$T = \frac{\phi_1 T_1 + (1 - \phi_1)\phi_2 T_2 + (1 - \phi_1)(1 - \phi_2)\phi_3 T_3}{\phi_1 + (1 - \phi_1)\phi_2 + (1 - \phi_1)(1 - k\phi_2)\phi_3}. \tag{13.49}$$

In these equations ρ_F denotes the fluid density with $\zeta_{12}, \zeta_{23}, h_{12}, h_{23}$ being interaction coefficients.

[227] analyse the onset of thermal convection in a horizontal layer of tripdispersive porous material saturated with a viscous, incompressible fluid, and heated from below. They find stationary convection thresholds. [227] present many numerical results and they certainly show that the critical Rayleigh number may be substantially larger or smaller than that found in a classical Darcy porous material, depending on the values of the appropriate parameters.

Chapter 14
Resonance in Thermal Convection

Resonance in thermal convection has many applications since interactions between the fluid layers may greatly increase the critical Rayleigh number threshold for the onset of convection, which in turn is of interest to the energy industry. In particular, with modern heat transfer devices being increasingly employed on a microscale there is much need to understand penetrative and resonant convection on a nanoscale, cf. [18, 118, 133, 300, 303, 402, 420, 425, 428, 433]. MEMS (micro-electro-mechanical-systems) will play an important part in future heat transfer technology, cf. [402], therefore, we believe an analysis of thermal convection influenced by resonance is important.

There have been some very interesting recent analyses of resonance in thermal convection. For clarification, by resonance we mean where instability in one part of a fluid layer may simultaneously occur with instability in another part of the layer. Such resonance can lead to unusually high Rayleigh numbers at the onset of thermal convection and so this is very much of interest to the heat transfer industry. Typically, a high Rayleigh number may be synonymous with delaying or prohibiting heat transfer and is thus important in insulation, while low Rayleigh numbers may be desirable when one requires rapid heat transfer such as in cooling pipes used in many modern devices such as computers. The implications of resonance in the energy sector may be important, especially with nano devices, see e.g. [118], or with application in building design and heat loss, see e.g. [289]. [303] investigated the possibility of thermal resonance in a bidispersive porous medium where resonance is between small scale perturbations in the porous blocks and large scale behaviour in the open fluid.

14.1 Resonant Penetrative Convection in a Fluid

In this section we recall some interesting results of [325]. These writers produce a resonance situation in a layer of water by having a constant heat source Q in the layer but simultaneously assuming that water in part of the layer is below $4°C$,

© Springer International Publishing Switzerland 2015

B. Straughan, *Convection with Local Thermal Non-Equilibrium and Microfluidic Effects*,
Advances in Mechanics and Mathematics 32, DOI 10.1007/978-3-319-13530-4_14

the temperature where water has a density maximum. This combination of physical effects leads to a scenario where a two cell convection structure may form in the fluid layer exhibiting resonance between the two cells.

[325] commence with a horizontal layer of water of height h bounded by two rigid surfaces. They assume that the upper boundary, $z = h$, is held at fixed temperature 4°C. They suppose that the lower boundary, $z = 0$, is maintained at constant temperature $T_0 \neq T_m = 4°C$. Therefore, the water in the vicinity of the plane $z = 0$ is less dense. [325] suppose the density in the buoyancy term is quadratic in temperature T so they employ the [468] (6.2) equation for the density–temperature relationship. In addition they assume a constant heat source is acting throughout the fluid.

The steady solution whose stability is investigated by [325] is the one where $\bar{v}_i \equiv 0$ while the steady temperature is

$$\bar{T}(z) = T_0 + \frac{\Delta T_1}{h^2}(1 - \mu)z^2 + \frac{\Delta T_1}{h}\mu z. \tag{14.1}$$

The coefficients appearing in (14.1) are

$$\Delta T_1 = T_m - T_0, \quad \mu = (2/\gamma)[1 + \gamma + (1 + \gamma)^{1/2}], \quad \text{and} \quad \gamma = \Delta T_1/\Delta T_2.$$

The quantity ΔT_2 is given by $\Delta T_2 = T_{\text{ex}} - T_m$ where T_{ex} is the temperature where $d\bar{T}/dz = 0$ at a depth d. For this configuration we have a steady state three layer situation in which there is an unstable layer of depth $d_2 = h - d$ lying above a stable layer, which in turn lies above an unstable layer of depth $d_1 = 1/(\mu - 1)$. The lowest of the three layers commences at $z = 0$. This means there there is a stable layer of fluid bounded above and below by two potentially unstable layers. Thermal convection may, therefore, commence in one or other of the layers and penetrate into the rest of the layer. The possibility arises that for appropriate layer depths thermal convection will switch from one layer to the other or essentially commence in both layers simultaneously like a resonance effect.

This resonance effect was discovered by [325]. These writers perform a linear instability analysis and they demonstrate that for values of μ between $\mu = 4.6$ and $\mu = 4.7$ the instability curve switches from one mode to another and there is a region in between where oscillatory convection occurs.

The momentum equation, continuity equation, and balance of energy equation, governing the [325] problem take the form

$$v_{i,t} + v_j v_{i,j} = -\frac{1}{\rho_m}p_{,i} + \nu \Delta v_i - g k_i \left[1 - \alpha(T - T_m)^2\right],$$

$$v_{i,i} = 0, \tag{14.2}$$

$$T_{,t} + v_i T_{,i} = \kappa \Delta T + Q.$$

In these equations v_i is velocity field, p is the pressure, and T is the temperature. The perturbation equations which arise from this are non-dimensionalized with length scale h, time scale h^2/ν, velocity scale ν/h, pressure scale $\nu\rho_m/h$, and temperature scale $U\sqrt{\nu/\kappa\alpha g h}$. The resulting non-dimensional perturbation equations become

$$\frac{\partial u_i}{\partial t} + u_j\frac{\partial u_i}{\partial x_j} = -\frac{\partial \pi}{\partial x_i} + \Delta u_i + 2RF(z)\theta k_i + Pr\theta^2 k_i,$$

$$\frac{\partial u_i}{\partial x_i} = 0, \tag{14.3}$$

$$Pr\left(\frac{\partial \theta}{\partial t} + u_i\frac{\partial \theta}{\partial x_i}\right) = -RF'(z)w + \Delta\theta,$$

where the function $F(z)$ is given by

$$F(z) = (1-\mu)z^2 + \mu z - 1. \tag{14.4}$$

As [414, p. 337, 338], observes, one may develop an unconditional nonlinear energy stability analysis for these equations by employing a weighted energy of form

$$E(t) = \frac{1}{2}\|\mathbf{u}\|^2 + <(\delta - 2z)\theta^2>, \tag{14.5}$$

for $\delta > 2$ a coupling parameter. One may show that this leads to an energy equation of the form

$$\frac{dE}{dt} = RI - D, \tag{14.6}$$

where the functions I and D are given by

$$I = <(2F - [\delta - 2z]F')w\theta>,$$
$$D = \|\nabla\mathbf{u}\|^2 + <(\delta - 2z)|\nabla\theta|^2>.$$

One may develop an unconditional nonlinear stability analysis from equation (14.6). The nonlinear critical Rayleigh number is determined by calculating $R_E^{-1} = \max_H I/D$. Numerical solution of the Euler-Lagrange equations which arise from this variational problem yields the critical nonlinear stability threshold.

The resonance behaviour found by [325] is discovered by solving the eigenvalue problem arising from the linearized instability problem. From equations (14.3) this means solving the eigenvalue system

$$\sigma u_i = -\pi_{,i} + \Delta u_i + 2RF(z)\theta k_i + Pr\theta^2 k_i,$$
$$u_{i,i} = 0, \tag{14.7}$$
$$Pr\sigma\theta = -RF'(z)w + \Delta\theta.$$

[325] find that for $\mu \in (4.6, 4.7)$ there are values such that the neutral stability curves have a shape like that shown in figure 14.1. The Rayleigh number Ra is given by $Ra = R^2$ and a is the wavenumber associated to (14.7). In figure 14.1 the branches of curve marked S indicate stationary convection whereas O indicates oscillatory convection. The minimum value of Ra is found on the O branch and this is where resonance is obtained. Precise numerical details are given in [325].

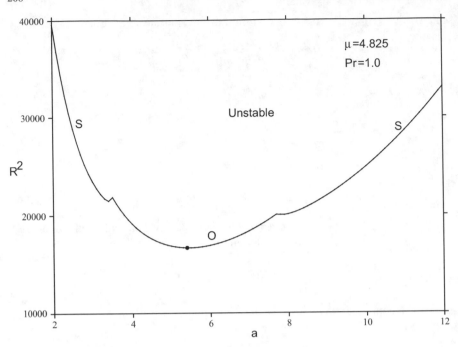

Fig. 14.1 A picture of a neutral curve found from equations (14.7), for $\mu = 4.825, Pr = 1$. The branch marked O indicates oscillatory convection whereas S indicates stationary convection. The minimum is at $a = 5.4$ where $Ra = 16681$.

14.2 Resonant Penetrative Convection in a Porous Medium

14.2.1 Nonlinear Density, Heat Source Model

[415] extended the work of [325] to an analogous convection problem in a porous medium. In addition to a linearized instability analysis [415] was interested in obtaining a global nonlinear stability threshold. Since the momentum equation in the porous case contains a θ^2 term like the last term in equation $(14.3)_1$ he used a quadratic Forchheimer theory to control the cubic nonlinearity which arises in the energy stability analysis. In this section we analyse the analogous porous convection problem but we employ a cubic Forchheimer equation rather than a quadratic one. That is, we employ equations (1.60) rather than equations (1.59) which were utilized by [415].

We follow [415] and commence with a layer of saturated porous material bounded by the horizontal planes $z = 0$ and $z = h(> 0)$. The layer is assumed infinite in horizontal extent so that $(x, y) \in \mathbb{R}^2$. The upper boundary $z = h$ is held at the constant temperature 4°C. The lower boundary $z = 0$ is held at a fixed temperature T_0 which is either in the range $0 \le T_0 < 4$, or $T_0 > 4$. In either case the density of water at $z = 0$ is smaller than that at $z = h$. The equation governing the temperature field in the porous medium is equation (1.35). However, we now allow for a constant value heat source or sink Q. Thus, the equation for the temperature field is

$$\frac{\partial T}{\partial t} + v_i \frac{\partial T}{\partial x_i} = \kappa \Delta T + Q. \tag{14.8}$$

We look for a steady solution of form $\bar{v}_i \equiv 0$ and $\bar{T} = \bar{T}(z)$. The presence of the heat source or heat sink means $\bar{T}(z)$ is a quadratic function when Q is constant. The constant Q is chosen such that the temperature field has a maximum greater than $4°C$ in $(0, h)$ if $0 \le T_0 < 4$, and has a minimum less than $4°C$ in $(0, h)$ if $T_0 > 4$. This leads to a situation in which there are effectively three layers in $(0, h)$, one stable with two potentially unstable. The steady solution may then be found as in [415]. Define ΔT_1, ΔT_2 and γ by

$$\Delta T_1 - T_m - T_0, \qquad \Delta T_2 - T_{ex} - T_m, \qquad \gamma = \frac{\Delta T_1}{\Delta T_2}.$$

It turns out that a key variable influencing convective instability is the parameter μ, given by

$$\mu = \frac{2}{\gamma} \left[1 + \gamma + \sqrt{1 + \gamma} \right]. \tag{14.9}$$

One may then show that the steady temperature field $\bar{T}(z)$, corresponding to the conduction solution, may be written as, cf. [415],

$$\bar{T}(z) - T_m = \frac{\Delta T_1}{h^2(1 - \mu)} z^2 - \frac{\mu \Delta T_1}{h(1 - \mu)} z - \Delta T_1, \tag{14.10}$$

where $z \in (0, h)$.

To complete the problem we add an appropriate form of the cubic Forchheimer equations (1.60). In this case the density in the buoyancy force depends on temperature and so we include the equations of momentum and continuity as

$$\frac{\hat{\mu}}{K} v_i + \hat{\lambda} |\mathbf{v}|^2 v_i = -\frac{\partial p}{\partial x_i} - \rho(T) g k_i, \tag{14.11}$$

$$\frac{\partial v_i}{\partial x_i} = 0, \tag{14.12}$$

where $\mathbf{k} = (0, 0, 1)$. The density in equation (14.11) is taken to be a quadratic function of temperature, so

$$\rho = \rho_0 \left(1 - \alpha [T - T_m]^2 \right), \tag{14.13}$$

where α is an expansion coefficient and ρ_0 is the density of water at $T = T_m = 4°C$. The nonlinear relation is necessary since we have a porous layer saturated with water whose temperature is in the $4°C$ range, i.e. the maximum temperature range. Thus, the full system of equations governing the problem of this section is

$$\frac{\hat{\mu}}{K} v_i + \hat{\lambda} |\mathbf{v}|^2 v_i = -\frac{\partial p}{\partial x_i} - \rho(T) g k_i,$$

$$\frac{\partial v_i}{\partial x_i} = 0, \tag{14.14}$$

$$\frac{\partial T}{\partial t} + v_i \frac{\partial T}{\partial x_i} = \kappa \Delta T + Q.$$

Perturbation variables (u_i, θ, π) are introduced to the steady solution to (14.14) via the relations

$$v_i = \bar{v}_i + u_i, \qquad T = \bar{T} + \theta, \qquad p = \bar{p} + \pi,$$

and then perturbation equations are derived. Upon using h as a length scale, the time, velocity, pressure and temperature scales are chosen as

$$\mathscr{T} = h^2/\kappa, \qquad U = \kappa/h, \qquad P = \hat{\mu} U h/K, \qquad T^\sharp = U \sqrt{\hat{\mu} h/\kappa g \alpha K \rho_0}.$$

The Rayleigh number Ra and the parameter R are defined as

$$Ra = R^2 = (\Delta T_1)^2 h \left(\frac{g \alpha k \rho_0}{\hat{\mu} \kappa} \right). \tag{14.15}$$

It may be shown that the non-dimensional version of the temperature field $\bar{T}$ given by (14.10) may be written as

$$F(z) = (1 - \mu) z^2 + \mu z - 1, \tag{14.16}$$

where $F(z)$ is the non-dimensional steady temperature field. Then, one may show that the the non-dimensional perturbation equations which arise from the equations (14.14) are

$$u_i + \lambda u_i |\mathbf{u}|^2 = -\frac{\partial \pi}{\partial x_i} + 2k_i RF \theta + k_i \theta^2,$$

$$\frac{\partial u_i}{\partial x_i} = 0, \tag{14.17}$$

$$\frac{\partial \theta}{\partial t} + u_i \frac{\partial \theta}{\partial x_i} = \Delta \theta - RF' w.$$

In these equations $F' = dF/dz$, λ is a non-dimensional Forchheimer coefficient, and $w = u_3$. Equations (14.17) are defined on the domain $\mathbb{R}^2 \times (0,1) \times \{t > 0\}$. The boundary conditions which the solution to (14.17) must satisfy are

$$w = 0, \quad \theta = 0, \qquad z = 0, 1, \tag{14.18}$$

with u_i, θ, π satisfying a plane tiling periodicity in the $x, y-$plane.

To investigate thermal convection and possible resonance behaviour which may arise from the solution to equations (14.17) and (14.18) we analyse linearized instability and global nonlinear stability theory for this model.

14.2.2 Linear Instability Analysis

Equations (14.17) are linearized and a time dependence like

$$u_i = e^{\sigma t} u_i(\mathbf{x}), \quad \theta = e^{\sigma t} \theta(\mathbf{x}), \quad \pi = e^{\sigma t} \pi(\mathbf{x}),$$

is assumed. Upon removing the pressure perturbation the linearized instability equations may be shown to reduce to

$$\begin{aligned} \Lambda w &= 2RF\Delta^*\theta, \\ \sigma\theta &= -RF'w + \Delta\theta, \end{aligned} \tag{14.19}$$

where $\Delta^* = \partial^2/\partial x^2 + \partial^2/\partial y^2$ is the horizontal Laplacian. Equations (14.19) are the same as those encountered by [415] and so the deductions drawn there for linearized instability apply also to the present problem.

We may thus argue as [415] and denote the depths of the "fictitious" layers as d_1 (lowest) with the two layers of depth d_2 above. Then one shows that with $n = d_1/d_2$,

$$\mu = \frac{2(n+1)}{n} \quad \text{and} \quad \gamma = n(n+2). \tag{14.20}$$

One might expect resonant-like behaviour when $d_1 = d_2$ and this corresponds to $\mu = 4$. In the porous medium context $Ra_1 \propto (\Delta T_1)^2 d_1$ and $Ra_2 \propto (\Delta T_2)^2 d_2$ and these will be equal when $d_1/d_2 = (\Delta T_2/\Delta T_1)^2$, i.e. when $n = 1/\gamma^2$. Since $\gamma = n(n+2)$ the equation $\gamma^2 n = 1$ means that we must solve the equation

$$n^5 + 4n^4 + 4n^3 = 1. \tag{14.21}$$

One finds $n \approx 0.54$ for which $\mu \approx 5.6$. Hence, we might expect complex growth rates in the region $\mu \in [4, 5.7]$ and possible oscillatory convection there. Numerical computation does find a range of μ where $\sigma \in \mathbb{C}$. However, the situation is very different in the porous case to what is found in the analogous fluid problem by [325]. This is discussed below.

To progress from equations (14.19) introduce a plane tiling form f, put $w = W(z)f(x,y)$, $\theta = \Theta(z)f(x,y)$, and introduce the wavenumber a by $\Delta^*f = -a^2 f$. The linear instability equations (14.19) then reduce to

$$\begin{aligned} (D^2 - a^2)W &= -2Ra^2 F\Theta, \\ (D^2 - a^2)\Theta - RF'W &= \sigma\Theta, \end{aligned} \tag{14.22}$$

where $D = d/dz$, $z \in (0,1)$. The boundary conditions are

$$W = \Theta = 0, \quad z = 0, 1. \tag{14.23}$$

System (14.22), (14.23) is solved numerically in [415].

14.2.3 Global Nonlinear Stability Analysis

In this section we investigate the problem of deriving a threshold for global nonlinear stability for the stationary solution (14.10) via the perturbation equations (14.17) and (14.18).

Let V be a period cell for the solution to (14.17). Energy identities are derived by multiplying $(14.17)_1$ by u_i and integrating over V, and by multiplying $(14.17)_3$ by θ and integrating over V, to find

$$\|\mathbf{u}\|^2 + \lambda \|\mathbf{u}\|_4^4 = 2R(F\theta, w) + (\theta^2, w), \tag{14.24}$$

and

$$\frac{d}{dt} \frac{1}{2} \|\theta\|^2 = -R(F'\theta, w) - \|\nabla\theta\|^2, \tag{14.25}$$

where $\|\cdot\|_4$ is the norm on $L^4(V)$. We wish to control the cubic term on the right of (14.24) and to utilize the $\|\mathbf{u}\|_4$ term in order to do so. One way to do this is to introduce an energy functional which contains more than the L^2 norm $\|\theta\|$. Because of the nature of the cubic Forchheimer term and the resulting L^4 integral of u_i in (14.24) a natural way to dominate the nonlinearity is to introduce into the energy function a term involving $\|\theta\|_4$. Thus, we now derive an equation for $\|\theta\|_4$. One may find

$$\frac{d}{dt} \frac{1}{4} \|\theta\|_4^4 = -R(F'w, \theta^3) - \frac{3}{4} \|\nabla\theta^2\|^2. \tag{14.26}$$

At this point one could form the combination $(14.24) + \lambda_1(14.25) + \lambda_2(14.26)$ and try to find an optimal nonlinear stability bound by varying the coupling parameters λ_1 and λ_2. To minimize on technical details we here simply add equations (14.24), (14.25) and (14.26) together. Thus, we define our Lyapunov functional $E(t)$ by

$$E(t) = \frac{1}{2} \|\theta\|^2 + \frac{1}{4} \|\theta\|_4^4. \tag{14.27}$$

By following the above procedure we thus obtain

$$\begin{aligned}
\frac{dE}{dt} = &-\|\mathbf{u}\|^2 - \lambda \|\mathbf{u}\|_4^4 - \|\nabla\theta\|^2 - \frac{3}{4} \|\nabla\theta^2\|^2 \\
&+ 2R(F\theta, w) - R(F'\theta, w) + (\theta^2, w) - R(F'w, \theta^3).
\end{aligned} \tag{14.28}$$

Let F_m be the maximum value of $|F'(z)|$ in $[0,1]$. Then employ Young's inequality on the last term in (14.28) to find for a constant $\varepsilon > 0$,

$$\begin{aligned}
-R(F'w, \theta^3) &\le \frac{RF_m}{4\varepsilon^4} \|w\|_4^4 + \frac{3RF_m\varepsilon^{4/3}}{4} \|\theta\|_4^4, \\
&\le \frac{RF_m}{4\varepsilon^4} \|\mathbf{u}\|_4^4 + \frac{3RF_m\varepsilon^{4/3}}{4} \|\theta\|_4^4.
\end{aligned} \tag{14.29}$$

We now employ inequality (14.29) in equation (14.28) and select ε to cancel out the $\|\mathbf{u}\|_4$ terms. Thus, we select $\varepsilon = (RF_m/4\lambda)^{1/4}$. This then leads from (14.28) to the inequality

$$
\begin{aligned}
\frac{dE}{dt} \leq & -\|\mathbf{u}\|^2 - \|\nabla\theta\|^2 - \frac{3}{4}\|\nabla\theta^2\|^2 + 2R(F\theta, w) \\
& - R(F'\theta, w) + (\theta^2, w) + \frac{3}{\lambda^{1/3}}\left(\frac{RF_m}{4}\right)^{4/3}\|\theta\|_4^4.
\end{aligned}
\tag{14.30}
$$

From Poincaré's inequality we know $\|\nabla\theta^2\|^2 \geq \pi^2\|\theta\|_4^4$ and we now require F_m and λ to be such that

$$
\lambda > \frac{(RF_m)^4}{4\pi^2}.
\tag{14.31}
$$

Note that inequality (14.31) ensures $\pi^2 > (RF_m)^{4/3}/\pi^2 4^{1/3}$ and so we may select a constant $\beta \in (0,1)$ such that $(1-\beta)\pi^2 = (RF_m)^{4/3}/(4\lambda)^{1/3}$. We do this and then regard $\phi = \theta^2$ as independent in (14.30) to derive an energy inequality of form

$$
\frac{dE}{dt} \leq I - D.
\tag{14.32}
$$

In inequality (14.32) the terms I and D are given by

$$
I = 2R(F\theta, w) - R(F'\theta, w) + (\phi, w),
\tag{14.33}
$$

and

$$
D = \|\mathbf{u}\|^2 + \|\nabla\theta\|^2 + \frac{3\beta}{4}\|\nabla\phi\|^2.
\tag{14.34}
$$

From inequality (14.32) we may deduce

$$
\frac{dE}{dt} \leq -D\left(1 - \frac{1}{R_E}\right).
\tag{14.35}
$$

where R_E is defined by

$$
\frac{1}{R_E} = \max_H \frac{I}{D},
\tag{14.36}
$$

with H being the space of admissible solutions.

Provided $R_E > 1$ we may employ Poincaré's inequality in (14.35) to show there is a constant $c > 0$ such that

$$
\frac{dE}{dt} \leq -cE.
\tag{14.37}
$$

One may obtain exponential decay of $E(t)$ from (14.37) and hence global nonlinear stability follows. The conditions which must hold are that $R_E > 1$ and λ satisfies the restriction (14.31).

The optimum value of $R_E(\mu, \zeta)$ is found by calculating the Euler-Lagrange equations from (14.36) and solving these numerically.

14.2.4 Oscillatory Behaviour Observed

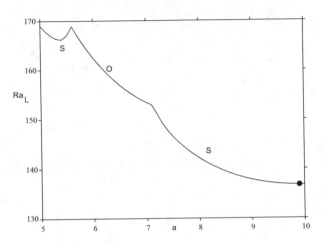

Fig. 14.2 Ra_L vs. a, for system (14.22), (14.23). The S branches indicate stationary convection, whereas the O branch represents oscillatory convection. Here $\mu = 5.1$ and the minimum value of Ra_L is shown on the right branch marked with a black dot.

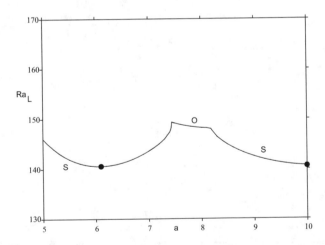

Fig. 14.3 Ra_L vs. a, for system (14.22), (14.23). The S branches indicate stationary convection, whereas the O branch represents oscillatory convection. Here $\mu = 5.2311$. The minimum value of Ra_L occurs on both S branches as shown by the black dots.

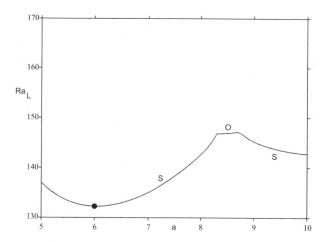

Fig. 14.4 Ra_L vs. a, for system (14.22), (14.23). The S branches indicate stationary convection, whereas the O branch represents oscillatory convection. Here $\mu = 5.3$ and the minimum value of Ra_L is shown on the left branch marked with a black dot.

The linear instability problem explained in section 14.2.2 gives rise to the same differential equations and boundary conditions, (14.22) and (14.23), as those of [415]. Thus, the conclusions drawn by [415] for the linear problem apply also to the problem of this section. [415] finds that there is a critical value of μ, $\mu_c = 5.2311$. For μ in the neighbourhood of μ_c there is a switch of convection from one arising in the lower part of the layer for $\mu < \mu_c$ to one commencing in the upper part of the layer for $\mu > \mu_c$. At μ_c a resonant-like behaviour may occur and convection may arise in the lower or upper layer.

Figures 14.2–14.4 display the neutral curves for values of $\mu = 5.1$, $\mu = 5.2311$ and $\mu = 5.3$. These curves display the linear instability critical Rayleigh number, Ra_L, against the critical wave number a. In each curve the parts marked S indicate stationary convection, whereas O indicates where oscillatory convection occurs. In figure 14.2 where $\mu = 5.1$ the minimum is found on the right branch with $a_c = 9.9$, as shown on the figure with a black dot. Figure 14.3 is where the instability switches from one branch to the other and the minimum occurs simultaneously on both S branches with a_c as shown by the black dots in this figure. Then, in figure 14.4, we see that the minimum has switched again but this time onto the left hand S branch where $a_c = 6.0$. It is interesting to note that even though there is a parameter region for μ such that in each of figures 14.2–14.4 σ is complex for a range of a values, the instability always occurs by stationary convection. Thus, for the problem of convection in a porous medium with a constant heat source and quadratic density, convection is by stationary convection. This is in contrast with the analogous problem in a clear fluid as studied by [325] who discovered a similar region of complex σ, but there oscillatory convection dominates.

14.3 Resonance Linear Heat Source

In this section we describe work of [433] who develops a model for penetrative convection in a fluid involving a heat source which varies linearly with vertical height across the layer. This allows one to obtain very strong resonance between sublayers. The mathematical analysis in [433] involves a linear instability technique which yields a definite instability boundary coupled with a global nonlinear energy stability analysis which yields a definite stability threshold. [433] points out that such analyses for a constant heat source have been performed previously, cf. [198, 377, 380], although the richness of the results for resonance obtained in [433] are not found with a constant heat source or sink. It is worth observing that [377] did allow both gravity and the heat source to depend on the vertical coordinate, but they did not investigate resonance.

The model begins with an incompressible fluid contained between the planes $z = 0$ and $z = d$, where $(x, y) \in \mathbb{R}^2$.

The relevant equations are the momentum equation

$$\rho_0 \left(\frac{\partial v_i}{\partial t} + v_j \frac{\partial v_i}{\partial x_j} \right) = -\frac{\partial p}{\partial x_i} + \mu \Delta v_i - g k_i \rho(T), \tag{14.38}$$

the balance of mass equation

$$\frac{\partial v_i}{\partial x_i} = 0, \tag{14.39}$$

and the energy balance equation

$$\frac{\partial T}{\partial t} + v_i \frac{\partial T}{\partial x_i} = \kappa \Delta T + Q(z). \tag{14.40}$$

In these equations v_i, T and p are velocity, temperature and pressure, with μ being dynamic viscosity and κ thermal conductivity. The function $Q(z)$ is a heat source or sink and is assumed to be a linear function of z of form

$$Q(z) = \tilde{Q}_0 + \tilde{Q}_1 z \tag{14.41}$$

where $\tilde{Q}_0$ and $\tilde{Q}_1$ are constants we control.

In equation (14.38) $\rho(T)$ is linear in temperature, and has form

$$\rho(T) = \rho_0 (1 - \alpha[T - T_0]). \tag{14.42}$$

The boundary conditions are no slip on $z = 0, d$ so that $v_i = 0$ there and the temperature is kept constant on the boundaries with $T = T_L$ at $z = 0$, and $T = T_U$ at $z = d$, where $T_L > T_U$.

The steady solution is assumed to be of form

$$\bar{v}_i \equiv 0, \qquad \bar{T} = \bar{T}(z), \qquad \bar{p} = \bar{p}(z).$$

The values of $\tilde{Q}_0$ and $\tilde{Q}_1$ are chosen in order that $\bar{T}(z)$ has the shape shown in figure 14.5.

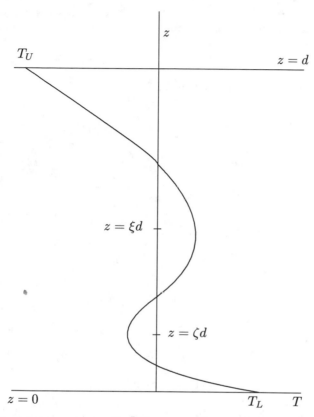

Fig. 14.5 Steady state temperature profile $\bar{T}(z)$, corresponding to solution (14.43).

The steady solution found by [433] has form

$$\bar{T}(z) = -\frac{Q_0}{2}z^2 - \frac{Q_1}{6}z^3 + \left(\frac{Q_0 d}{2} + \frac{Q_1 d^2}{6} - \beta\right)z + T_L, \qquad (14.43)$$

where β, Q_0 and Q_1 are given by

$$\beta = \frac{T_L - T_U}{d} > 0, \qquad Q_0 = \frac{\tilde{Q}_0}{\kappa}, \qquad Q_1 = \frac{\tilde{Q}_1}{\kappa},$$

cf. figure 14.5. In keeping with the physics of the problem convective instability may arise in the sublayer $(0, \zeta d)$ and in the sublayer $(\xi d, d)$ where ζ and ξ are calculated in [433], see equations (14.47). The steady pressure is found from equation (14.38).

To investigate stability of the steady solution $(\bar{v}_i, \bar{T}, \bar{p})$ we let (u_i, θ, π) be perturbations so that $v_i = \bar{v}_i + u_i$, $T = \bar{T} + \theta$ and $p = \bar{p} + \pi$. Equations are derived for (u_i, θ, π) from (14.38) to (14.43) and these are non-dimensionalized with length, time and velocity scales of $d, d^2/v$, and v/d, where $v = \mu/\rho_0$ is the kinematic viscosity. The temperature scale and pressure scales are

$$T^\sharp = Ud\sqrt{Q_1 v/\kappa\alpha g}, \qquad P = vU\rho_0/d,$$

and [433] introduces the numbers A_1 and B by

$$A_1 = \left|\frac{Q_0}{Q_1 d}\right|, \qquad B = \frac{\beta}{Q_1 d^2}.$$

Finally the Prandtl and Rayleigh numbers, Pr and Ra, are defined in [433] as

$$Pr = \frac{v}{\kappa}, \qquad Ra = \frac{d^6 Q_1 \alpha g}{\kappa v}. \tag{14.44}$$

The non-dimensional perturbation equations then become

$$\begin{aligned} u_{i,t} + u_j u_{i,j} &= -\pi_{,i} + R\theta k_i + \Delta u_i, \\ u_{i,i} &= 0, \\ Pr(\theta_{,t} + u_i \theta_{,i}) &= \Delta\theta + RF(z)w, \end{aligned} \tag{14.45}$$

where $R = \sqrt{Ra}$, $u_3 = w$, and where F is the negative of the non-dimensional temperature gradient given by

$$F(z) = -A_1\left(z - \frac{1}{2}\right) + \frac{1}{2}\left(z^2 - \frac{1}{3}\right) + B.$$

Equations (14.45) hold on the domain $\mathbb{R}^2 \times \{z \in (0,1)\} \times \{t > 0\}$. On the boundaries one has

$$u_i = 0, \quad \theta = 0, \qquad z = 0, 1, \tag{14.46}$$

and (u_i, θ, π) satisfy a plane tiling periodic form in the (x, y) directions.

[433] shows

$$\begin{aligned} \xi &= A_1 + \sqrt{A_1^2 - 2B - A_1 + \frac{1}{3}}, \\ \zeta &= A_1 - \sqrt{A_1^2 - 2B - A_1 + \frac{1}{3}}. \end{aligned} \tag{14.47}$$

In addition [433] shows that A_1 and B may be recast as functions of ζ and ξ such that

$$A_1 = \frac{\xi + \zeta}{2}, \qquad B = \frac{\xi\zeta}{2} - \left(\frac{\xi + \zeta}{4}\right) + \frac{1}{6}. \tag{14.48}$$

For the important case where $\xi + \zeta = 1$ it is shown in [433] that ξ is restricted to satisfy

$$0.211325 \approx \frac{1}{2} - \frac{1}{\sqrt{12}} < \xi < \frac{1}{2} + \frac{1}{\sqrt{12}} \approx 0.788675. \tag{14.49}$$

One may define different Rayleigh numbers Ra_1 and Ra_2 in terms of the depths $d_1 = \zeta$ and $d_2 = 1 - \xi$ and since $Ra \propto d^6$ this suggests resonance may occur with $d_1 = d_2$. This is investigated in detail in [433].

14.3.1 Linear Instability

[433] shows that the linearized instability eigenvalue problem arising from equations (14.15) and (14.46) reduces to solving the system

$$(D^2 - a^2)^2 W - Ra^2 \Theta = \sigma(D^2 - a^2)W$$
$$(D^2 - a^2)\Theta + RFW = \sigma Pr\Theta, \tag{14.50}$$

where $D = d/dz$ and $z \in (0,1)$. The boundary conditions for two fixed surfaces are

$$W = DW = \Theta - 0, \qquad z = 0,1. \tag{14.51}$$

In these equations W and Θ arise from $u_3 = w$ and θ where $w = W(z)f(x,y), \theta = \Theta(z)f(x,y)$, where f is a planform which tiles the plane and satisfies the equation $\Delta^* f = -a^2 f$, for a wavenumber a. System (14.50) together with the boundary conditions (14.51) is solved numerically by the D^2–Chebyshev tau method, cf. [115]. [433] calculates numerically various combinations of W, Θ for a variety of ξ, ζ values. These are displayed in [433]. In section 14.3.3 we do not show the eigenfunctions W, Θ but instead display some streamfunction and isotherm values which arise from W and Θ.

14.3.2 Global Nonlinear Stability

[433] develops a global nonlinear energy stability analysis based on the energy function E given by

$$E(t) = \frac{1}{2}\|\mathbf{u}\|^2 + \frac{\lambda Pr}{2}\|\theta\|^2. \tag{14.52}$$

He calculates the energy stability critical Rayleigh number, R_E, defined by

$$\frac{1}{R_E} = \max_{\mathcal{H}} \frac{I}{D} \tag{14.53}$$

where $\mathscr{H}$ is the space of admissible functions, i.e. $u_i \in H^1(V), \theta \in H^1(V)$, with u_i solenoidal and u_i, θ satisfying the boundary conditions. In the maximum problem (14.53) the functions I and D are defined by

$$I(t) = \big((1+\lambda F)w, \theta\big), \tag{14.54}$$

and

$$D(t) = \|\nabla \mathbf{u}\|^2 + \lambda \|\nabla \theta\|^2. \tag{14.55}$$

[433] shows that the nonlinear stability threshold R_E follows from the Euler-Lagrange equations

$$2\Delta u_i + R_E \mathscr{F} \phi k_i = -\frac{\partial \omega}{\partial x_i},$$

$$\frac{\partial u_i}{\partial x_i} = 0, \tag{14.56}$$

$$2\Delta \phi + R_E \mathscr{F} w = 0,$$

where ω is a Lagrange multiplier, $\phi = \lambda^{1/2} \theta$, and

$$\mathscr{F} = \frac{1 + \lambda F}{\sqrt{\lambda}}. \tag{14.57}$$

Remove ω from equation $(14.56)_1$ then seek a solution of form

$$w = W(z)f(x,y), \qquad \phi = \Phi(z)f(x,y).$$

To find R_E one then has to solve the eigenvalue problem

$$(D^2 - a^2)^2 W - R_E \frac{\mathscr{F}}{2} a^2 \Phi = 0,$$

$$(D^2 - a^2)\Phi + R_E \frac{\mathscr{F}}{2} W = 0. \tag{14.58}$$

System (14.58) is subject to the boundary conditions (14.51). This eigenvalue problem is solved numerically by the D^2–Chebyshev tau method, in [433], where many numerical results may be found.

14.3.3 Cellular Instability Structure

[433] finds resonance like behaviour for (ζ, ξ) in the range $(0.42, 0.58)$ to $(0.22, 0.78)$. Several tables of critical Rayleigh number results are displayed there for values of (ζ, ξ) in the above ranges. In addition, several figures displaying the critical functions $W(z)$ and $\Theta(z)$ are included for (ζ, ξ) in the same ranges. In particular, a rapid rise in Ra is found when $\xi = 0.76, 0.77$ and $\xi = 0.78$ as ζ decreases from 0.27 to 0.22. For example, with $\xi = 0.77$ when $\zeta = 0.27$, $Ra = 2412610$ and

when $\zeta = 0.22$, $Ra = 5232867$. For $\xi = 0.78$ when $\zeta = 0.27$, $Ra = 2380197$ whereas when $\zeta = 0.22$, $Ra = 6460507$. Such a large Ra increase is striking and should be of interest in insulation studies.

We now report on some of the results given in [433] and refer to this publication for further details, although our interpretation is very different. Differently from [433] we do not include figures of W, Θ but instead include streamline figures and isotherms.

Figures 14.6 and 14.7, respectively, display some streamlines and isotherms when $\zeta = 0.42$, $\xi = 0.58$ for which $Ra_c = 169565$ and $a_c = 3.18$. Throughout this discussion it is assumed $Pr = 5$. In figures 14.6 and 14.7 the cell width, L, is not 1 but $L = 0.987922$. One sees that the streamlines display the velocity structure to be that of a fairly regular single convection cell. However, the isotherms show that the temperature perturbation is already displaying signs of a two cell structure. The isotherms are all for either $\Theta > 0$ or $\Theta < 0$ and there is no change of sign.

In figures 14.8 and 14.9 we have $\zeta - 0.329$, $\xi = 0.671$ with $Ra_c - 1043618$ and $a_c = 5.53$. The convection cells are much narrower with width $L = 0.568100$. It is observed that while the streamlines in figure 14.8 mostly indicate one cell there is a tendency for a two cell structure to form. The isotherms as shown in figure 14.9 already display a three cell structure. The inner cell is where Θ has already changed sign.

The values of ζ and ξ in figures 14.10 and 14.11 change to $\zeta = 0.328$ and $\xi = 0.672$. Here $Ra_c = 1062452$ and $a_c = 5.66$. Thus, the cells are narrower than in figures 14.8 and 14.9 with width $L = 0.555052$. The streamlines in figure 14.10 display a strong change from those in figure 14.8. Indeed a two cell structure has definitely formed with circulation in opposite directions. The isotherms in figure 14.11 likewise show a clear two cell structure.

When we keep $\xi + \zeta = 1$ and increase ξ we find another change as shown in figures 14.12 and 14.13. The values of ξ and ζ are $\xi = 0.756$ and $\zeta = 0.244$ where $Ra_c = 4043926$ and $a_c = 7.96$. The streamlines in figure 14.12 still display a two layer structure with counter rotating cells. However, the flow rate is stronger toward the upper and lower boundaries. The isotherms in figure 14.13 display a strong change from those of lower ξ values. Here we find a four cell structure with Θ changing sign from cell to cell as we vary z from 0 to 1. Observe that the cell is narrower than in figures 14.10 and 14.11 with $L = 0.394672$.

We display in figures 14.14 and 14.15 the final streamlines and isotherms we can compute for $\xi + \zeta = 1$ because ξ is restricted by inequalities (14.49). In these figures $\xi = 0.78$, $\zeta = 0.22$, $Ra_c = 6460507$ and $a_c = 8.85$. The cell width is the narrowest of those we display and $L = 0.354982$. The streamlines in figure 14.14 display a three cell structure with a weak cell in the centre rotating in a different direction to the cells above and below. The isotherms in figure 14.15 show a four cell structure but one which is different from that of figure 14.13. In figure 14.15 the values of Θ are the same in the two outer cells and change sign in the inner two cells where Θ also has the same sign.

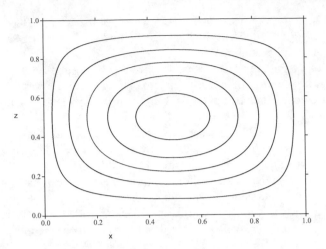

Fig. 14.6 Linear instability streamfunctions at criticality, $\psi = \sin(a_c x)W(z)$. Here $Ra = 169565$, $a_c = 3.18$, $\zeta = 0.42$, $\xi = 0.58$, $L = 0.987922$, $Pr = 5$. Note the cell width is L and not 1. The innermost streamfunction curve is for $\psi = 0.9$, with ψ taking values 0.7, 0.5, 0.3 until one reaches the outer streamfunction curve which is for $\psi = 0.1$.

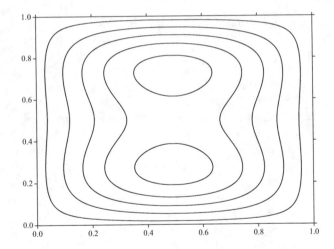

Fig. 14.7 Linear instability perturbation isotherms at criticality, $A = \sin(a_c x)\Theta(z)$. Here $Ra = 169565$, $a_c = 3.18$, $\zeta = 0.42$, $\xi = 0.58$, $L = 0.987922$, $Pr = 5$. Note the cell width is L and not 1. The two innermost isotherm curves are for $A = 0.9$. As one moves out from the centre of the cell the curves are for A taking values 0.7, 0.5, 0.3 until one reaches the outer isotherm curve which is for $A = 0.1$.

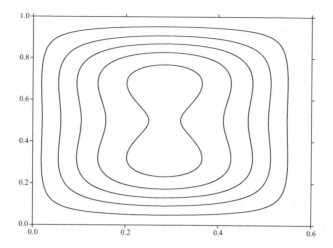

Fig. 14.8 Linear instability streamfunctions at criticality, $\psi = \sin(a_c x)W(z)$. Here $Ra = 1043618$, $a_c = 5.53$, $\zeta = 0.329$, $\xi = 0.671$, $L = 0.568100$, $Pr = 5$. Note the cell width is L and not 0.6. The innermost streamfunction curve is for $\psi = 0.9$, with ψ taking values 0.7, 0.5, 0.3 until one reaches the outer streamfunction curve which is for $\psi = 0.1$.

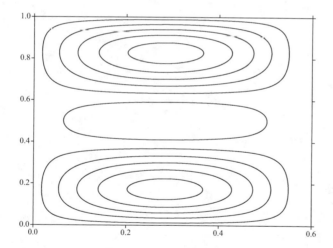

Fig. 14.9 Linear instability perturbation isotherms at criticality, $A = \sin(a_c x)\Theta(z)$. Here $Ra = 1043618$, $a_c = 5.53$, $\zeta = 0.329$, $\xi = 0.671$, $L = 0.568100$, $Pr = 5$. Note the cell width is L and not 0.6. The innermost isotherm curves are for $A = 0.9$. As one moves out from the centre of the cell the curves are for A taking values 0.7, 0.5, 0.3 until one reaches the outer isotherm curve which is for $A = 0.1$, and there is also a third cell for this value with the temperature perturbation changing sign.

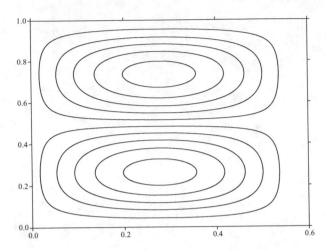

Fig. 14.10 Linear instability streamfunctions at criticality, $\psi = \sin(a_c x)W(z)$. Here $Ra = 1062452$, $a_c = 5.66$, $\zeta = 0.328$, $\xi = 0.672$, $L = 0.555052$, $Pr = 5$. Note the cell width is L and not 0.6. The innermost streamfunction curves are for $\psi = 0.9$, with ψ taking values 0.7, 0.5, 0.3 until one reaches the outer streamfunction curves which are for $\psi = 0.1$.

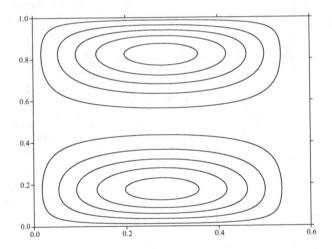

Fig. 14.11 Linear instability perturbation isotherms at criticality, $A = \sin(a_c x)\Theta(z)$. Here $Ra = 1062452$, $a_c = 5.66$, $\zeta = 0.328$, $\xi = 0.672$, $L = 0.555052$, $Pr = 5$. Note the cell width is L and not 0.6. The innermost isotherm curves are for $A = 0.9$. As one moves out from the centre of the cell the curves are for A taking values 0.7, 0.5, 0.3 until one reaches the outer isotherm curves which are for $A = 0.1$.

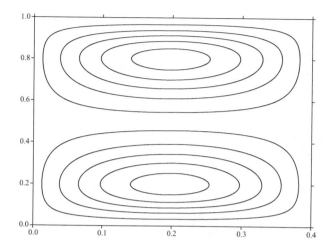

Fig. 14.12 Linear instability streamfunctions at criticality, $\psi = \sin(a_c x)W(z)$. Here $Ra = 4043926$, $a_c = 7.96$, $\zeta = 0.244$, $\xi = 0.756$, $L = 0.394672$, $Pr = 5$. Note the cell width is L and not 0.4. The innermost streamfunction curves are for $\psi = 0.9$, with ψ taking values 0.7, 0.5, 0.3 until one reaches the outer streamfunction curves which are for $\psi = 0.1$.

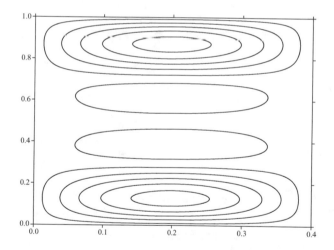

Fig. 14.13 Linear instability perturbation isotherms at criticality, $A = \sin(a_c x)\Theta(z)$. Here $Ra = 4043926$, $a_c = 7.96$, $\zeta = 0.244$, $\xi = 0.756$, $L = 0.394672$, $Pr = 5$. Note the cell width is L and not 0.4. The innermost isotherm curves are for $A = 0.9$. As one moves out from the centre of the cell the curves are for A taking values 0.7, 0.5, 0.3 until one reaches the outer isotherm curve which is for $A = 0.1$, and there are also two separate cells for this value with the temperature perturbation changing sign.

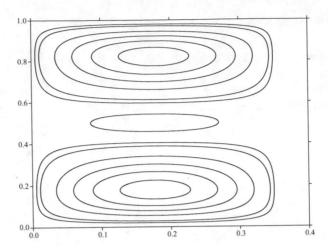

Fig. 14.14 Linear instability streamfunctions at criticality, $\psi = \sin(a_c x)W(z)$. Here $Ra = 6460507$, $a_c = 8.85$, $\zeta = 0.22$, $\xi = 0.78$, $L = 0.354982$, $Pr = 5$. Note the cell width is L and not 0.4. The innermost streamfunction curves are for $\psi = 0.9$, with ψ taking values 0.7, 0.5, 0.3, 0.1 until one reaches the outer streamfunction curves which are for $\psi = 0.05$, and we notice a third separate cell for this value.

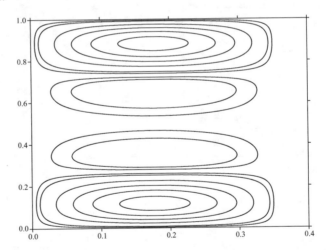

Fig. 14.15 Linear instability perturbation isotherms at criticality, $A = \sin(a_c x)\Theta(z)$. Here $Ra = 6460507$, $a_c = 8.85$, $\zeta = 0.22$, $\xi = 0.78$, $L = 0.354982$, $Pr = 5$. Note the cell width is L and not 0.4. The innermost isotherm curves are for $A = 0.9$. As one moves out from the centre of the cell the curves are for A taking values 0.7, 0.5, 0.3, 0.1 until one reaches the outer isotherm curves which are for $A = 0.05$. Observe that there are also two separate cells for the valueis 0.1 and 0.05 with the temperature perturbation changing sign.

14.4 Triple Resonance

[427] investigated a situation of thermal convection in a fluid layer where resonance may possibly occur simultaneously in three distinct sublayers. To achieve this he assumes the density-temperature relation in the buoyancy term is quadratic and he allows the heat source/sink to vary linearly with vertical height z. It transpires that the actual heat source/sink necessary is a heat sink in part of the layer and a heat source in the rest. This has some resemblance to the experimental situation of [217], although it is different, and appropriate experiments would be appreciated. [427] derives linear instability results as well as global nonlinear stability ones and he finds that there is a very interesting parameter range where the Rayleigh number for instability rises very rapidly. The basic steady state configuration of interest is one where there are three potential sub-layers which may give rise to instability which can then penetrate into the other layer(s).

14.4.1 The Model

The balance of linear momentum equation has form

$$\frac{\partial v_i}{\partial t} + v_j \frac{\partial v_i}{\partial x_j} = -\frac{1}{\rho_0} \frac{\partial p}{\partial x_i} + \nu \Delta v_i - g k_i \frac{\rho(T)}{\rho_0}, \tag{14.59}$$

where $\mathbf{v}$ and p denote velocity and pressure, with T the temperature field. The density function $\rho(T)$ is quadratic in T so,

$$\rho = \rho_0(1 - \alpha[T - T_m]^2), \tag{14.60}$$

where α is a thermal expansion coefficient and T_m is that value of temperature where the density ρ achieves a maximum.

The balance of mass equation is the incompressibility condition

$$\frac{\partial v_i}{\partial x_i} = 0, \tag{14.61}$$

while the balance of energy equation is

$$\frac{\partial T}{\partial t} + v_i \frac{\partial T}{\partial x_i} = \kappa \Delta T + Q(z). \tag{14.62}$$

Equations (14.61)–(14.62) hold on the domain $\mathbb{R}^2 \times \{z \in (0,h)\} \times \{t > 0\}$.

[427] constructs a heat source function Q which is linear in z and produces a steady state solution to equations (14.59)–(14.62) where the velocity $\bar{v}_i \equiv 0$ and the temperature $\bar{T}(z)$ is a cubic in z which yields the profile in figure 14.16. He considers $T_0 < T_m$ and shows that in this situation there are three potentially unstable layers

where thermal convection may arise, namely $(0, d_1)$, (d_2, d_3) and (d_4, h), this being due to relation (14.60) which ensures the fluid density is maximum at $z = d_1, d_3$ and h. Thus, in principle, there is a potential for resonant convection between each or all three of these layers.

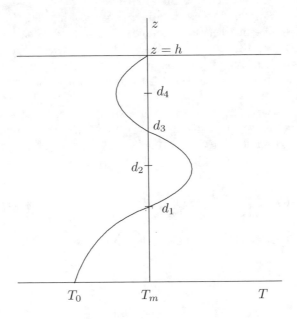

Fig. 14.16 Steady state temperature profile corresponding to solution (14.64).

Denote by $\Delta T_1 = T_m - T_0$, then [427] shows the function Q which yields the scenario of figure 14.16 is given by

$$Q = -\frac{6\kappa\Delta T_1}{h^3\xi\zeta}z + \frac{2\kappa\Delta T_1(1+\xi+\zeta)}{h^2\xi\zeta}, \qquad (14.63)$$

where $\zeta h = d_1$ and $\xi h = d_3$. This choice of Q yields the temperature profile

$$\bar{T}(z) = \Delta T_1\frac{z^3}{\xi\zeta h^3} - \Delta T_1\frac{(1+\xi+\zeta)z^2}{\xi\zeta h^2} + \Delta T_1\left[1+\frac{1}{\zeta}+\frac{1}{\xi}\right]\frac{z}{h} + T_0. \qquad (14.64)$$

The turning points d_2 and d_4 are given by

$$d_4 = \frac{h}{3}(1+\xi+\zeta) + \frac{h}{3}\sqrt{1+\xi^2+\zeta^2-\xi-\zeta-\xi\zeta},$$

$$d_2 = \frac{h}{3}(1+\xi+\zeta) - \frac{h}{3}\sqrt{1+\xi^2+\zeta^2-\xi-\zeta-\xi\zeta}. \qquad (14.65)$$

To study stability of the basic steady state solution [427] introduces perturbations u_i, θ, π by $v_i = \bar{v}_i + u_i$, and non-dimensionalizes with length, velocity and temperature scales $h, v/h = U, U\sqrt{v/\kappa g \alpha h}$ together with the Prandtl and Rayleigh numbers defined by

$$Pr = \frac{v}{\kappa}, \qquad Ra = (\Delta T_1)^2 \frac{g\alpha h^3}{\kappa v}. \tag{14.66}$$

If one puts $R = \sqrt{Ra}$ and defines the non-dimensional temperature function $F(z)$ by

$$F = \frac{z^3}{\xi\zeta} - \frac{(1+\zeta+\xi)z^2}{\xi\zeta} + \left(1 + \frac{1}{\zeta} + \frac{1}{\xi}\right)z - 1, \tag{14.67}$$

for $0 \le z \le 1$, the non-dimensional perturbation equations may be shown to be

$$\frac{\partial u_i}{\partial t} + u_j \frac{\partial u_i}{\partial x_j} = -\frac{\partial \pi}{\partial x_i} + \Delta u_i + 2RF\theta k_i + Prk_i\theta^2,$$

$$\frac{\partial u_i}{\partial x_i} = 0, \tag{14.68}$$

$$Pr\left(\frac{\partial \theta}{\partial t} + u_i \frac{\partial \theta}{\partial x_i}\right) = -RF'w + \Delta\theta,$$

where equations (14.68) hold on the domain $\mathbb{R}^2 \times \{z \in (0,1)\} \times \{t > 0\}$, with $w = u_3$, and where $F' = dF/dz$. The boundary conditions to be satisfied are

$$u_i = 0, \quad \theta = 0, \quad z = 0, 1, \tag{14.69}$$

with u_i, θ, π satisfying a plane-tiling periodicity in the horizontal plane.

14.4.2 Instability

[427] linearizes equations (14.68) and shows this results in dealing with the eigenvalue problem

$$\sigma \Delta w = \Delta^2 w + 2RF\Delta^*\theta,$$

$$\sigma Pr\theta = -RF'w + \Delta\theta, \tag{14.70}$$

where $\Delta^* = \partial^2/\partial x^2 + \partial^2/\partial y^2$, and equations (14.70) hold on the spatial domain $\mathbb{R}^2 \times (0,1)$. He argues that in looking for resonance one might expect such behaviour when the three potentially unstable layers have the same depths, i.e. when

$$d_1 = d_3 - d_2 = h - d_4. \tag{14.71}$$

He also argues that similar resonance may be expected if the Rayleigh numbers pertaining to each layer have approximately the same values. He uses these arguments to conclude that resonance is likely to be found with values of ζ and ξ in the vicinity of $\zeta = 0.128, \xi = 0.532$ and $\zeta = 0.191, \xi = 0.531$.

[427] writes $w = W(z)f(x,y)$, $\theta = \Theta(z)f(x,y)$ where f is a planform satisfying the equation $\Delta^* f = -a^2 f$, where a is the wavenumber. This leads to the eigenvalue problem

$$(D^2 - a^2)^2 W - 2RFa^2\Theta = \sigma(D^2 - a^2)W$$
$$(D^2 - a^2)\Theta - RF'W = \sigma Pr\Theta, \tag{14.72}$$

where $D = d/dz$, $z \in (0,1)$, and the boundary conditions are

$$W = DW = \Theta = 0, \qquad z = 0, 1. \tag{14.73}$$

System (14.72) together with (14.73) is solved by a D^2–Chebyshev tau method, cf. Dongarra *et al.* (1996), and detailed numerical results may be found in [427].

14.4.3 Global Nonlinear Stability

[427] employs a weighted energy method on system (14.68) with an energy function $E(t)$ of form

$$E(t) = \frac{1}{2}\|\mathbf{u}\|^2 + \frac{Pr}{2}\int_V \hat{\mu}(z)\theta^2 dx, \tag{14.74}$$

where $\hat{\mu} = \mu - 2z$, $\mu > 2$ is a constant to be selected optimally. We do not give details but note that this choice of energy functional leads to unconditional nonlinear stability results.

In addition [427] argues that one could employ a momentum equation which has a viscosity which depends on the velocity gradient. For such a choice he derives the nonlinear perturbation equations

$$\frac{\partial u_i}{\partial t} + u_j\frac{\partial u_i}{\partial x_j} = -\frac{\partial \pi}{\partial x_i} + \frac{\partial}{\partial x_j}\left[\left(1 + \nu_1|\nabla\mathbf{u}|\right)\frac{\partial u_i}{\partial x_j}\right]$$
$$+ 2RF\theta k_i + Prk_i\theta^2,$$

$$\frac{\partial u_i}{\partial x_i} = 0, \tag{14.75}$$

$$Pr\left(\frac{\partial \theta}{\partial t} + u_i\frac{\partial \theta}{\partial x_i}\right) = -RF'w + \Delta\theta,$$

with the same boundary conditions as in (14.69). He shows that one may employ a generalized energy of form

$$\mathscr{E}(t) = E + \frac{a_1 Pr}{3}\|\theta\|_3^3, \tag{14.76}$$

where the function E is given by

$$E(t) = \frac{1}{2}\|\mathbf{u}\|^2 + \frac{\lambda Pr}{2}\|\theta\|^2, \tag{14.77}$$

for coupling parameters λ and a_1. [427] shows that such an energy in connection with equations (14.75) also leads to an unconditional energy stability analysis and the critical Rayleigh number threshold is sharper than that obtained with the previous energy.

[427] computed critical Rayleigh numbers and wave numbers for many combinations of ζ and ξ, $0 < \zeta < \xi < 1$ to solve equations (14.72) and (14.73) for the linear instability thresholds. He solved equations for the weighted energy thresholds as well as equations for the generalized energy thresholds. For the linear instability thresholds he concentrates mainly on reporting values for ζ near 0.16 and 0.17 with ξ near 0.51 to 0.53. This is the range of values in (Ra, ζ, ξ) space where he found a very rapid rise in Rayleigh number, and the possibility of oscillatory convection with the major component of convection switching between the three basic layers, and penetrative convection ensuing.

[427] finds that for ζ, ξ in the ranges $[0.1626, 0.1628]$ and $[0.51, 0.53]$ the Ra vs. a curves display behaviour similar to that of [325]. Full details are given in [427] but we observe that the possibility of triple resonance is predicted with a much increased critical Rayleigh number which is likely to be of interest in insulation theory.

14.5 Resonance with Variable Gravity

In this chapter we have concentrated on situations involving thermal convection in a horizontal layer which have led to the possibility of resonance in the sense that convective fluid motion may commence simultaneously in two or more sublayers. This has been achieved in a fluid layer by having a constant heat source coupled with a density in the buoyancy term which has a quadratic dependence on temperature, by a heat source linear in the vertical coordinate coupled with a density linear in temperature, or by a heat source linear in the vertical coordinate coupled with a density quadratic in temperature. In a fluid saturated porous medium we only analysed the case of a constant heat source coupled with a density linear in temperature in the buoyancy force term, although the other combinations of heat source and density are possible in the porous case.

In the present section we wish to consider another possibility which may possibly lead to convective fluid motion in sublayers and so possible resonance and we do this by taking a density quadratic in the temperature field, but we allow the gravity field to vary linearly in z. This situation may possibly arise in an experimental set-up, or even possibly in stellar convection.

Suppose we have a layer of water which we assume is an incompressible viscous fluid contained in the horizontal layer bounded by the planes $z = 0$ and $z = d$ with a variable gravity field $g(z)$ acting in the z-direction. The boundaries are held at fixed constant temperatures $T_U > 0$, for $z = d$, and $T_L = 0°C$, for $z = 0$. The density in the buoyancy force term is the [468] one, namely,

$$\rho = \rho_0 \left(1 - \alpha[T - T_m]^2\right), \tag{14.78}$$

where ρ_0 is a constant, α is a thermal expansion coefficient, and $T_m \approx 4°C$. The gravity field varies linearly with z and has value $g_U < 0$ at $z = d$ whereas it is $g_L > 0$ at $z = 0$. The situation we wish to construct is as in figure 14.17, which represents the fluid layer in the stationary state.

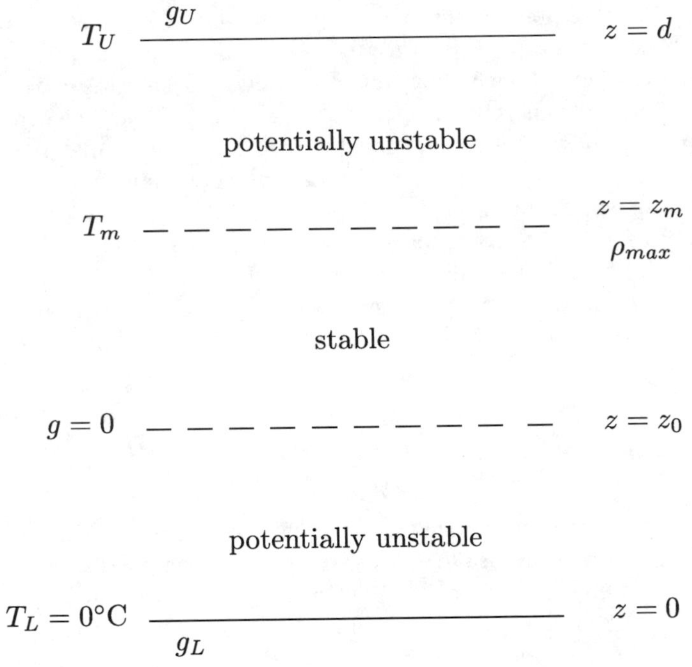

$$T_U \quad \underline{\quad g_U \quad \rule{6cm}{0.4pt} \quad} \quad z = d$$

potentially unstable

$$T_m \; -\,-\,-\,-\,-\,-\,-\,-\,-\,-\,- \quad \begin{array}{l} z = z_m \\ \rho_{max} \end{array}$$

stable

$$g = 0 \; -\,-\,-\,-\,-\,-\,-\,-\,-\,-\,- \quad z = z_0$$

potentially unstable

$$T_L = 0°C \quad \underline{\quad \rule{5cm}{0.4pt} \quad} \quad z = 0$$
$$g_L$$

Fig. 14.17 Configuration for thermal convection with variable gravity and quadratic density.

The gravity field across the layer is given by

$$g = g_L - \frac{\Delta g}{d} z,$$

where $\Delta g = g_L - g_U > 0$. The temperature gradient in the steady state is

$$\bar{T}(z) = \beta z = \frac{T_U}{d} z. \tag{14.79}$$

Then

$$z_m = \frac{T_m d}{T_U}.$$

Furthermore, since $g = 0$ at $z = z_0$,

$$z_0 = \frac{g_L d}{\Delta g}.$$

We require

$$0 < z_0 < z_m < d$$

and so if we put $z_m = \xi d$, $z_0 = \zeta d$, where $\xi = T_m/T_U$ and $\zeta = g_L/\Delta g$, then this is equivalent to needing

$$0 < \zeta < \xi < 1.$$

Thus, the upper temperature and gravity field are restricted by requiring $T_U > T_m$ and $g_L/\Delta g < T_m/T_U$.

Assuming the above to be the case we then have a situation, see figure 14.17, where there may potentially be convective instability in the sub-layers $(0, \zeta d)$ and $(\xi d, d)$. The equations governing the behaviour of the linear viscous fluid in this scenario are

$$\frac{\partial v_i}{\partial t} + v_j \frac{\partial v_i}{\partial x_j} = -\frac{1}{\rho_0} \frac{\partial p}{\partial x_i} + \nu \Delta v_i - \left(1 - \alpha[T - T_m]^2\right) g k_i,$$

$$\frac{\partial v_i}{\partial x_i} = 0, \tag{14.80}$$

$$\frac{\partial T}{\partial t} + v_i \frac{\partial T}{\partial x_i} = \kappa \Delta T.$$

To study stability of the basic solution $\bar{v}_i \equiv 0$ and $\bar{T}$ given by (14.79) we introduce perturbations (u_i, π, θ) to the basic velocity, pressure and temperature $(\bar{v}_i, \bar{p}, \bar{T})$ and then we non-dimensionalize with the scales

$$x_i = x_i^* d, \qquad t = t^* \mathcal{T}, \qquad u_i = u_i^* U, \qquad \theta = T^\sharp \theta^*,$$
$$\pi = \pi^* P, \qquad \mathcal{T} = d^2/\nu, \qquad U = \nu/d, \qquad P = \nu U \rho_0/d,$$

and select

$$T^\sharp = U \sqrt{\frac{\nu}{\kappa \alpha d \Delta g}}$$

and pick the Rayleigh number $Ra = R^2$ such that

$$R = T_U \sqrt{\frac{d^3 \Delta g \alpha}{\kappa \nu}}.$$

Then, dropping stars, the non-dimensional perturbation equations take the form

$$\frac{\partial u_i}{\partial t} + u_j \frac{\partial u_i}{\partial x_j} = -\frac{\partial \pi}{\partial x_i} + \Delta u_i$$

$$- 2R(\xi - z)(\zeta - z)\theta k_i + Pr(\zeta - z)\theta^2 k_i,$$

$$\tag{14.81}$$

$$\frac{\partial u_i}{\partial x_i} = 0,$$

$$Pr\left(\frac{\partial \theta}{\partial t} + u_i \frac{\partial \theta}{\partial x_i}\right) = -Rw + \Delta\theta,$$

where $Pr = \nu/\kappa$ is the Prandtl number.

One might employ a heuristic argument to estimate for what values of ζ, ξ one might expect resonance. In this regard equal sub-layer depths is a candidate so that $1 - \xi = \zeta$. Also if we denote by R_ξ^2 and R_ζ^2 the Rayleigh numbers associated with each sub-layer $(\xi d, d)$ and $(0, \zeta d)$ then apart from the constants in R_ξ^2 and R_ζ^2 this leads to

$$d_1^3 \Delta g_1 = d_2^3 \Delta g_2 \tag{14.82}$$

where $d_1 = d(1-\xi), d_2 = d\zeta$, and $\Delta g_1 = |g_U - g_m|, \Delta g_2 = g_L - 0$, with g_m denoting the value of $g(z)$ at $z = z_m$. Since

$$g_m = g_L + \left(\frac{g_U - g_L}{d}\right)\frac{T_m}{T_U}d$$

(14.82) leads to

$$(g_L - g_U)(1-\xi)^4 = g_L\zeta^3$$

and since $\zeta = g_L/\Delta g$ this means $(1-\xi)^4 = \zeta^4$, or $1-\xi = \zeta$. Therefore, when searching numerically it may be worth considering especially the case $\xi + \zeta = 1$.

Linear instability is governed by the eigenvalue problem,

$$\sigma u_i = -\pi_{,i} - 2R(\xi - z)(\zeta - z)\theta k_i + \Delta u_i,$$

$$u_{i,i} = 0,$$

$$Pr\sigma\theta = -Rw + \Delta\theta,$$

and upon removing the pressure π we find one needs to solve the eigenvalue problem

$$\sigma\Delta w = \Delta^2 w - 2R(\xi - z)(\zeta - z)\Delta^*\theta,$$

$$\sigma Pr\theta = \Delta\theta - Rw,$$

$$\tag{14.83}$$

where Δ^* is the horizontal Laplacian.

To my knowledge the numerical calculation of R^2 from (14.83) has not been performed, but it could be done by using a D^2–Chebyshev tau method, cf. [115], to find the critical Rayleigh numbers of linear instability theory.

To determine a global nonlinear stability threshold from equations (14.81) is not entirely straightforward. One could do so by means of a weighted energy method using as energy the function

$$E(t) = \frac{1}{2}\|\mathbf{u}\|^2 + \frac{Pr}{2}\int_V \mu\theta^2 dx, \tag{14.84}$$

where V is a period cell for the solution and $\mu(z)$ is a suitable weight. One such weight is the function

$$\mu = \mu_1 + z^2 - 2\zeta z, \tag{14.85}$$

where $\mu_1 > 0$ is a constant to be optimally selected with $\mu_1 > \zeta^2$ so that $\mu > 0$. If one employs this energy then the cubic terms arising from the Pr term in $(14.81)_1$ disappear and one finds an energy equation of form

$$\frac{dE}{dt} = I - D,$$

where

$$I = \|\theta\|^2 - R\int_V \left[\mu(z) + 2(\xi - z)(\zeta - z)\right]\theta w\, dx,$$

and

$$D = \|\nabla\mathbf{u}\|^2 + \int_V \mu|\nabla\theta|^2 dx.$$

To find the energy stability critical Rayleigh numbers one then needs to find the Euler-Lagrange equations for the maximum $\max_H I/D$ and then solve these numerically.

Chapter 15
Thermal Convection in Nanofluids

15.1 Heat Transfer Enhancement in Nanofluids

Nanofluids consist of a stable suspension of very small metallic or metallic-like particles suspended in a carrier fluid. The particles are called nanoparticles and as [321] note the stability of the suspension and prevention of particle settlement has been achieved in the laboratory. There are many types of nanofluids but typical examples might involve a suspension of copper, Cu, copper oxide, CuO, or aluminium oxide, Al_2O_3, in water or ethylene glycol, cf. [135, 237, 341, 463]. Other types of nanofluids are made by suspending carbon nanotubes in an appropriate oil, cf. [463].

Research on nanofluids and their use in heat transfer devices is very much in vogue in the engineering and industrial communities. One reason for this is that nanofluids appear to have highly desirable properties for greatly increasing heat transfer by comparison with ordinary fluids. Thermal conductivities of metals like copper, or their oxides, are usually much greater than those of a typical carrier fluid. Because of this the nanofluid suspension may have an increased thermal conductivity over that of the pure fluid, although the effects of changes to both the thermal conductivity and the viscosity should be considered together, cf. [230, 237, 274, 341] and [480]. One belief is that an increased thermal conductivity may have a pronounced effect on heat transfer.

While some believe the thermal conductivity of a nanofluid suspension is very much increased, [463] question the basis of measurement leading to the assertion of increased thermal conductivity. They argue that interpretation of experimental measurements on the thermal conductivity usually employs theory which is based on the Fourier law of heat conduction. They also argue that factors influencing thermal properties of nanofluids may be ballistic rather than diffusive. In view of this [463] propose that thermal wave effects should be taken into account when interpreting experimental results. [463] propose six possible reasons for the increased effective thermal conductivity of a nanofluid. Among these we note effects of hyperbolic or phase-lagging thermal waves, particle driven, or thermally driven, fluid convection, and hyperbolic thermal convection.

© Springer International Publishing Switzerland 2015

B. Straughan, *Convection with Local Thermal Non-Equilibrium and Microfluidic Effects,*
Advances in Mechanics and Mathematics 32, DOI 10.1007/978-3-319-13530-4_15

In this chapter we review some recent work on models to describe the behaviour of a nanofluid. We also present work on hyperbolic thermal convection in a fluid incorporating one possibility suggested by [463].

Convective transport in nanofluids is investigated by [458] and by [65]. [454] produces a model for thermal convection in a nanofluid and analyses this. [390] investigate a model which allows the fluid to be compressible. [224, 225] develop a theory and analysis for thermal convection in a nanofluid which saturates a porous medium. [321] and [323] further develop models for thermal convection in a nanofluid employing a single temperature. In [321] they allow for slip boundary conditions whereas [323] analyses a boundary condition connecting the temperature gradient to the gradient of nanoparticle concentration.

Of particular relevance to the present book is the work of [320] who allowed the nanoparticles and carrier fluid to have different temperatures, thereby adapting an LTNE theory. At about the same time [420] employed an LTNE theory to describe a nanofluid by suitably adapting the thermodynamic analysis of [168]. The theory developed in [420] incorporates higher derivatives of velocity in the momentum equation. This is consistent with what is found experimentally for a suspension where, for example, a non-parabolic profile is found in Couette or Poiseuille flow.

15.2 The Tzou Nanofluid Model

The [454] nanofluid model is reviewed in some depth in [425, pp. 224–228]. However, for completeness we include a very brief description here. This model employs a conservation equation for the nanoparticle concentration and such an equation frequently occurs in many other models for nanofluid suspensions. [454] employs an incompressible fluid with a Boussinesq approximation and his theory involves differential equations for the velocity of a point in the suspension, v_i, the pressure, p, the temperature, T, and the concentration of nanoparticles, $\phi(\mathbf{x}, t)$. The model of [454] is based on the momentum equation,

$$\rho_0 \left(\frac{\partial v_i}{\partial t} + v_j \frac{\partial v_i}{\partial x_j} \right) = -\frac{\partial p}{\partial x_i} + \mu \Delta v_i - \rho g k_i, \qquad (15.1)$$

the equation of continuity of mass,

$$\frac{\partial v_i}{\partial x_i} = 0, \qquad (15.2)$$

the balance of energy equation,

$$\rho_0 c_F \left(\frac{\partial T}{\partial t} + v_i \frac{\partial T}{\partial x_i} \right) = -\frac{\partial q_i}{\partial x_i} + h_S \frac{\partial J_i^S}{\partial x_i}, \qquad (15.3)$$

together with an equation describing the conservation of nanoparticles,

$$\rho_S\left(\frac{\partial \phi}{\partial t} + v_i \frac{\partial \phi}{\partial x_i}\right) = -\frac{\partial J_i^S}{\partial x_i}. \tag{15.4}$$

In these equations F or S refer to (bulk) fluid or (nanoparticle) solid components, ρ is the density at a point $\mathbf{x}$, ρ_0 is the constant fluid density at a reference temperature T_0, μ is the dynamic viscosity of the bulk fluid, g is gravity, $\mathbf{k} = (0,0,1)$, c_F or c_S denote the specific heat (at constant pressure) of the fluid or solid particles, $\mathbf{q}$ is the heat flux vector, h_S is the enthalpy of the solid, ρ_S is the solid density, and $\mathbf{J}^S$ is a flux vector which is associated to the nanoparticle density. The constitutive theory [454] adopts for $\mathbf{J}^S$ is that

$$J_i^S = -\rho_S D_B \frac{\partial \phi}{\partial x_i} - \rho_S \frac{D_T}{T_B} \frac{\partial T}{\partial x_i}, \tag{15.5}$$

where T_B is a constant (bulk fluid temperature) and D_B and D_T are effectively constants.

To describe the heat flux [454] writes it as a linear combination of the classical Fourier theory plus a term linear in $\mathbf{J}^S$. This is thus a generalized Fourier law which takes account of the flux due to nanoparticle density. In fact, for the heat flux [454] writes

$$q_i = -k\frac{\partial T}{\partial x_i} + h_S J_i^S, \tag{15.6}$$

where k is the thermal conductivity of the bulk fluid.

In the momentum equation (15.1) [454] splits the density ρ into a term linear in ϕ, which expresses the nanoparticle component, plus a term in $(1 - \phi)$ to account for the fluid itself. In the fluid part he adopts a Boussinesq approximation and expresses the density as a linear function of temperature. Thus, in equation (15.1) [454] puts

$$\rho = \phi \rho_S + \rho_0 (1 - \phi)\big[1 - \alpha(T - T_0)\big], \tag{15.7}$$

where α is the thermal expansion coefficient of the fluid. [454] also writes

$$\frac{\partial h_S}{\partial x_i} = c_S \frac{\partial T}{\partial x_i}. \tag{15.8}$$

This is analogous to [243, p. 8], in his treatment of enthalpy in a fluid. The energy balance equation (15.3) is rewritten by substituting from (15.6) for the heat flux. One must employ (15.5) and (15.8) in the resulting term of form $-h_{S,i} J_i^S$. In this way one arrives at

$$\rho_0 c_F \left(\frac{\partial T}{\partial t} + v_i \frac{\partial T}{\partial x_i}\right) = k\Delta T + \rho_S c_S D_B \frac{\partial \phi}{\partial x_i} \frac{\partial T}{\partial x_i} + \frac{\rho_S c_S D_T}{T_B} \frac{\partial T}{\partial x_i} \frac{\partial T}{\partial x_i}. \tag{15.9}$$

Collecting equations together one finds that the equations of [454] for a nanofluid suspension may be written as

$$\frac{\partial v_i}{\partial t} + v_j \frac{\partial v_i}{\partial x_j} = -\frac{1}{\rho_0} \frac{\partial p}{\partial x_i} + \frac{\mu}{\rho_0} \Delta v_i$$

$$- gk_i \left\{ \phi \frac{\rho_S}{\rho_0} + (1-\phi)(1 - \alpha[T - T_0]) \right\},$$

$$\frac{\partial v_i}{\partial x_i} = 0, \tag{15.10}$$

$$\frac{\partial T}{\partial t} + v_i \frac{\partial T}{\partial x_i} = \kappa \Delta T + k_1 \frac{\partial \phi}{\partial x_i} \frac{\partial T}{\partial x_i} + k_2 \frac{\partial T}{\partial x_i} \frac{\partial T}{\partial x_i},$$

$$\frac{\partial \phi}{\partial t} + v_i \frac{\partial \phi}{\partial x_i} = D_B \Delta \phi + \frac{D_T}{T_B} \Delta T,$$

where the constants k_1 and k_2 and the thermal diffusivity κ are given by

$$k_1 = \frac{\rho_S c_S D_B}{\rho_0 c_F}, \qquad k_2 = \frac{\rho_S D_T c_S}{\rho_0 c_F T_B}, \qquad \kappa = \frac{k}{\rho_0 c_F}.$$

Generalizations of equations (15.10) are presented by [321] and by [323]. [321] employ physically realistic slip boundary conditions and [323] address the important question of what are the correct boundary conditions for the nanoparticle concentration. [323] propose a boundary condition involving $\partial \phi / \partial z$ and $\partial T / \partial z$. Thermal convection instability thresholds are provided by these writers.

15.3 Convection with Heat Wave Theories

The propagation of thermal waves (heat waves) is a topic of great current interest and much of the development is reviewed in the book by [425]. It is now realised that the propagation of heat waves is not simply a low temperature phenomenon but is something which has application in everyday life. For example, thermal waves may be responsible for flash burns of skin, [101]. Thermal waves are important in medical treatments such as in laser heating in the cornea or in radiofrequency heating for elimination of cardiac arrhythmias, destruction of tumours, treatment of gastroesophageal reflux disease, corneal heating, [47, 251, 453]. They are also believed to be important in general laser interactions with matter, [270]. Recent work recognises the importance of heat waves in phase change mechanisms, see e.g. [99, 100, 246, 292, 293]. Thermal wave propagation is also believed to be important in biological materials, [384], and in planetary and stellar evolution, [30, 128, 178]. The importance of heat transfer by a thermal wave mechanism in nanofluids has been recognised by [420, 425, 457, 458, 463], and also in a graphene nanoribbon, [471].

The heat flux law of [79] may be regarded as a generalization of Fourier's law and is, therefore, a constitutive equation. In view of this [92] has suggested a Lie derivative invariant form of time derivative in Cattaneo's heat flux equation. The appropriate derivative to employ has also been suggested earlier by [276]. The history of this derivative as explained lucidly by [277] is worth noting. [421, 423, 426] developed thermal convection in a fluid and in a fluid saturated porous medium using the theory which follows from the work of [79] and [92] (see also [276, 277]). The appropriate system of equations which arises has been called in the literature Cattaneo–Christov theory. We point out that [79] theory has been very well examined in other areas of fluid and continuum mechanics, cf. [75, 93, 94, 354, 355], although they did not employ the objective derivative of [92]. Much of the work in fluid mechanics using Cattaneo theory is reviewed in [425]. Certainly the work of [421, 423, 426, 428] on thermal convection with an incompressible fluid indicates that Cattaneo–Christov theory has the potential to yield interesting novel results in thermal convection.

We now briefly review some pertinent theories of viscous fluid motion coupled with heat transfer via a Cattaneo-like law since we believe they may be relevant to heat transfer in nanofluid theory. We include an account of such thermal convection using a linear viscous fluid coupled with suitable Cattaneo-like laws for heat transfer. Even though we describe models derived already for the Cattaneo–Fox and Cattaneo–Christov theories we include several new numerical results not available elsewhere. In particular we examine the (strong) effect variation of the Prandtl number has on the thermal convection instability threshold.

15.3.1 Cattaneo–Fox Law

The Cattaneo law, [79], was introduced in thermal convection in fluid mechanics by [434], with later work by [148, 236] and [105].

[434] used the [147] invariant form of derivative for the heat flux in the Cattaneo law. They only treated the case of two free surfaces. [419] continued the work of [434], but he used boundary conditions appropriate to two fixed surfaces. [419] employed a Prandtl number of value 6. One of the things we do in this section is investigate numerically what happens when the Prandtl number is varied.

To study thermal convection with a Cattaneo–Fox theory we commence with an incompressible linear viscous fluid contained in the layer between the planes $z = 0, d$ with $(x, y) \in \mathbb{R}^2$. The governing equations are given by [434] as

$$\frac{\partial v_i}{\partial t} + v_j \frac{\partial v_i}{\partial x_j} = -\frac{1}{\rho} \frac{\partial p}{\partial x_i} + \nu \Delta v_i + g_i [1 - \alpha (T - T_R)],$$

$$\frac{\partial v_i}{\partial x_i} = 0,$$

$$\frac{\partial T}{\partial t} + v_i \frac{\partial T}{\partial x_i} = -\frac{\partial Q_i}{\partial x_i}, \qquad (15.11)$$

$$\tau \left(\frac{\partial Q_i}{\partial t} + v_j \frac{\partial Q_i}{\partial x_j} - \varepsilon_{ijk} \omega_j Q_k \right) = -Q_i - \kappa \frac{\partial T}{\partial x_i}.$$

The variables v_i, T, Q_i and p are the velocity, temperature, heat flux and pressure. In addition $\omega = \operatorname{curl} \mathbf{v}/2$, $\mathbf{g} = (0,0,-g)$ the gravity vector, T_R is a reference temperature, and τ is a constant with the dimensions of time, τ being referred to as a relaxation time.

The boundary conditions adopted are

$$v_i = 0, \ z = 0, d, \qquad T = T_L, \ z = 0, \qquad T = T_U, \ z = d, \qquad (15.12)$$

with constants T_L and T_U such that $T_L > T_U$. Our interest is in the instability of the motionless solution

$$\bar{v}_i = 0, \qquad \bar{T} = -\beta z + T_L, \qquad \bar{q}_i = (0, 0, \kappa\beta) \qquad (15.13)$$

where β is the temperature gradient,

$$\beta = \frac{T_L - T_U}{d}.$$

Perturbations (u_i, θ, π, q_i) are introduced to the basic solution $(\bar{v}_i, \bar{T}, \bar{p}, \bar{q}_i)$. One then derives *linearized* perturbation equations from (15.11). The resulting equations are non-dimensionalized using the length, time, pressure, heat flux, and temperature scales

$$d, \ \frac{d^2}{v}, \ \frac{vU}{d}, \ \frac{\kappa T^{\sharp}}{d}, \ T^{\sharp} = U\sqrt{\frac{\beta v}{\alpha \kappa g}}.$$

The Rayleigh number, Ra, Prandtl number, Pr, and the parameter Sg are introduced by

$$Ra = R^2 = \frac{g\alpha\beta d^4}{v\kappa}, \qquad Pr = \frac{v}{\kappa}, \qquad Sg = \frac{\tau v}{d^2}.$$

The non-dimensional parameter, Sg, was introduced by [330].

The non-dimensional linearized perturbation equations which arise from (15.11) may be shown to be

$$\frac{\partial u_i}{\partial t} = -\frac{\partial \pi}{\partial x_i} + R\theta k_i + \Delta u_i,$$

$$\frac{\partial u_i}{\partial x_i} = 0,$$

$$Pr\frac{\partial \theta}{\partial t} = Rw - \frac{\partial q_i}{\partial x_i}, \qquad (15.14)$$

$$Sg\frac{\partial q_i}{\partial t} = \frac{SgR}{2Pr}\left(\frac{\partial u_i}{\partial z} - \frac{\partial w}{\partial x_i}\right) - q_i - \frac{\partial \theta}{\partial x_i},$$

where we have used the notation $\mathbf{u} = (u, v, w)$, $\mathbf{x} = (x, y, z)$.

By removing the pressure term π and introducing the variable $\xi = q_{i,i}$ system (15.14) is reduced to solving

$$
\begin{aligned}
\Delta^2 w + R\Delta^* \theta &= \sigma \Delta w, \\
Pr\sigma\theta &= Rw - \xi, \\
\sigma Sg\xi &= -\frac{SgR}{2Pr}\Delta w - \xi - \Delta\theta,
\end{aligned}
\tag{15.15}
$$

where $\Delta^* = \partial^2/\partial x^2 + \partial^2 \partial y^2$, and a time dependence like $e^{\sigma t}$ has been introduced. One supposes w, θ, ξ satisfy a plane tiling form $f(x,y)$ with $\Delta^* f = -a^2 f$, where a is a wavenumber, cf. [85, pp. 43–52, 414, p. 51]. Thus, to find the Rayleigh number instability threshold one must solve equations (15.15) numerically in the form

$$
\begin{aligned}
\Delta W - \chi &= 0, \\
\Delta\chi - Ra^2\Theta &= \sigma\chi, \\
\Delta\Theta + \Xi + \frac{SgR}{2Pr}\chi &= -Sg\,\Xi\,\sigma, \\
RW - \Xi &= \sigma Pr\Theta,
\end{aligned}
\tag{15.16}
$$

where W, Θ and Ξ are the z–parts of w, θ, ξ (e.g. $w = W(z)f(x,y)$), and χ is defined as ΔW. The boundary conditions for two *fixed* surfaces are

$$
W = DW = \Theta = 0, \qquad z = 0, 1,
\tag{15.17}
$$

where $D = d/dz$.

The Chebyshev tau D^2 numerical method is now used to solve equations (15.16) subject to the boundary conditions (15.17).

Table 15.1 Critical values of Ra and a against Sg, with $Pr = 5$, for Cattaneo–Fox theory. The quantity σ_1 is the imaginary part of the growth rate σ

Sg	a	Ra	σ_1
0.14	2.80	2321.775	0
0.15	2.77	2378.814	0
0.152	2.77	2390.490	0
0.154	2.76	2402.252	0
0.156	2.76	2414.108	0
0.1564	2.76	2416.493	0
0.1564	5.04	2416.931	±4.756
0.1565	5.04	2415.194	±4.757
0.157	5.04	2406.550	±4.764
0.158	5.03	2389.443	±4.758
0.16	5.03	2355.940	±4.784
0.17	5.02	2201.493	±4.866

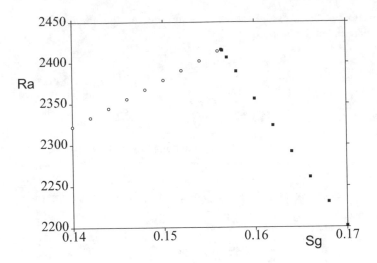

Fig. 15.1 Critical values of Ra vs. Sg, with $Pr = 5$. Cattaneo–Fox model.

Table 15.2 Critical values of Ra and a against Sg, with $Pr = 10$, for Cattaneo–Fox theory. The quantity σ_1 is the imaginary part of the growth rate σ

Sg	a	Ra	σ_1
0.3	2.77	2378.814	0
0.304	2.77	2390.490	0
0.3045	2.77	2391.958	0
0.3045	4.91	2391.508	± 2.331
0.305	4.91	2387.350	± 2.333
0.306	4.91	2379.078	± 2.337
0.308	4.91	2362.705	± 2.344
0.312	4.90	2330.624	± 2.349
0.316	4.90	2299.403	± 2.362
0.32	4.90	2269.009	± 2.375

[419, table 1], gives instability values arising from (15.16) and (15.17) when the Prandtl number is taken to have the numerical value of 6. We present Ra and a^2 at criticality for various values of Sg when the Prandtl number has values of 5 and 10. In this way we are able to witness the effect of changing the Prandtl number. Table 15.1 tabulates values when $Pr = 5$ whereas table 15.2 does likewise when $Pr = 10$. Figures 15.1 and 15.2 display graphically the values of Ra when

Sg increases for $Pr = 5$ and $Pr = 10$ concentrating on the changeover point where stationary convection turns to oscillatory convection.

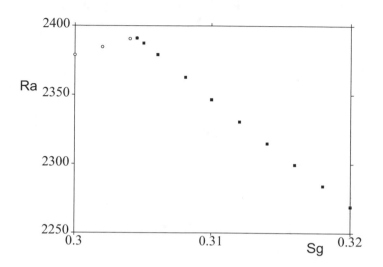

Fig. 15.2 Critical values of Ra vs. Sg, with $Pr = 10$. Cattaneo–Fox model.

Tables 15.1 and 15.2 together with figures 15.1 and 15.2 show that for small values of Sg stationary convection is preferred. When Sg is below a transition value the critical Rayleigh number increases as Sg increases. However, there is a transition value such that once Sg exceeds this value the onset of thermal convection is by oscillatory convection. After this transition value an increase in Sg leads to the threshold value of Ra decreasing. Thus, in figures 15.1 and 15.2 the circles to the left of the peak value of Ra denote stationary convection whereas the black squares to the right indicate oscillatory convection.

It would appear that before the transition value the instability mechanism is governed by thermal diffusion as in classical convection. Once Sg exceeds the transition value the hyperbolic nature of the energy balance equation is having an effect. The increasing of Sg corresponds to increasing τ the relaxation time.

It is noticeable that the Prandtl number has an effect on the value of Sg at the transition from stationary to oscillatory convection. For example, when $Pr = 5$ the transition value is $Sg = 0.1564$, when $Pr = 6$ it is $Sg = 0.1860$, [425], whereas when $Pr = 10$ the transition value is $Sg = 0.3045$. The values of the Rayleigh number at the transitions display less change and are 2416, 2408 and 2392 when $Pr = 5, 6$ and 10, respectively.

In all cases the switch from stationary to oscillatory convection is made with a marked change of wavenumber. This means the convection cells change to ones which are narrower. The critical wavenumber changes are 2.76 to 5.04, 2.76 to 4.99, and 2.77 to 4.91, according to whether $Pr = 5, 6$ or 10.

15.3.2 Cattaneo–Christov Law

The Cattaneo–Fox model used in section 15.3.1 can lead to instability when heating from above. This is likely to be physically unreasonable. To overcome this [423] appealed to work of [92] who proposed a Lie derivative form of invariant time derivative for the heat flux when dealing with a Cattaneo type theory for a fluid. This objective derivative has been employed in thermal convection studies by [329, 330] and by [421, 423]. The paper of [330] studies thermal convection in a two-dimensional domain. This is an interesting piece of work which solves directly the two-dimensional eigenvalue problem instead of employing normal modes.

The Cattaneo–Christov equations have form, cf. [423],

$$\frac{\partial v_i}{\partial t} + v_j \frac{\partial v_i}{\partial x_j} = -\frac{1}{\rho} \frac{\partial p}{\partial x_i} + \alpha g k_i T + \nu \Delta v_i, \tag{15.18}$$

$$\frac{\partial v_i}{\partial x_i} = 0, \tag{15.19}$$

$$\rho c_p \left(\frac{\partial T}{\partial t} + v_i \frac{\partial T}{\partial x_i} \right) = -\frac{\partial Q_i}{\partial x_i}, \tag{15.20}$$

$$\tau \left(\frac{\partial Q_i}{\partial t} + v_j \frac{\partial Q_i}{\partial x_j} - Q_j \frac{\partial v_i}{\partial x_j} \right) = -Q_i - \kappa \frac{\partial T}{\partial x_i}, \tag{15.21}$$

in which v_i, p, T, Q_i are the velocity, pressure, temperature and heat flux.

For the thermal convection instability problem we assume the fluid occupies the horizontal layer $(x, y) \in \mathbb{R}^2$, $z \in (0, d)$ and so equations (15.18)–(15.21) hold in the domain $\mathbb{R}^2 \times (0, d) \times \{t > 0\}$. The boundary conditions we use are those of no-slip and temperatures prescribed, so

$$\begin{aligned} v_i &= 0 \quad \text{on} \quad z = 0, d, \\ T &= T_L, \quad z = 0, \qquad T = T_U, \quad z = d, \end{aligned} \tag{15.22}$$

where T_L, T_U are constants with $T_L > T_U$.

The motionless solution whose instability is under investigation is

$$\bar{v}_i \equiv 0, \qquad \bar{T} = -\beta z + T_L, \qquad \bar{Q} = (0, 0, \kappa \beta), \tag{15.23}$$

with β being the temperature gradient, $\beta = (T_L - T_U)/d$.

To investigate instability of the basic solution (15.23) perturbations (u_i, θ, π, q_i) are introduced such that $v_i = \bar{v}_i + u_i$, $T = \bar{T} + \theta$, $p = \bar{p} + \pi$, $Q_i = \bar{Q}_i + q_i$. Equations for (u_i, θ, π, q_i) are derived from equations (15.18)–(15.21) and are non-dimensionalized as in section 15.3.1. The nonlinear, non-dimensional equations for

(u_i, θ, π, q_i) are found to be, cf. [423]

$$\frac{\partial u_i}{\partial t} + u_j \frac{\partial u_i}{\partial x_j} = -\frac{\partial \pi}{\partial x_i} + Rk_i \theta + \Delta u_i,$$

$$\frac{\partial u_i}{\partial x_i} = 0,$$

$$Pr\left(\frac{\partial \theta}{\partial t} + u_i \frac{\partial \theta}{\partial x_i}\right) = Rw - \frac{\partial q_i}{\partial x_i},$$

$$Sg\left(\frac{\partial q_i}{\partial t} + u_j \frac{\partial q_i}{\partial x_j} - q_j \frac{\partial u_i}{\partial x_j}\right) = -q_i + \frac{SgR}{Pr} \frac{\partial u_i}{\partial z} - \frac{\partial \theta}{\partial x_i}.$$

(15.24)

To analyse linear instability we remove the nonlinear terms from equations (15.24). Since the resulting system of equations is linear we may employ an exponential time dependence such that

$$u_i(\mathbf{x},t) = e^{\sigma t} u_i(\mathbf{x}), \quad \theta(\mathbf{x},t) = e^{\sigma t} \theta(\mathbf{x}),$$
$$q_i(\mathbf{x},t) = e^{\sigma t} q_i(\mathbf{x}), \quad \pi(\mathbf{x},t) = e^{\sigma t} \pi(\mathbf{x}).$$

In this way one obtains the non-dimensional linearized equations

$$\sigma u_l = -\frac{\partial \pi}{\partial x_i} + Rk_i \theta + \Delta u_i,$$

$$\frac{\partial u_i}{\partial x_i} = 0,$$

$$\sigma Pr\theta = Rw - \frac{\partial q_i}{\partial x_i},$$

$$\sigma Sgq_i = -q_i + \frac{SgR}{Pr} \frac{\partial u_i}{\partial z} - \frac{\partial \theta}{\partial x_i}.$$

(15.25)

Next eliminate the pressure and define $Q = q_{i,i}$. This leads one to solve the equations

$$\sigma \Delta w = R\Delta^* \theta + \Delta^2 w$$
$$\sigma Pr\theta = Rw - Q$$
$$\sigma SgQ = -Q - \Delta\theta,$$

(15.26)

with $\Delta^* = \partial^2/\partial x^2 + \partial^2/\partial y^2$ being the horizontal Laplacian.

To investigate stationary convection, put $\sigma = 0$, and then one derives from equations (15.26) the single equation for w,

$$\Delta^3 w = R^2 \Delta^* w.$$

(15.27)

For two fixed surfaces we obtain the stationary convection instability values

$$Ra = R^2 = 1707.762, \qquad a_c = 3.117,$$

cf. Chandrasekhar [85].

To analyse oscillatory convection we suppose in (15.26) $w = W(z)f(x,y)$, $\theta = \Theta(z)f(x,y)$, and $Q = Q(z)f(x,y)$, where f is a plane tiling planform. In fact, f satisfies the equation $\Delta^* f + a^2 f = 0$ where a is the wavenumber. After doing this we employ a Chebyshev-tau D^2 numerical technique and solve the system of equations

$$
\begin{aligned}
(D^2 - a^2)\chi - a^2 R\Theta &= \sigma\chi, \\
(D^2 - a^2)W - \chi &= 0, \\
Rw - Q &= \sigma Pr\Theta, \\
(D^2 - a^2)\Theta + Q &= -\sigma Sg Q.
\end{aligned}
\tag{15.28}
$$

The relevant boundary conditions are

$$
W = DW = \Theta = 0, \qquad z = 0, 1.
\tag{15.29}
$$

Equations (15.28) and (15.29) are solved numerically by [423] for a fixed value of Prandtl number, $Pr = 6$. In this section we solve the same equations numerically by a D^2 Chebyshev tau method but when Pr takes values 1, 5 and 10. In this way we examine the effect of Pr and Sg on the stationary-oscillatory convection boundary when the Cattaneo–Christov theory is utilized.

Table 15.3 Critical values of Ra and a against Sg, with $Pr = 1$, Cattaneo–Christov theory

Sg	a	Ra	σ_1
$0 - 0.064283$	3.12	1707.765	0
0.064283	5.18	1707.734	± 12.128
0.0643	5.18	1707.122	± 12.131
0.065	5.17	1682.249	± 12.195
0.066	5.17	1647.866	± 12.356
0.067	5.16	1614.767	± 12.449
0.068	5.16	1582.888	± 12.584
0.069	5.15	1552.160	± 12.655
0.070	5.15	1522.530	± 12.769

Tables 15.3–15.5 display critical Rayleigh number values against critical wavenumber values for various values of Sg when Pr takes the values 1, 5 and 10. The value σ_1 is these tables is the imaginary part of σ at the onset of convection. Figures 15.3–15.5 show the transition from stationary convection to oscillatory convection for $Pr = 1, 5$ and 10. Note that the scales for Sg in figures 15.3–15.5 change appreciably.

As in section 15.3.1 we again witness stationary convection when Sg is small. However, differently from the Cattaneo–Fox model of section 15.3.1 when we

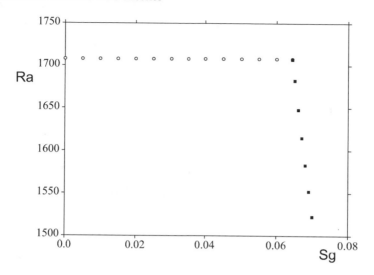

Fig. 15.3 Critical values of *Ra* vs. *Sg*, with *Pr* = 1, Cattaneo–Christov model.

Table 15.4 Critical values of *Ra* and *a* against *Sg*, with *Pr* = 5, Cattaneo–Christov theory

Sg	a	Ra	σ_1
0 – 0.22624	3.12	1707.765	0
0.22624	4.89	1707.702	±2.750
0.22625	4.89	1707.619	±2.750
0.228	4.89	1693.287	±2.770
0.230	4.89	1677.198	±2.792
0.235	4.89	1638.273	±2.843
0.240	4.89	1601.104	±2.887
0.245	4.88	1565.572	±2.913
0.250	4.88	1531.575	±2.948

employ the Cattaneo–Christov theory the critical Rayleigh number for stationary convection remains fixed at the value 1707.765 as *Sg* varies. We do witness a transition to oscillatory convection in each of the cases *Pr* = 1, 5 and 10 once the parameter *Sg* is sufficiently large. The transition values are available for *Pr* = 1, 5 and 10 from tables 15.3–15.5 and when *Pr* = 6 from [425], section 8.3.2. We see that the transition to oscillatory convection commences when *Pr* = 1 for *Sg* = 0.06428, when *Pr* = 5 for *Sg* = 0.2262, when *Pr* = 6 for *Sg* = 0.2669, and when *Pr* = 10 the transition value is *Sg* = 0.4298.

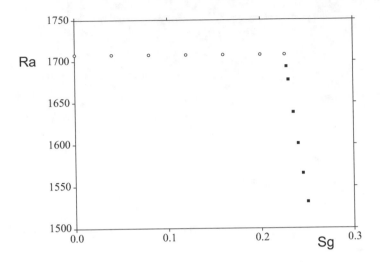

Fig. 15.4 Critical values of *Ra* vs. *Sg*, with *Pr* = 5, Cattaneo–Christov model.

Table 15.5 Critical values of *Ra* and *a* against *Sg*, with *Pr* = 10, Cattaneo–Christov theory

Sg	a	Ra	σ_1
0 − 0.4298	3.12	1707.765	0
0.4298	4.83	1707.748	±1.394
0.4299	4.83	1707.332	±1.394
0.43	4.83	1706.917	±1.394
0.44	4.83	1666.338	±1.424
0.45	4.83	1627.642	±1.450
0.46	4.83	1590.703	±1.473
0.47	4.83	1555.402	±1.493
0.48	4.83	1521.633	±1.510

Stationary convection is always with wavenumber $a = 3.12$. The wavenumber for oscillatory convection is always larger than that of stationary convection which means that the convection cells are narrower when oscillatory convection occurs. Also as *Pr* increases the critical wavenumber at the transition decreases as *Sg* increases. In fact, at the transition values the wavenumbers of oscillatory convection are $a = 5.18$ when $Pr = 1$, $a = 4.89$ when $Pr = 5$, $a = 4.87$ when $Pr = 6$, and $a = 4.83$ when $Pr = 10$.

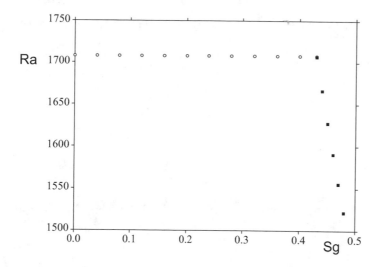

Fig. 15.5 Critical values of *Ra* vs. *Sg*, with $Pr = 10$, Cattaneo–Christov model.

Once again, in all Prandtl number cases we have studied there is a strong transition from stationary to oscillatory convection once *Sg* is sufficiently large. We stress that *Sg* being sufficiently large corresponds to the relaxation time τ being large enough. We believe this indicates that the hyperbolic nature of the Cattaneo–Christov system dominates for τ (or *Sg*) large enough and oscillatory convection is commensurate with wavelike behaviour. A detailed analysis of the effect of different Prandtl numbers on thermal convection with the Cattaneo–Christov model is given by [53].

15.3.3 Cattaneo Theories and Porous Materials

[421] investigated analogous thermal convection problems to those of sections 15.3.1 and 15.3.2 in a porous medium of Darcy type and this material is reviewed in some depth in [425], section 8.3.3. Apart from the fact that such analysis might prove valuable in a practical convection situation in a star or planet, the Darcy equations are of lower order in spatial derivatives than the Navier–Stokes equations and this enables one to proceed further with calculations analytically without having to resort to numerical solution of the relevant equations. For details of the actual Rayleigh number thresholds we refer to [421], or [425], section 8.3.3.

For completeness we record the equations one may employ to study thermal convection in a porous material of Darcy type. [421] and [425] include an isotropic inertia in the Darcy momentum equation. We follow this approach here and denote by v_i, p, T and Q_i the velocity, pressure, temperature and heat flux vector. Then for a porous material with Cattaneo–Fox theory the governing differential equations are

$$\frac{\partial v_i}{\partial t} = -\frac{1}{\rho}\frac{\partial p}{\partial x_i} + \alpha g k_i T - \frac{\mu}{\rho K} v_i,$$

$$\frac{\partial v_i}{\partial x_i} = 0,$$

$$\frac{1}{M}\frac{\partial T}{\partial t} + v_i\frac{\partial T}{\partial x_i} = -\frac{\partial Q_i}{\partial x_i},$$

$$\tau\frac{\partial Q_i}{\partial t} + \tau_f\left(v_j\frac{\partial Q_i}{\partial x_j} - \frac{1}{2}Q_j\frac{\partial v_i}{\partial x_j} + \frac{1}{2}Q_j\frac{\partial v_j}{\partial x_i}\right) = -Q_i - \kappa\frac{\partial T}{\partial x_i}. \tag{15.30}$$

When one employs a Cattaneo–Christov theory then instead the governing differential equations are

$$\frac{\partial v_i}{\partial t} = -\frac{1}{\rho}\frac{\partial p}{\partial x_i} + \alpha g k_i T - \frac{\mu}{\rho K} v_i,$$

$$\frac{\partial v_i}{\partial x_i} = 0,$$

$$\frac{1}{M}\frac{\partial T}{\partial t} + v_i\frac{\partial T}{\partial x_i} = -\frac{\partial Q_i}{\partial x_i},$$

$$\tau\frac{\partial Q_i}{\partial t} + \tau_f\left(v_j\frac{\partial Q_i}{\partial x_j} - Q_j\frac{\partial v_i}{\partial x_j}\right) = -Q_i - \kappa\frac{\partial T}{\partial x_i}. \tag{15.31}$$

In these equations ρ, α, g, μ and K are density, thermal expansion coefficient, gravity, dynamic viscosity and permeability, respectively. As usual $\mathbf{k} = (0,0,1)$, and

$$M = (\rho_0 c_p)_f / (\rho_0 c)_m,$$

where

$$(\rho_0 c)_m = \varepsilon(\rho_0 c_p)_f + (1 - \varepsilon)(\rho_0 c)_s,$$

with ε being the porosity. The f and s denote fluid and solid components, respectively, while m denotes the porous medium average.

Extensions of the work of section 15.3 to include extra effects like those of Brinkman, Guyer–Krumhansl, magnetic fields, dielectric fluids, phase changes, fractional derivatives and objective derivatives may be found in articles by [49, 95, 126, 127, 171–174, 212, 235, 272, 294, 301, 302, 324, 329, 330, 349], and [495].

The effects of Guyer–Krumhansl in some ways add extra stability to the system. To include Guyer–Krumhansl effects we may modify equations (15.30) and (15.31) in the following way. The equations for thermal convection in a Darcy porous material incorporating the Cattaneo–Fox theory with Guyer–Krumhansl effects are

$$\frac{\partial v_i}{\partial t} = -\frac{1}{\rho}\frac{\partial p}{\partial x_i} + \alpha g k_i T - \frac{\mu}{\rho K} v_i,$$

$$\frac{\partial v_i}{\partial x_i} = 0,$$

$$\frac{1}{M}\frac{\partial T}{\partial t} + v_i\frac{\partial T}{\partial x_i} = -\frac{\partial Q_i}{\partial x_i}, \tag{15.32}$$

$$\tau\frac{\partial Q_i}{\partial t} + \tau_f\left(v_j\frac{\partial Q_i}{\partial x_j} - \frac{1}{2}Q_j\frac{\partial v_i}{\partial x_j} + \frac{1}{2}Q_j\frac{\partial v_j}{\partial x_i}\right)$$
$$= -Q_i - \kappa\frac{\partial T}{\partial x_i} + \hat{\tau}\left(\Delta Q_i + 2\frac{\partial^2 Q_k}{\partial x_k \partial x_i}\right).$$

The equations for thermal convection in a Darcy porous material incorporating the Cattaneo–Christov theory with Guyer–Krumhansl effects are

$$\frac{\partial v_i}{\partial t} = -\frac{1}{\rho}\frac{\partial p}{\partial x_i} + \alpha g k_i T - \frac{\mu}{\rho K} v_i,$$

$$\frac{\partial v_i}{\partial x_i} = 0,$$

$$\frac{1}{M}\frac{\partial T}{\partial t} + v_i\frac{\partial T}{\partial x_i} = -\frac{\partial Q_i}{\partial x_i}, \tag{15.33}$$

$$\tau\frac{\partial Q_i}{\partial t} + \tau_f\left(v_j\frac{\partial Q_i}{\partial x_j} - Q_j\frac{\partial v_i}{\partial x_j}\right)$$
$$= -Q_i - \kappa\frac{\partial T}{\partial x_i} + \hat{\tau}\left(\Delta Q_i + 2\frac{\partial^2 Q_k}{\partial x_k \partial x_i}\right).$$

In equations (15.32) and (15.33) the coefficient $\hat{\tau}$ is a relaxation time. Within heat conduction theory this relaxation time is discussed in section 1.3 of [425]. We observe that the analogous equations to (15.32) and (15.33) but in the context of a pure fluid are given in sections 3.1.3 and 3.1.4 of [425].

15.4 LTNE Cattaneo Solid, Fourier Fluid

In this section we continue with the idea of heat propagating as a thermal wave although we now wish to combine this with LTNE effects where the fluid and solid temperatures in a porous material may be different. Certainly it is believed that finite speed heat propagation is important in certain metallic material situations and we believe it is worthwhile considering this aspect in thermal convection flows in porous metallic foams, especially if the device dimensions are small.

Thermal convection in a fluid saturated porous material allowing for a second sound effect has already been reviewed in section 15.3, with this work extended to thermosolutal convection by [426], cf. also [329, 330] and section 15.5. However, this work employs only one temperature field and does not allow for the possibility of local thermal non-equilibrium. As mentioned in the introduction there are many potential applications of LTNE theory such as those in heat exchangers, cf. [107]. Due to potential applications we describe work of [428] who combines second sound and LTNE. Since the second sound effect appears greater in a solid, especially those involved in porous metallic foams, [428] restricts attention to this effect in the solid whilst retaining the Fourier heat transfer law in the fluid. [428] observes some novel results and discovers that when an interaction coefficient is sufficiently large simultaneously with a relaxation time having sufficient magnitude, then commencement of convective motion according to linear instability theory is via oscillatory convection. This is *not* the case with standard LTNE theory, see section 2.2 and chapter 2. One notable feature of the work of [428] is that he is able to determine a global nonlinear stability threshold which is interesting since previous work on thermal convection instability incorporating heat wave effects did not appear to allow use of an energy method. As [428] observes the use of an energy method is novel since the system of partial differential equations involves Darcy's law, a parabolic equation for the fluid temperature, and a hyperbolic system for the solid temperature. [428] finds that for certain parameter ranges the global stability threshold is close to the linear instability one which greatly restricts the region where sub-critical instabilities might arise. We now describe work of [428].

15.4.1 LTNE–Cattaneo Model

We commence with the basic equations for thermal convection in a porous medium with LTNE effects employing Darcy's law, cf. section 2.2, but modify the equation for the solid temperature to allow the heat flux to satisfy a Cattaneo law. Thus, the basic system of equations has form, cf. [428],

$$v_i = -\frac{K}{\mu}\frac{\partial p}{\partial x_i} + \frac{\rho_f g \alpha K}{\mu} T_f k_i,$$

$$\frac{\partial v_i}{\partial x_i} = 0,$$

$$(1-\varepsilon)(\rho c)_s \frac{\partial T^s}{\partial t} = -\frac{\partial Q_i}{\partial x_i}(1-\varepsilon) - h(T_s - T_f), \qquad (15.34)$$

$$\tau_s \frac{\partial Q_i}{\partial t} = -Q_i - k_s \frac{\partial T}{\partial x_i},$$

$$\varepsilon(\rho c)_f \frac{\partial T^f}{\partial t} + (\rho c)_f v_i \frac{\partial T^f}{\partial x_i} = \varepsilon k_f \Delta T^f + h(T_s - T_f).$$

In these equations x_i and t denote space and time, v_i, p, T^s, Q_i, T^f denote fluid (pore averaged) velocity, pressure, solid temperature, heat flux in the solid, and fluid temperature, respectively. The quantities $K, \mu, g, \alpha, \varepsilon, \rho, c, h, k_s, k_f, \tau_s$ denote permeability, fluid dynamic viscosity, gravity, fluid expansion coefficient, porosity, density, specific heat at constant pressure, a thermal interaction coefficient, thermal conductivity of the solid, thermal conductivity of the fluid, and solid thermal relaxation time, respectively. A sub or superscript s or f refers to solid or fluid.

Equation $(15.34)_1$ is the balance of momentum via Darcy's law, with equation $(15.34)_2$ being the conservation of mass. Equation $(15.34)_3$ represents energy balance equation in the solid recalling $\mathbf{Q}$ is the heat flux vector. Equation $(15.34)_4$ is Cattaneo's law for the solid heat flux, whilst equation $(15.34)_5$ is the energy balance equation for the fluid. The system of equations (15.34) hold in the layer $\mathbb{R}^2 \times \{z \in (0,d)\}$ for $t > 0$, with gravity acting in the negative $z-$direction. The boundary conditions considered are

$$v_3 = 0 \text{ on } z = 0, d \quad \text{and}$$
$$T_f = T_s = T_L, z = 0; \quad T_f = T_s = T_U, z = d; \tag{15.35}$$

where T_L, T_U are constants with $T_L > T_U$.

The steady solution to equations (15.34) subject to boundary conditions (15.35) in whose stability we are interested, is

$$\bar{v}_i \equiv 0, \quad \bar{T}_f = \bar{T}_s = -\beta z + T_L, \quad \bar{\mathbf{Q}} = (0, 0, k_s \beta), \tag{15.36}$$

where β is the temperature gradient given by

$$\beta = \frac{T_L - T_U}{d}.$$

The steady pressure field, $\bar{p}(z)$, follows from $(15.34)_1$ which reduces to

$$\frac{\partial \bar{p}}{\partial z} = \rho_f g \alpha (T_L - \beta z).$$

To study instability of the steady solution (15.36) we let $(u_i, \pi, \theta, q_i^s, \phi)$ be perturbations to $(\bar{v}_i, \bar{p}, \bar{T}^f, \bar{Q}_i, \bar{T}^s)$, so that

$$v_i = \bar{v}_i + u_i, \quad p = \bar{p} + \pi, \quad T^f = \bar{T}^f + \theta, \quad Q_i = \bar{Q}_i + q_i^s, \quad T^s = \bar{T}^s + \phi.$$

One then derives from equations (15.34) and (15.36) the equations governing the perturbation quantities. [428] non-dimensionalizes these equations with the length, time, velocity, pressure, and heat scales $L = d, \mathcal{T} = (\rho c)_f d^2 / k_f, U = \varepsilon k_f / (\rho c)_f d, P = \mu dU / K$, and $Q^s = k_s T^\sharp / d$ where $T^\sharp$ is the temperature scale, $T^\sharp = U \sqrt{c_f \beta d^2 \mu / \varepsilon k_f g \alpha K}$, and further introduces the Rayleigh number $Ra = R^2$, and the non-dimensional coefficients H, A, γ and $\hat{\tau}$ by

$$Ra = R^2 = \frac{\rho_f^2 c_f \beta d^2 K g \alpha}{\mu \varepsilon k_f}, \quad H = \frac{h d^2}{\varepsilon k_f},$$

$$A = \frac{(\rho c)_s}{(\rho c)_f}, \quad \gamma = \frac{\varepsilon}{1 - \varepsilon}, \quad \hat{\tau} = \frac{\tau_s \kappa_f}{d^2}$$

where $\kappa_f = k_f/\rho_f d^2$. Then, the non-dimensional nonlinear system of perturbation equations takes form

$$u_i = -\frac{\partial \pi}{\partial x_i} + R\theta\, k_i,$$

$$\frac{\partial u_i}{\partial x_i} = 0,$$

$$\frac{\partial \theta}{\partial t} + u_i\frac{\partial \theta}{\partial x_i} = Rw + \Delta\theta + H(\phi - \theta), \tag{15.37}$$

$$A\frac{\partial \phi}{\partial t} = -\frac{\partial q_i^s}{\partial x_i} - \gamma H(\phi - \theta),$$

$$\hat{\tau}\frac{\partial q_i^s}{\partial t} = -q_i^s - \frac{\partial \phi}{\partial x_i}.$$

Equations (15.37) hold on the domain $\{(x,y) \in \mathbb{R}^2\} \times \{z \in (0,1)\} \times \{t > 0\}$, and the boundary conditions are

$$w = 0, \theta = 0, \phi = 0, \qquad z = 0, 1, \tag{15.38}$$

and $(u_i, \pi, \theta, \phi, q_i^s)$ satisfy a plane tiling periodicity in the (x,y) plane.

15.4.2 Linear Instability

To find the instability boundary [428] discards the nonlinear term in equation $(15.37)_3$, and removes the pressure term. One then seeks a time dependence like $e^{\sigma t}$ and uses the boundary conditions (15.38) and so may develop a $\sin n\pi z$ series solution. [428] shows the stationary convection boundary, $\sigma = 0$, is given by

$$R_{stat}^2 = \frac{\Lambda^3 + \Lambda^2(1+\gamma)H}{a^2(\Lambda + \gamma H)}, \tag{15.39}$$

where a is the wavenumber and $\Lambda = \pi^2 + a^2$.

The oscillatory convection boundary is found as

$$R_{osc}^2 = \frac{1}{a^2}\left(\frac{b - \sqrt{b^2 - 4Xc}}{2X}\right), \tag{15.40}$$

where the coefficients X, b and c are given by

$$X = \frac{k_1}{\Lambda}, \qquad b = 2k_1\Lambda + k_2,$$

$$c = k_1\Lambda^3 + \left(k_2 + \frac{k_1}{A\hat{\tau}}\right)\Lambda^2 + \Lambda H\left[2\gamma H + \frac{\gamma^2 H}{A} + AH + \frac{(\gamma + A)}{\hat{\tau}}\right],$$

in which the coefficients k_1 and k_2 have form

$$k_1 = A + \hat{\tau}\gamma H, \qquad k_2 = \frac{k_1^2}{\hat{\tau}A} + k_1 H + HA.$$

The imaginary part of the growth rate σ, σ_1 is found to be given by the expression

$$\sigma_1^2 = \frac{H}{\hat{\tau}}\left(\frac{\gamma}{A}+1\right) + \frac{\Lambda}{A\hat{\tau}}(1+k_1) - R^2 a^2 \frac{k_1}{A\hat{\tau}\Lambda}. \tag{15.41}$$

The oscillatory convection threshold is found by minimizing R^2_{osc} in a^2, comparing with the minimum of the stationary convection boundary R^2_{stat}, and checking with expression (15.41) whether $\sigma_1^2 > 0$ or not.

15.4.3 Global Nonlinear Stability

[428] applied an energy method to derive global nonlinear stability results. He observes that while the energy method is well known the application to system (15.37) is non-standard. In particular with other systems involving thermal waves in convection it does not seem easy to obtain global nonlinear stability bounds, cf. Straughan [423, 425, 426]. This point is discussed in more detail on pages 193–195 of Straughan [425], and is due to the fact that the difficulty with Cattaneo systems is the lack of dissipation in the temperature field.

To develop a global nonlinear stability analysis from equations (15.37) and (15.38) let $\| \cdot \|$ and $(\cdot, \cdot)$ denote the norm and inner product on $L^2(V)$. Firstly multiply (15.37)$_4$ by ϕ and integrate over the cell V and then multiply (15.37)$_5$ by q_i^s and likewise integrate over V, to obtain

$$\frac{d}{dt}\frac{A}{2}\|\phi\|^2 = -\gamma H(\phi - \theta, \phi) + (q_i^s, \phi_{,i}),$$
$$\frac{d}{dt}\frac{\hat{\tau}}{2}\|\mathbf{q}^s\|^2 = -\|\mathbf{q}^s\|^2 - (q_i^s, \phi_{,i}), \tag{15.42}$$

where we have integrated by parts and used the boundary conditions (15.38) to arrive at (15.42)$_1$. The key to the energy stability analysis is to now add the resulting equations in (15.42) which removes the $(q_i^s, \phi_{,i})$ terms. This is essential because there is no dissipation term in ϕ to control the $\phi_{,i}$ terms. Addition of equations (15.42) leads to the result

$$\frac{d}{dt}\left(\frac{A}{2}\|\phi\|^2 + \frac{\hat{\tau}}{2}\|\mathbf{q}^s\|^2\right) = -\|\mathbf{q}^s\|^2 - \gamma H(\phi - \theta, \phi). \tag{15.43}$$

Next multiply each of (15.37)$_1$ and (15.37)$_3$ by u_i and θ, respectively, and integrate over V to find

$$\|\mathbf{u}\|^2 = R(\theta, w), \tag{15.44}$$

and

$$\frac{d}{dt}\frac{1}{2}\|\theta\|^2 = R(w,\theta) - \|\nabla\theta\|^2 + H(\phi - \theta, \theta). \tag{15.45}$$

Let now $\lambda_1, \lambda_2 > 0$ be coupling parameters to be judiciously selected and form the combination (15.45)+λ_1(15.44)+λ_2(15.43). In this way one obtains an energy equation of form

$$\frac{dE}{dt} = I - D - \lambda_2\|\mathbf{q}^s\|^2, \tag{15.46}$$

where the energy function is

$$E(t) = \frac{1}{2}\|\theta\|^2 + \frac{A\lambda_2}{2}\|\phi\|^2 + \frac{\hat{\tau}\lambda_2}{2}\|\mathbf{q}^s\|^2, \tag{15.47}$$

the production term I is given by

$$I(t) = (1 + \lambda_1)R(w,\theta) + H(1 + \lambda_2\gamma)(\phi, \theta), \tag{15.48}$$

and the dissipation function D is

$$D(t) = \|\nabla\theta\|^2 + H\|\theta\|^2 + \lambda_1\|\mathbf{u}\|^2 + \lambda_2\gamma H\|\phi\|^2. \tag{15.49}$$

Next define R_E by putting

$$\frac{1}{R_E} = \max_{\mathscr{H}} \frac{I}{D} \tag{15.50}$$

where $\mathscr{H}$ is the space of admissible solutions. We now require $R_E > 1$ and set $\alpha = 1 - R_E^{-1} > 0$, so that from equation (15.46) we may obtain

$$\frac{dE}{dt} \leq -\alpha D - \lambda_2\|\mathbf{q}^s\|^2. \tag{15.51}$$

Use Poincaré's inequality and the definition of D to see that

$$D \geq (H + \pi^2)\|\theta\|^2 + \lambda_1\|\mathbf{u}\|^2 + \lambda_2\gamma H\|\phi\|^2. \tag{15.52}$$

We combine inequality (15.52) together with inequality (15.51) to see that there is a constant $k > 0$ such that

$$\frac{dE}{dt} \leq -kE.$$

From this inequality one finds $E(t) \leq E(0)\exp(-kt)$ and thus E decays at least exponentially which yields global stability in θ, ϕ and q_i^s. Decay and global stability of u_i then follows from (15.44) since we may derive the inequality

$$\|\mathbf{u}\|^2 \leq R^2\|\theta\|^2.$$

The global nonlinear stability theshold follows upon resolution of the variational problem (15.50). At the threshold value $R_E = 1$, the maximum problem (15.50) leads to the Euler-Lagrange equations

$$(1+\lambda_1)R\theta k_i - 2\lambda_1 u_i = -\frac{\partial \omega}{\partial x_i}; \qquad \frac{\partial u_i}{\partial x_i} = 0;$$

$$(1+\lambda_1)Rw + H(1+\lambda_2\gamma)\phi - 2H\theta + 2\Delta\theta = 0; \qquad (15.53)$$

$$H(1+\lambda_2\gamma)\theta - 2\lambda_2\gamma H\phi = 0,$$

where $\omega(\mathbf{x})$ is a Lagrange multiplier. The energy stability Rayleigh number R^2 follows from equations (15.53) as

$$R^2 = \frac{1}{a^2} \frac{2\lambda_1}{(1+\lambda_1)^2} \left[2\Lambda^2 + 2H\Lambda - \frac{H\Lambda}{2\gamma} \frac{(1+\lambda_2\gamma)^2}{\lambda_2} \right], \qquad (15.54)$$

where $\Lambda = \pi^2 + a^2$. The critical energy stability Rayleigh number Ra_E follows from the maximization-minimization problem

$$Ra_E = \max_{\lambda_1,\lambda_2} \min_{a^2} R^2.$$

[428] observes that the maximum of R^2 in λ_1 is achieved when $\lambda_1 = 1$, and then a similar calculation involving λ_2 yields the maximum with $\lambda_2 = 1/\gamma$. Hence,

$$R^2 = \frac{\Lambda^2}{a^2}$$

which yields the critical energy Rayleigh number maximum as $Ra_E = R_{en}^2 = 4\pi^2$.

15.4.4 Conclusions for LTNE–Cattaneo Convection

[428] reports numerical computations based on the linear instability critical Rayleigh numbers obtained from (15.39) and (15.40), and on the global nonlinear stability Rayleigh number. He employs H values which are in line with those suggested by [364, 365] and varies the non-dimensional relaxation time $\hat{\tau}$. Unlike the classical case of LTNE thermal convection in a Darcy material studied by [27], see section 2.2, where only stationary convection occurs, [428] discovers that in the present scenario with Cattaneo theory for the solid skeleton, there is the possibility of convective instability arising by oscillatory convection. One of his findings is that the critical value of the Rayleigh number at which the transition to oscillatory convection takes place decreases as $\hat{\tau}$ increases.

We here include several new graphs, not in [428], in figures 15.6–15.12. The numerical calculations for figures 15.6–15.12 are produced using relations (15.39) and (15.40). We base the computations on the solid skeleton being of two types. The first is a skeleton with a porosity found in everyday porous materials having a porosity in the range 0.2–0.4. The second type of porous material is a highly porous

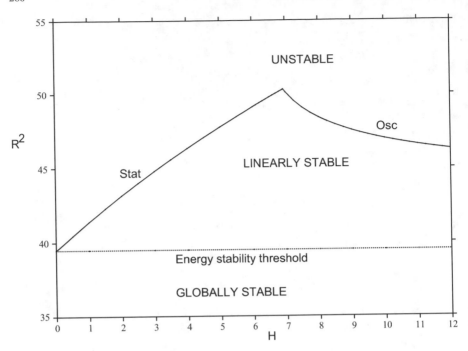

Fig. 15.6 Critical values of R^2 vs. H, $A = 0.81853$, $\gamma = 0.66667$, $\hat{\tau} = 1$. The A value is that for a porous skeleton made of Al_2O_3 which is saturated with water. The porosity value is $\varepsilon = 0.4$. The linear instability threshold is comprised of the curves marked Stat and Osc. The global non-linear stability threshold is the line indicated by "Energy stability threshold". The transition from stationary to oscillatory convection is at $H_{trans} = 6.93$.

foam where the porosities are greater than 0.9. The actual values of parameters we employ are based on the real materials aluminium oxide, Al_2O_3, a sandstone known as Sander sandstone, or on an aluminium foam AL1050. The aluminium oxide and the aluminium foam have practical use in heat exchangers and sandstone is a well known building material. The saturating fluid in all cases is water. We have performed computations when the saturating fluid is a 50/50 mixture of water and ethylene glycol and the results are very similar to those we present for water alone.

We need values for the coefficient $A = \rho_s c_s / \rho_f c_f$. For Sander sandstone we find from [1] that $\rho_s = 2.081$ g cm^{-3}, $c_s = 0.795$ kJ kg^{-1} $^\circ$K^{-1}, and the porosity has value $\varepsilon = 0.17$. This yields a value for $\gamma = \varepsilon/(1 - \varepsilon)$ for Sander sandstone as $\gamma = 0.204819$. For water we use [124] to find $\rho_f = 0.9992$g cm^{-3}, and $c_f = 4.1855$ kJ kg^{-1} $^\circ$K^{-1}. For Al_2O_3 we use [2] to see that $\rho_s = 3.89$g cm^{-3}, and $c_s = 0.880$ kJ kg^{-1} $^\circ$K^{-1}. This yields a value for A with Al_2O_3-water as $A = 0.81853$. The combination of Sander sandstone and water has $A = 0.395585$. When we deal with the Al_2O_3-water combination we use two values for $\hat{\tau}$. The first is to use a porosity of $\varepsilon = 0.4$ which is consistent with loosely packed gravel. The next is $\varepsilon = 0.94$ which is appropriate for a high porosity foam, cf. [4]. These porosity values lead to values of γ as $\gamma = 0.6667$ (to 4 dp) and $\gamma = 15.6667$ (to 4 dp). For the Al_2O_3-water

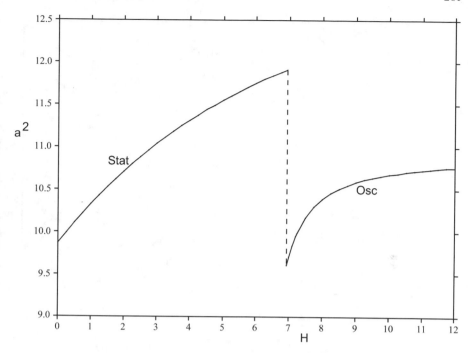

Fig. 15.7 Critical values of the linear instability critical wavenumbers squared, a^2 vs. H. The constant A has value $A = 0.81853$, and is that for a porous skeleton made of Al_2O_3 which is saturated with water. The porosity value is $\varepsilon = 0.4$. The constant $\hat{\tau} = 1$. The branch marked Stat is for stationary convection whereas that marked Osc denotes oscillatory convection. The jump in the wavenumbers from stationary to oscillatory convection is denoted by a dashed line. The transition from stationary to oscillatory convection is at $H_{trans} = 6.93$.

combination we choose the non-dimensional relaxation time $\hat{\tau}$ to be $\hat{\tau} = 1$. When we deal with the sandstone-water combination we select $\hat{\tau} = 0.1$.

For the aluminium foam AL1050 we find from [189] that $\rho_s = 2.719$ g cm^{-3} and $c_s = 0.871$ kJ kg^{-1} $^\circ$K^{-1}. [189] report porosity values of 0.913 - 0.967 and we select the value of $\varepsilon = 0.913$. This leads to a value of $\gamma = 10.49425$. For water in this case we employ the values of [57] so that $\rho_f = 0.997$ g cm^{-3} and $c_f = 4.179$ kJ kg^{-1} $^\circ$K^{-1}. Thus we find that for a water saturated AL1050 aluminium foam $A = 0.568407543$. In this case we set $\hat{\tau} = 1$. When we computed the critical Rayleigh numbers and wavenumbers using the ethylene glycol-water mixture we used the density and specific heat values reported in [57].

The solid curves in figures 15.6–15.12 indicated by the *Stat* and *Osc* branches show the instability threshold. In figures 15.6, 15.8, 15.10 and 15.11 when R^2 is above the curve marked Stat-Osc then there is definitely instability. The transition value from stationary to oscillatory convection is where $R^2 = 50.282$ and $H_{trans} = 6.93$, in figure 15.6. For H greater than H_{trans} instability is by oscillatory convection whereas when H is less than H_{trans} instability is by stationary convection. It is evident from figure 15.7 that the wavenumber changes strongly from 11.91

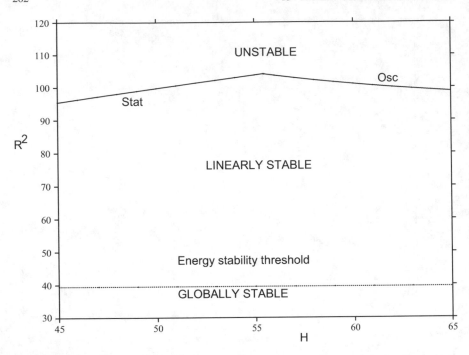

Fig. 15.8 Critical values of R^2 vs. H, $A = 0.395585$, $\hat{\tau} = 0.1$. The A value is that for a porous skeleton made of Sander sandstone which is satuated with water. The porosity value is $\varepsilon = 0.17$. The linear instability threshold is comprised of the curves marked Stat and Osc. The global non-linear stability threshold is the line indicated by "Energy stability threshold". The transition from stationary to oscillatory convection is at $H_{trans} = 55.43$.

to 9.61 as one passes from stationary to oscillatory convection. Thus, the convection cells become much wider as we cross through H_{trans}. For $H < H_{trans}$ we see that the wavenumber increases and then at H_{trans} it changes discontinuously and thereafter again increases.

In figure 15.8 the transition value from stationary to oscillatory convection is where $R^2 = 104.058$ and $H_{trans} = 55.43$. The wavenumber transition values are $a^2 = 16.90$ on the stationary convection branch and $a^2 = 14.34$ on the oscillatory convection branch, cf. figure 15.9.

The transition value from stationary to oscillatory convection is where $R^2 = 41.757$ and $H_{trans} = 11.94$, in figure 15.10. We do not include a figure of the wavenumber transition for the case of figure 15.10 because the wavenumber changes very little before and after the transition. The graph of the critical wavenumber corresponding to figure 15.10 is not dissimilar to figure 15.12. For the case of figures 15.10–15.12 the wavenumber change at the transition is very interesting. For the figure 15.10 the critical wavenumber for the stationary convection branch at the transition is $a^2 = 9.921$ which jumps to the value of $a^2 = 10.379$ on the oscillatory convection branch. For the figure 15.12 the critical wavenumber for the stationary convection branch at the transition is $a^2 = 10.044$ which jumps to the value

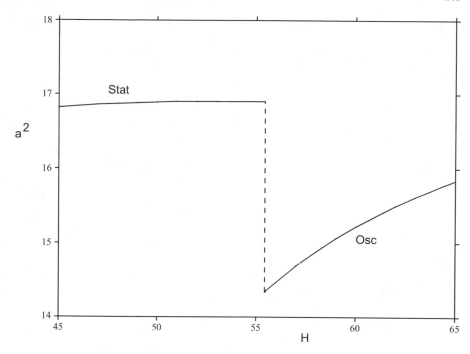

Fig. 15.9 Critical values of the linear instability critical wavenumbers squared, a^2 vs. H. The constant A has value $A = 0.395585$, and is that for a porous skeleton made of Sander sandstone which is saturated with water. The porosity value is $\varepsilon = 0.17$. The constant $\hat{\tau} = 0.1$. The branch marked Stat is for stationary convection whereas that marked Osc denotes oscillatory convection. The jump in the wavenumbers from stationary to oscillatory convection is denoted by a dashed line. The transition from stationary to oscillatory convection is at $H_{trans} = 55.43$.

of $a^2 = 10.373$ on the oscillatory convection branch. Thus, the behaviour is different from either of the cases in figures 15.7 and 15.9 in that the wavenumber here increases during the transition from stationary to oscillatory convection. The much greater porosity appears to be having a major effect on the cell sizes at transition. Observe that the variation in R^2 in figures 15.10 and 15.11 is much less than that of figures 15.6 and 15.8. The greater porosity in the situation of figures 15.10 and 15.11 would appear to be lowering the Rayleigh number substantially.

The curve marked "Energy stability threshold" in figures 15.6, 15.8, 15.10 and 15.11 indicates the global nonlinear stability threshold. If R^2 is below the value of the energy stability threshold, namely 39.478, there is definitely global nonlinear stability. We cannot make any judgement about stability in the region between the curves marked "Stat-Osc" and "Energy stability threshold" in figures 15.6, 15.8, 15.10 and 15.11, and this may represent a region of sub-critical instability. In figures 15.6, 15.8, 15.10 and 15.11 the region of possible sub-critical instability is marked as LINEARLY STABLE.

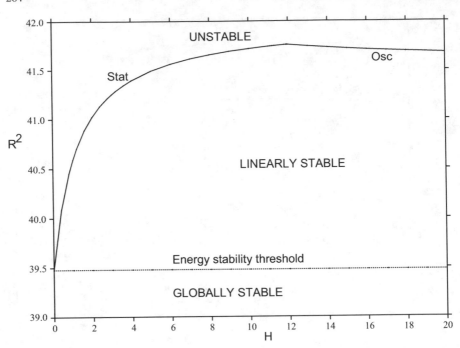

Fig. 15.10 Critical values of R^2 vs. $H, A = 0.81853, \hat{\tau} = 1$. The A value is that for a porous skeleton made of Al_2O_3 which is saturated with water. The porosity value is $\varepsilon = 0.94$. The linear instability threshold is comprised of the curves marked Stat and Osc. The global nonlinear stability threshold is the line indicated by "Energy stability threshold". The transition from stationary to oscillatory convection is at $H_{trans} = 11.94$.

15.5 Cattaneo Double Diffusion

We now describe work of [426] who analyses the problem of heated and salted below double diffusive convection, but allows the heat flux to be of Cattaneo–Christov type. While the motivation of [426] was based on work of [178] and of [128] we believe the general modelling analysis may well be of use in other thermodiffusive microfluidic situations. This viewpoint is also advanced by [15] who despite the title of their article give other relevant non-relativisitic examples and a very interesting account of the role of the relaxation time. [178] proposed the idea to employ the theory of [79] to study thermohaline convection and concluded that it could have important implications in stellar atmospheres, for example, in a close binary system where helium rich material may be transferred to a main sequence star, or in a neutron star, or in the core of a collapsed supernovae. [128] proposed analysing cooling of white dwarfs and neutron stars by employing a Cattaneo theory. [178] do not produce a precise model for thermohaline convection employing a [79] heat flux theory as was done by [426]. He derives a model for thermohaline convection in a porous medium saturated with a linear viscous fluid when the heat

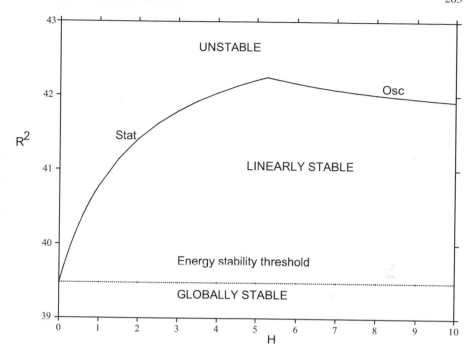

Fig. 15.11 Critical values of R^2 vs. H, $A = 0.568407543$, $\hat{\tau} = 1$. The A value is that for a highly porous foam skeleton made of AL1050 which is saturated by water. The porosity value is $\varepsilon = 0.913$. The linear instability threshold is comprised of the curves marked Stat and Osc. The global nonlinear stability threshold is the line indicated by "Energy stability threshold". The transition from stationary to oscillatory convection is at $H_{trans} = 5.265$.

flux law is one of Cattaneo–Christov type and it is this we now describe. [426] finds that in the Cattaneo–Christov theory of thermohaline convection in a porous material the thermal relaxation effect plays a key role. He discovers that an instability threshold is established which depends strongly on the Cattaneo number (relaxation time), but which is influenced by the porous properties and by the salt concentration.

15.5.1 Cattaneo–Christov Model

The momentum, mass balance, and salt concentration equations are

$$\tilde{p}_{,i} = -\frac{\mu}{K} v_i - \rho g k_i, \qquad (15.55)$$

and

$$v_{i,i} = 0, \qquad (15.56)$$

and

$$\varepsilon C_{,t} + v_i C_{,i} = \varepsilon k_C \Delta C, \qquad (15.57)$$

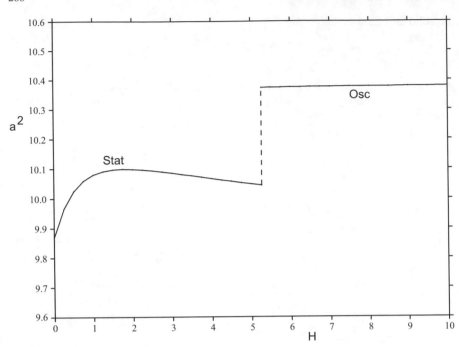

Fig. 15.12 Critical values of the linear instability critical wavenumbers squared, a^2 vs. H. The constant A has value $A = 0.568407543$, and is that for a highly porous foam skeleton made of AL1050 which is saturated with water. The porosity value is $\varepsilon = 0.913$. The constant $\hat{\tau} = 1$. The branch marked Stat is for stationary convection whereas that marked Osc denotes oscillatory convection. The jump in the wavenumbers from stationary to oscillatory convection is denoted by a dashed line. The transition from stationary to oscillatory convection is at $H_{trans} = 5.265$.

where $\tilde{p}, \mu, K, \rho, g$ are pressure, dynamic viscosity, permeability, density and gravity, $\mathbf{k} = (0,0,1)$, and v_i is the pore averaged velocity, T, C are the temperature and salt concentration, and k_C is the salt diffusion coefficient. Let V_i be the actual velocity in the fluid with ε the porosity, so that $v_i = \varepsilon V_i$. A Boussinesq approximation is adopted and in the buoyancy term we employ as density

$$\rho = \rho_0\left(1 - \alpha[T - T_0] + \alpha_s[C - C_0]\right). \tag{15.58}$$

Using equation (15.58) in equation (15.55) allows us to replace (15.55) by the equation

$$p_{,i} = -\frac{\mu}{K}v_i + \rho_0\alpha g T k_i - \rho_0\alpha_s g C k_i, \tag{15.59}$$

in which $p = \tilde{p} + \rho_0(1 + \alpha T_0 - \alpha_s C_0)gz$ is a modified pressure, and z is the vertical coordinate.

The next step is to write energy balances for the solid and fluid parts of the porous medium separately, and likewise write Cattaneo–Christov heat flux laws for each of the solid and fluid components. Then the energy balance and Cattaneo law for the solid are,

$$(\rho_0 c)_s T_{,t} = -\tilde{Q}_{i,i}, \tag{15.60}$$

$$\tau_s \tilde{Q}_{i,t} = -\tilde{Q}_i - k_s T_{,i}, \tag{15.61}$$

where c is the specific heat, τ_s is the relaxation time, k_s is thermal conductivity, and $\tilde{Q}_i$ is heat flux. For the fluid, the energy balance and [92] heat flux laws are

$$(\rho_0 c_p)_f (T_{,t} + V_i T_{,i}) = -\tilde{Q}_{i,i}, \tag{15.62}$$

$$\tau_f (\tilde{Q}_{i,t} + V_j \tilde{Q}_{i,j} - \tilde{Q}_j V_{i,j}) = -\tilde{Q}_i - k_f T_{,i}, \tag{15.63}$$

where c_p is the specific heat at constant pressure, k_f is the thermal conductivity, and τ_f is a relaxation time.

[426] shows that with the variables

$$Q_i = \frac{\tilde{Q}_i}{(\rho_0 c_p)_f}, \qquad (\rho_0 c)_m = \varepsilon (\rho_0 c_p)_f + (1 - \varepsilon)(\rho_0 c)_s,$$

$$k_m = \varepsilon k_f + (1 - \varepsilon) k_s,$$

$$\kappa = \frac{k_m}{(\rho_0 c_p)_f}, \qquad M = \frac{(\rho_0 c_p)_f}{(\rho_0 c)_m}, \qquad \tau = \varepsilon \tau_f + (1 - \varepsilon)\tau_s,$$

then equations (15.60)–(15.63) may be replaced by

$$\frac{1}{M} T_{,t} + v_i T_{,i} = -Q_{i,i}, \tag{15.64}$$

$$\tau Q_{i,t} + \tau_f (v_j Q_{i,j} - Q_j v_{i,j}) = -Q_i - \kappa T_{,i}. \tag{15.65}$$

The complete system of equations for the model then becomes

$$\frac{\partial p}{\partial x_i} = -\frac{\mu}{K} v_i + \rho_0 \alpha g T k_i - \rho_0 \alpha_s g C k_i,$$

$$\frac{\partial v_i}{\partial x_i} = 0,$$

$$\frac{1}{M} \frac{\partial T}{\partial t} + v_i \frac{\partial T}{\partial x_i} = -\frac{\partial Q_i}{\partial x_i}, \tag{15.66}$$

$$\tau \frac{\partial Q_i}{\partial t} + \tau_f \left(v_j \frac{\partial Q_i}{\partial x_j} - Q_j \frac{\partial v_i}{\partial x_j} \right) = -Q_i - \kappa \frac{\partial T}{\partial x_i},$$

$$\varepsilon \frac{\partial C}{\partial t} + v_i \frac{\partial C}{\partial x_i} = \varepsilon k_C \Delta C.$$

We have only employed Cattaneo–Christov theory in the temperature and heat flux equations. One could propose a model which involves hyperbolic diffusion, cf. [425, pp. 252–254]. Hyperbolic diffusion is a topic of much current interest, see e.g. [60, 136, 152, 166, 194, 381, 483], and [296], and the references therein.

To study stability of a solution to equations (15.66) let us suppose the fluid saturated porous layer occupies the region $\{(x,y) \in \mathbb{R}^2\} \times \{z \in (0,d)\}$. On the boundaries $z = 0, d$ we pose the boundary conditions

$$
\begin{aligned}
T = T_L, \quad C = C_L, \quad z = 0; \\
T = T_U, \quad C = C_U, \quad z = d;
\end{aligned}
\tag{15.67}
$$

where T_L, T_U, C_L, C_U are constants with $T_L > T_U, C_L > C_U$. The steady solution in
which we are interested is

$$
\bar{v}_i = 0, \qquad \bar{T} = -\beta z + T_L, \qquad \bar{C} = -\beta_s z + C_L,
\tag{15.68}
$$

where the temperature gradient, β, and the salt gradient, β_s, are defined by

$$
\beta = \frac{T_L - T_U}{d},
$$

and

$$
\beta_s = \frac{C_L - C_U}{d}.
$$

There is a non-zero heat flux in the steady state of form $\bar{\mathbf{Q}} = (0,0,\bar{Q}_3)$ where

$$
\bar{Q}_3 = \kappa\beta.
$$

To complete the base solution we note that the steady pressure $\bar{p}(z)$ may then be
found from equation (15.66)$_1$.

The instability analysis introduces perturbations $(u_i, \pi, \theta, q_i, \varphi)$ to the steady
solution $(\bar{v}_i, \bar{p}, \bar{T}, \bar{Q}_i, \bar{C})$ by

$$
v_i = \bar{v}_i + u_i, \quad p = \bar{p} + \pi, \quad T = \bar{T} + \theta, \quad Q_i = \bar{Q}_i + q_i, \quad C = \bar{C} + \varphi.
$$

Then with the non-dimensional variables

$$
x_i = dx_i^*, \quad t = \mathscr{T}t^*, \quad \mathscr{T} = \frac{d^2}{M\kappa},
$$

$$
T^\sharp = \sqrt{\frac{\beta d\mu U}{\rho_0 \alpha g K}}, \qquad C^\sharp = Ud\sqrt{\frac{\mu\beta_s}{\rho_0 \alpha_s g K \phi k_C}},
$$

$$
U = \frac{\kappa}{d}, \qquad P = \frac{\mu Ud}{K}, \qquad Q^* = \frac{\kappa T^\sharp}{d},
$$

$$
Ca = \frac{\kappa\tau}{2d^2}, \quad \hat{\tau}_f = \frac{\tau_f U Q^*}{\kappa T^\sharp}, \quad C_f = \frac{\kappa\tau_f}{2d^2}, \quad Le = \frac{Ud}{\varepsilon k_C},
$$

$$
R^2 = \frac{\alpha g\beta K d^2}{(\mu/\rho_0)\kappa}, \qquad R_s^2 = \frac{\alpha_s g\beta_s K d^2}{(\mu/\rho_0)\varepsilon k_C},
$$

one may derive the following system of non-dimensional perturbation equations

$$\frac{\partial \pi}{\partial x_i} = -u_i + R\theta k_i - R_s \varphi k_i,$$

$$\frac{\partial u_i}{\partial x_i} = 0,$$

$$\varepsilon Le \frac{\partial \varphi}{\partial t} + Le\, u_i \frac{\partial \varphi}{\partial x_i} = R_s w + \Delta \varphi, \qquad (15.69)$$

$$\frac{\partial \theta}{\partial t} + u_j \frac{\partial \theta}{\partial x_j} = Rw - \frac{\partial q_i}{\partial x_i},$$

$$2MCa \frac{\partial q_i}{\partial t} + \hat{\tau}_f \left(u_j \frac{\partial q_i}{\partial x_j} - q_j \frac{\partial u_i}{\partial x_j} \right) = 2C_f R \frac{\partial u_i}{\partial z} - q_i - \frac{\partial \theta}{\partial x_i},$$

where $\varepsilon = \varepsilon M$ and $Ra = R^2$ and R_s^2 are the Rayleigh and salt Rayleigh numbers. These equations are to be solved together with the boundary conditions

$$u_i n_i = 0, \quad \theta = \varphi = 0, \qquad \text{on} \quad z = 0, 1, \qquad (15.70)$$

together with the fact that the (x, y) behaviour of the solution satisfies a plane tiling periodic pattern. In equations (15.70) the vector $\mathbf{n}$ denotes the unit outward normal to the boundary.

15.5.2 Linear Instability

[426] studies instability of the steady solution (15.68) by linearizing equations (15.69) and removes the pressure perturbation π. He further takes the divergence of the linear version of equation (15.69)$_5$ and sets $Q = q_{i,i}$. He writes $w = e^{\sigma t} w(\mathbf{x})$, $\theta = e^{\sigma t} \theta(\mathbf{x})$, $\varphi = e^{\sigma t} \varphi(\mathbf{x})$, $Q = e^{\sigma t} Q(\mathbf{x})$ and defines

$$\mathscr{C} = 2MCa \qquad \text{and} \qquad \varepsilon_1 = \varepsilon Le.$$

In this way one may show that the relevant equations to solve to obtain the linear instability boundary are

$$\Delta w - R\Delta^* \theta + R_s \Delta^* \varphi = 0,$$
$$\Delta \theta + Q = -\sigma \mathscr{C} Q, \qquad (15.71)$$
$$\Delta \varphi + R_s w = \sigma \varepsilon_1 \varphi,$$

where Δ^* is the horizontal Laplacian, $\Delta^* = \partial^2/\partial x^2 + \partial^2/\partial y^2$. This is an eigenvalue problem for σ to be solved subject to the boundary conditions

$$w = 0, \theta = 0, \varphi = 0, \qquad z = 0, 1.$$

Introduce a plane tiling function f, and a wavenumber, a, with $\Delta^* f = -a^2 f$, and then with

$$w = W(z)f(x,y), \qquad \theta = \Theta(z)f(x,y),$$
$$\varphi = \Phi(z)f(x,y), \qquad Q = Q(z)f(x,y),$$

equations (15.71) reduce to solving

$$(D^2 - a^2)W + Ra^2\Theta - R_s a^2 \Phi = 0,$$
$$(D^2 - a^2)\Theta + Q = -\sigma\mathscr{C}Q,$$
$$(D^2 - a^2)\Phi + R_s W = \sigma\varepsilon_1\Phi, \qquad (15.72)$$
$$Q - RW = -\sigma\Theta,$$

on $z \in (0,1)$, where $D = d/dz$. These equations are to be solved subject to the boundary conditions

$$W = 0, \Theta = 0, \Phi = 0, \qquad z = 0, 1. \qquad (15.73)$$

[426] solves (15.72) and (15.73) exactly by analytical means, but also by using a D^2–Chebyshev tau numerical method.

He finds the stationary convection boundary ($\sigma = 0$) is given by

$$R^2 = R_s^2 + \frac{\Lambda^2}{a^2}$$

where $\Lambda = n^2\pi^2 + a^2$. Then one shows the critical Rayleigh number is given by

$$R^2 = R_s^2 + 4\pi^2, \qquad a_c = \pi, \qquad (15.74)$$

a_c being the critical value of a which yields a minimum in R^2. To find the oscillatory convection boundary put $\sigma = i\sigma_1$ in (15.72). [426] shows that this results in having to solve the equation

$$AR^4 a^4 - BR^2 a^2 + C_1 = 0, \qquad (15.75)$$

where the coefficients A, B and C_1 are functions of $\varepsilon_1, \mathscr{C}, R_s, n$ and the wavenumber. Precisely, A, B and C_1 have the forms

$$A = \varepsilon_1\mathscr{C}(\varepsilon_1 + \mathscr{C}\Lambda),$$

and

$$B = \mathscr{C}^2\Lambda^3 + \Lambda\varepsilon_1^2 + 2\mathscr{C}\varepsilon_1\Lambda^2 + \varepsilon_1^2\mathscr{C}\Lambda^2 + R_s^2 a^2\{\mathscr{C}^2\Lambda + 2\varepsilon_1\mathscr{C}\},$$

and

$$C_1 = R_s^2 a^2[2\mathscr{C}\Lambda^2 + \Lambda\varepsilon_1 + \mathscr{C}R_s^2 a^2] + \Lambda^3(\mathscr{C}\Lambda + \varepsilon_1^2 + \varepsilon_1).$$

The critical Rayleigh number for oscillatory convection is then found by minimizing

$$R^2 = \frac{B \pm \sqrt{B^2 - 4AC_1}}{2Aa^2} \tag{15.76}$$

in a, n with $\mathscr{C} \neq 0$.

The instability surface in $R^2, R_s^2, \mathscr{C}$ space is determined in [425]. It is an interesting surface which is formed by eigenvalues changing places as the Cattaneo number increases. The instabilities may be stationary or oscillatory, depending on the parameter values, R_s^2 and Ca.

15.6 Green–Naghdi Nanofluid Model

In this section we describe a system of equations proposed by [420] to model the behaviour of a nanofluid suspension. The model of [420] is based on a fluid flow model of [168] who were interested primarily in attempting to model turbulence. The thermodynamic theory of [168] hinges on use of an entropy balance law instead of an entropy inequality. They work with a temperature field T which they call the "empirical" temperature and they also have recourse to use a function α they call the "thermal displacement variable". The thermal displacement variable is defined as

$$\alpha = \int_{t_0}^{t} T(\mathbf{x}, s) ds + \alpha_0,$$

where α_0 is a suitable value for α at time t_0.

The model we are interested in is a reduced version of the general theory of [168] where only one equation remains for the mechanical differential equation. From the point of view of producing a model to describe a suspension the theory of [168] is appealing since it involves the vorticity and the spin of vorticity as constitutive variables. The introduction of these extra variables produces higher spatial gradients in the resulting equations when considered against those for a linear viscous fluid, i.e. Navier–Stokes theory. We believe this is important for a suspension which has some characteristics of a non-Newtonian fluid. [168] are motivated by attempting to describe turbulence and they introduce two temperatures. We believe this is a natural theory to adapt to nanofluid behaviour where we identify the two temperatures with the temperature of the fluid and with the temperature of the nanoparticles themselves.

The differential equations for the model of [168] are firstly given in the form they presented. We do not include details of the continuum thermodynamical development from the entropy balance equation since this may be found in [168]. The equations of [168] involve conservation laws for the balance of linear momentum, the balance of mass, and two further equations describing balances of entropy. The two equations for the balances of entropy are necessary since there are two temperatures θ_H and θ_T. [168] call θ_H the usual temperature, and they call θ_T a turbulent temperature. The basic equations of [168] for their incompressible fluid are thus the balance of linear momentum, balance of mass, and balances of entropy which have the form

$$\rho\left(\dot{v}_i - \frac{\mu_1}{\mu}\frac{d}{dt}\Delta v_i\right) = \rho b_i - \frac{\partial p}{\partial x_i} + \mu\Delta v_i - 2\mu_1\Delta^2 v_i,$$

$$\frac{\partial v_i}{\partial x_i} = 0,$$

$$\rho\dot{\eta}_H = \rho s_H + \rho\xi_H - \frac{\partial p_i^H}{\partial x_i},$$

$$\rho\dot{\eta}_T = \rho s_T + \rho\xi_T - \frac{\partial p_i^T}{\partial x_i}.$$

(15.77)

In equations (15.77) we use either a superposed dot or d/dt to denote the material derivative $d/dt = \partial/\partial t + v_i\partial/\partial x_i$. The quantities ρ, v_i, b_i, p are the density, velocity, prescribed body force, and the pressure. The coefficient μ is the kinematic viscosity of the fluid whereas μ_1 is another constant which depends on the geometry of the particles in the suspension and on their interaction with the fluid.

The presence of the higher derivative term in the inertia on the left hand side of $(15.77)_1$ is one which is important when dealing with a suspension. Its presence may be justified by appealing to work in the appendices to [54] and [169]. These writers derive an expression for the kinetic energy of a system of particles as a function of the velocity of the centroid and the derivative of this velocity. The method of [169] and [54] produces a kinetic energy which can be likened to the kinetic energy of the fluid theory of [168].

In equations (15.77) the variables η_H and η_T are entropies which correspond to the temperatures θ_H and θ_T. The quantities s_H and s_T are *external* supplies of entropy, while the quantities ξ_H and ξ_T are *intrinsic* supplies of entropy which depend on the constitutive variables of the theory. The terms p_i^H and p_i^T are entropy flux vectors.

To make equations (15.77) determinate and therefore useful one needs to propose a specific form for the Helmholtz free energy function ψ in terms of the constitutive variables. The Helmholtz free energy chosen by [168] is the following

$$\psi = c_H(\theta_H - \theta_H\ln\theta_H) + c_T(\theta_T - \theta_T\ln\theta_T).$$

(15.78)

In equation (15.78) c_H and c_T are positive constants. The entropies and entropy fluxes selected by [168] are

$$\eta_H = c_H\ln\theta_H, \qquad \eta_T = c_T\ln\theta_T,$$

(15.79)

and

$$p_i^H = -\frac{\kappa_H}{\theta_0}\frac{\partial\theta_H}{\partial x_i}, \qquad p_i^T = -\frac{\kappa_T}{\theta_0}\frac{\partial\theta_T}{\partial x_i}.$$

(15.80)

In equations (15.80) κ_H and κ_T are positive constants and $\theta_0 > 0$ is a reference temperature.

The instrinsic entropy supply functions of [168] are

$$\xi_H = \frac{1}{\rho \theta_H} \left(\frac{\kappa_H}{\theta_0} \frac{\partial \theta_H}{\partial x_i} \frac{\partial \theta_H}{\partial x_i} + 2\mu d_{ij} d_{ij} + \phi \right), \tag{15.81}$$

and

$$\xi_T = \frac{1}{\rho \theta_T} \left(\frac{\kappa_T}{\theta_0} \frac{\partial \theta_T}{\partial x_i} \frac{\partial \theta_T}{\partial x_i} + 4\mu_1 d_{ij} P_{ij} + \frac{2\mu_1^2}{\mu} P_{ij} P_{ij} - \phi \right). \tag{15.82}$$

The quantity d_{ij} is the symmetric part of the velocity gradient, i.e. $d_{ij} = (v_{i,j} + v_{j,i})/2$, and P_{ij} is defined by $P_{ij} = -\Delta v_{i,j}$. [168] take the term ϕ to be constant although they do note that ϕ could depend on the temperatures θ_H and θ_T.

Upon employing expressions from equations (15.78)–(15.82) in the balance laws (15.77) one derives equations for an incompressible viscous fluid in the [168] extended theory as

$$\rho \frac{dv_i}{dt} - \frac{\rho \mu_1}{\mu} \frac{d}{dt} \Delta v_i = \rho b_i - \frac{\partial p}{\partial x_i} + \mu \Delta v_i - 2\mu_1 \Delta^2 v_i, \qquad \frac{\partial v_i}{\partial x_i} = 0,$$

$$\rho c_H \frac{d\theta_H}{dt} = \rho s_H \theta_H + \frac{\kappa_H}{\theta_0} \frac{\partial \theta_H}{\partial x_i} \frac{\partial \theta_H}{\partial x_i} + \frac{\kappa_H}{\theta_0} \theta_H \Delta \theta_H$$

$$+ 2\mu d_{ij} d_{ij} + \phi, \tag{15.83}$$

$$\rho c_T \frac{d\theta_T}{dt} = \rho s_T \theta_T + \frac{\kappa_T}{\theta_0} \frac{\partial \theta_T}{\partial x_i} \frac{\partial \theta_T}{\partial x_i} + \frac{\kappa_T}{\theta_0} \theta_T \Delta \theta_T$$

$$+ 4\mu d_{ij} P_{ij} + 2\frac{\mu_1^2}{\mu} P_{ij} P_{ij} - \phi.$$

One objection one may raise to equations (15.83) is that if we take $v_i \equiv 0$, $s_H = 0$, $s_T = 0$, and $\phi = 0$ then equations (15.83)$_3$ and (15.83)$_4$ do not reduce to the heat equation for each of θ_H and θ_T. This may be remedied by modifying equations (15.80) for the entropy fluxes $\mathbf{p}^H$ and $\mathbf{p}^T$, and by modifying equations (15.81) and (15.82) for the intrinsic entropy supply functions ξ_H and ξ_T. One way to do this is to replace θ_0 in equations (15.80), (15.81) and (15.82) by θ_H or θ_T as appropriate. This we now do to derive a model for the behaviour of a nanofluid suspension.

We now briefly describe work of [420] who adapts the Green–Naghdi theory outlined above to be applicable to a thermal convection in a nanofluid.

[420] begins by modifying the entropy flux vectors and intrinsic entropy supplies as outlined above by appropriate replacement of θ_0. Rather than employ H and T as sub or superscripts we now use F or S to denote the fluid component or the solid nanoparticles. The entropy flux vectors $\mathbf{p}^F$ and $\mathbf{p}^S$ are specified as

$$p_i^F = -\frac{k_F}{\theta_F} \frac{\partial \theta_F}{\partial x_i}, \tag{15.84}$$

and

$$p_i^S = -\frac{k_S}{\theta_S} \frac{\partial \theta_S}{\partial x_i}. \tag{15.85}$$

Note that θ_0 in equation (15.80)$_1$, has been replaced by θ_F in equation (15.84)$_1$ while θ_0 in equation (15.80)$_2$ becomes instead θ_S. In a similar manner the intrinsic entropy supply functions ξ_F and ξ_S are now written as

$$
\begin{aligned}
\xi_F &= \frac{1}{\rho\theta_F}\left(\frac{k_F}{\theta_F}\frac{\partial\theta_F}{\partial x_i}\frac{\partial\theta_F}{\partial x_i} + 2\mu d_{ij}d_{ij} + \phi\right), \\
\xi_S &= \frac{1}{\rho\theta_S}\left(\frac{k_S}{\theta_S}\frac{\partial\theta_S}{\partial x_i}\frac{\partial\theta_S}{\partial x_i} + 4\mu_1 d_{ij}P_{ij} + \frac{2\mu_1^2}{\mu}P_{ij}P_{ij} - \phi\right).
\end{aligned}
\tag{15.86}
$$

In equations (15.86) the coefficients k_F and k_S are thermal conductivities. [420] treats the ϕ terms in a manner akin to the development of a convection theory in a local thermal non-equilibrium porous medium. He chooses

$$
\phi = -h(\theta_F - \theta_S)
\tag{15.87}
$$

where $h > 0$ is a constant.

[420] considers two expressions for the buoyancy term, the ρb_i term, in equations (15.77). He writes either

$$
\rho b_i = -gk_i\rho_0\left[1 - \alpha_F(\theta_F - \theta_F^0) + \alpha_S(\theta_S - \theta_S^0)\right]
\tag{15.88}
$$

or

$$
\rho b_i = -gk_i\left[\rho_S\phi + (1 - \phi)\rho_F^0(\theta_F - \theta_F^0)\right].
\tag{15.89}
$$

In these equations g is gravity, $\mathbf{k} = (0, 0, 1)$, α_F, α_S are the thermal expansion coefficients of the fluid and solid, ρ_0 is a constant, θ_F^0, θ_S^0 are (constant) reference temperatures, and ϕ denotes the volume fraction of particles in the nanofluid suspension.

If we take into account equations (15.84), (15.85), (15.86), (15.87), and (15.88), then the system of equations governing nanofluid behaviour derived by [420] may be shown from equations (15.77) to reduce to

$$
\dot{v}_i - \frac{\mu_1}{\mu}\frac{d}{dt}\Delta v_i = -\frac{1}{\rho_0}\frac{\partial p}{\partial x_i} + \alpha_F gk_i\theta_F - \alpha_S gk_i\theta_S + v\Delta v_i - 2v_1\Delta^2 v_i,
\tag{15.90}
$$

and

$$
\frac{\partial v_i}{\partial x_i} = 0,
\tag{15.91}
$$

and

$$
\rho c_F\dot{\theta}_F = \rho s_F\theta_F + 2\mu d_{ij}d_{ij} + k_F\Delta\theta_F - h(\theta_F - \theta_S),
\tag{15.92}
$$

together with

$$
\rho c_S\dot{\theta}_S = \rho s_S\theta_S + k_S\Delta\theta_S + 4\mu_1 d_{ij}P_{ij} + \frac{2\mu_1^2}{\mu}P_{ij}P_{ij} + h(\theta_F - \theta_S),
\tag{15.93}
$$

in which $v = \mu/\rho_0$ and $v_1 = \mu_1/\rho_0$. An analogous system may be derived if equation (15.88) is replaced by (15.89).

Observe that if we set $v_i = 0, s_F = 0, s_S = 0$, and $h = 0$, then equations (15.92) and (15.93) reduce to the standard diffusion equations for θ_F and θ_S, namely

$$\rho c_F \frac{\partial \theta_F}{\partial t} = k_F \Delta \theta_F$$

and

$$\rho c_S \frac{\partial \theta_S}{\partial t} = k_S \Delta \theta_S.$$

Thus, we believe the approach taken here is consistent with thermodynamics and correct.

[420] analysed thermal convection on the basis of equations (15.90)–(15.93). He supposes the fluid occupies the horizontal layer $\{(x,y) \in \mathbb{R}^2\} \times \{z \in (0,d)\}$ with gravity acting downward. The boundaries are fixed and the temperatures maintained at constant values T_L at $z = 0$ and T_U at $z = d$, where $T_L > T_U$, i.e.

$$\theta_F = \theta_S = T_L, \quad z = 0, \qquad \theta_F = \theta_S = T_U, \quad z = d. \tag{15.94}$$

The motionless solution whose linearized instability is investigated is then

$$\bar{v}_i = 0, \qquad \bar{\theta}_F = \bar{\theta}_S = -\beta z + T_L, \tag{15.95}$$

where β is the temperature gradient, given by $\beta = (T_L - T_U)/d$. The steady pressure $\bar{p}(z)$ is determined from equation (15.90) assuming $\bar{v}_i$ and $\bar{\theta}_F$ and $\bar{\theta}_S$ are as given in equations (15.95).

[420] derives the nonlinear system of non-dimensional equations for the perturbations $(u_i, \pi, \theta_F, \theta_S)$,

$$\frac{\partial u_i}{\partial t} + u_j \frac{\partial u_i}{\partial x_j} - \hat{\mu} \Delta \frac{\partial u_i}{\partial t} - \hat{\mu} u_j \Delta \frac{\partial u_i}{\partial x_j} = -\frac{\partial \pi}{\partial x_i} + R k_i \theta_F - \frac{\alpha_S}{\alpha_F} R k_i \theta_S$$

$$+ \Delta u_i - 2\hat{\mu} \Delta^2 u_i,$$

$$\frac{\partial u_i}{\partial x_i} = 0,$$

$$Pr\left(\frac{\partial \theta_F}{\partial t} + u_j \frac{\partial \theta_F}{\partial x_j}\right) = Rw + \Delta \theta_F - H(\theta_F - \theta_S) + a_1 d_{ij} d_{ij},$$

$$\frac{\kappa_F}{\kappa_S} Pr\left(\frac{\partial \theta_S}{\partial t} + u_j \frac{\partial \theta_S}{\partial x_j}\right) = \frac{\kappa_F}{\kappa_S} Rw + \Delta \theta_S + \frac{k_F}{k_S} H(\theta_F - \theta_S)$$

$$+ b_1 d_{ij} p_{ij} + b_2 p_{ij} p_{ij}. \tag{15.96}$$

In these equations $Ra = \alpha_F g \beta d^4 / \nu \kappa_F$ is the Rayleigh number which is connected to the parameter R by the relation $Ra = R^2$. The other coefficients are non-dimensional versions of the coefficients in equations (15.90)–(15.93) and they are given explicitly in [420], while d_{ij} and p_{ij} are now derived from u_i.

The next step is to linearize equations (15.96) and then employ a time dependence like $e^{\sigma t}$. After removing the pressure field π this leads to the linearized system of equations in $w(\mathbf{x}) = u_3$, $\theta_F(\mathbf{x})$ and $\theta_S(\mathbf{x})$,

$$\sigma(\hat{\mu}\Delta^2 w - \Delta w) = -R\Delta^*\theta_F + \frac{\alpha_S}{\alpha_F}R\Delta^*\theta_S - \Delta^2 w + 2\hat{\mu}\Delta^3 w,$$

$$\sigma Pr\theta_F = Rw + \Delta\theta_F - H(\theta_F - \theta_S), \tag{15.97}$$

$$\sigma\frac{\kappa_F}{\kappa_S}Pr\theta_S = \frac{\kappa_F}{\kappa_S}Rw + \Delta\theta_S + \frac{k_F}{k_S}H(\theta_F - \theta_S),$$

where $\Delta^* = \partial^2/\partial x^2 + \partial^2/\partial y^2$ is the horizontal Laplacian.

The boundary conditions are important and not trivial because of the presence of higher spatial derivatives. Suppose we consider two fixed surfaces at constant temperatures. Then the usual arguments lead to the boundary conditions

$$w = \frac{\partial w}{\partial z} = 0, \quad \theta_F = \theta_S = 0, \quad \text{on} \quad z = 0, d. \tag{15.98}$$

A further boundary condition is necessary for w due to the higher spatial derivatives in equation $(15.97)_1$. This may be shown to be, see [420] or [425], section 8.4,

$$\frac{\partial^3 w}{\partial z^3} + \Delta^*\frac{\partial w}{\partial z} = 0. \tag{15.99}$$

Finally one sets $w = W(z)f(x,y)$, $\theta_F = \Theta_F(z)f(x,y)$, and $\theta_S = \Theta_S(z)f(x,y)$, where f is a plane tiling function such that

$$\Delta^* f + a^2 f = 0,$$

where a is a wavenumber.

Then from equations (15.97) the resolution of the critical Rayleigh number R^2 reduces to solving the system of equations and boundary conditions (15.100), (15.101),

$$2\hat{\mu}(D^2 - a^2)^3 W - (D^2 - a^2)^2 W + Ra^2\Theta_F - \alpha Ra^2\Theta_S$$
$$= \sigma[\hat{\mu}(D^2 - a^2)^2 W - (D^2 - a^2)W],$$
$$(D^2 - a^2)\Theta_F - H(\Theta_F - \Theta_S) + RW = \sigma Pr\Theta_F, \tag{15.100}$$
$$(D^2 - a^2)\Theta_S + kH(\Theta_F - \Theta_S) + \kappa RW = \sigma\kappa Pr\Theta_S,$$

and

$$W = DW = 0, \quad D^3 W - a^2 DW = 0,$$
$$\Theta_F = 0, \quad \Theta_S = 0, \quad \text{on } z = 0, 1. \tag{15.101}$$

In equations (15.100) the operator D is given by $D = d/dz$. System (15.100) and (15.101) is solved numerically by a D^2 Chebyshev tau method in [420], where details of the numerical method may be found.

Numerical results for the critical values of R^2 and a^2 for system (15.100) and (15.101) are reported in [420]. Realistic values of the coefficients in equations (15.100) were employed for nanofluid suspensions composed of water and copper oxide particles, water and aluminium oxide particles, ethylene glycol and copper oxide particles, and ethylene glycol and aluminium oxide particles. For the range of parameters employed in [420] the onset of instability was always found to be by stationary convection.

Since the thermal conductivity of a nanofluid may be considerably higher than that of the solvent and the effective viscosity of the nanofluid may also be significantly different from that of the base fluid, it is often argued that one should employ a Rayleigh number which will reflect the effective properties of the nanofluid suspension itself. This is done in [420] where for small values of $\hat{\mu}$ one obtains a large reduction in the Rayleigh number as compared to that for a classical Newtonian fluid with no particles, i.e. $Ra \approx 1707$. This is in agreement with the findings of other studies of thermal convection in nanofluids, e.g. [454]. A significant reduction of an appropriate Rayleigh number means convective motion occurs much more easily in the nanofluid suspension. This is useful because it means heat transfer occurs more readily which agrees with the perceived use of a nanofluid in a heat transfer device.

15.7 Nield–Kuznetsov Nanofluid Model

[320] develop a model for nanofluid behaviour which also allows for different temperatures for the fluid, T^f, and for the solid nanoparticles, T^s, as was described in section 15.6, although their approach is entirely different to that of section 15.6. Instead, [320] employ an equation for the concentration of nanoparticles as in [454] and [65].

If one denotes by $\phi(\mathbf{x}, t)$ the concentration of nanoparticles in the suspension then the model of [320] is based on the following equations of balance of linear momentum, balance of mass, balance of energy for the fluid phase, balance of energy for the solid phase, and the conservation equation for the nanoparticles. These equations are written in a notation consistent with that employed in section 15.6 whereas [320] use p rather than s for the solid particles in suspension. The governing equations of [320] are,

$$
\begin{aligned}
\big[\phi\rho_s + (1-\phi)\rho_f\big](v_{i,t} + v_j v_{i,j}) &= -p_{,i} + \mu\Delta v_i \\
&\quad - g k_i \big[\phi\rho_s + (1-\phi)\rho_f\{1 - \alpha(T^f - T_0)\}\big],
\end{aligned}
$$

$$
v_{i,i} = 0,
$$

$$
\begin{aligned}
(1-\phi)(\rho c)_f(T^f_{,t} + v_i T^f_{,i}) &= (1-\phi)k_f \Delta T^f + h_{fs}(T^s - T^f) \\
&\quad + (1-\phi)(\rho c)_s\Big[D_B \phi_{,i} T^f_{,i} + \frac{D_T}{T^F} T^f_{,i} T^f_{,i}\Big],
\end{aligned}
\tag{15.102}
$$

$$
\phi(\rho c)_s(T^s_{,t} + v_i T^s_{,i}) = \phi k_s \Delta T^s - h_{fs}(T^s - T^f),
$$

$$
\phi_{,t} + v_i \phi_{,i} = (D_B \phi_{,i})_{,i} + \Big(\frac{D_T}{T^f} T^f_{,i}\Big)_{,i},
$$

where h_{fs} is an interphase heat transfer coefficient, D_B and D_T are the Brownian diffusion coefficient and the thermophoretic diffusion coefficient, which have the forms

$$
D_B = \frac{k_B T^f}{3\pi\mu d_f}, \qquad D_T = \frac{0.26 k_f \mu}{\rho_f(2k_f + k_s)}\phi.
$$

The coefficient k_B is the Boltzmann constant, d_f is the diameter of the nanoparticles, μ is the dynamic viscosity of the fluid, k_f and k_s are the thermal conductivities of the fluid and solid particles, and other notation is as in the rest of this chapter.

Clearly the system (15.102) derived by [320] is a very interesting system of partial differential equations, but it is highly nonlinear due to the presence of the D_B and D_T terms.

[320] make some assumptions about terms being constant and then non-dimensionalize (15.102) to write their equations in the non-dimensional form

$$
\frac{1}{Pr}\Big(\frac{\partial v_i}{\partial t} + v_j \frac{\partial v_i}{\partial x_j}\Big) = -\frac{\partial p}{\partial x_i} + \Delta v_i - Rm\,k_i + Ra\,T^f k_i - Rn\,\phi\,k_i,
$$

$$
\frac{\partial v_i}{\partial x_i} = 0,
$$

$$
\frac{\partial T^f}{\partial t} + v_i \frac{\partial T^f}{\partial x_i} = \Delta T^f
\tag{15.103}
$$
$$
+ \frac{N_B}{Le}\frac{\partial \phi}{\partial x_i}\frac{\partial T^f}{\partial x_i} + \frac{N_A N_B}{Le}\frac{\partial T^f}{\partial x_i}\frac{\partial T^f}{\partial x_i} + N_H(T^s - T^f),
$$

$$
\frac{\partial T^s}{\partial t} + v_i \frac{\partial T^s}{\partial x_i} = \varepsilon_1 \Delta T^s - \gamma N_H(T^s - T^f),
$$

$$
\frac{\partial \phi}{\partial t} + v_i \frac{\partial \phi}{\partial x_i} = \frac{1}{Le}\Delta\phi + \frac{N_A}{Le}\Delta T^f,
$$

where Pr and Le are the Prandtl and Lewis numbers, N_A, N_B and N_H are non-dimensional coefficients, Rm and Rn are a density Rayleigh number and a concentration Rayleigh number, and Ra is the thermal Rayleigh number given by

$$Ra = \frac{\rho_f g \alpha d^3 (T_L - T_U)(\rho c)_f}{\mu k_f}.$$

Since [320] are primarily interested in thermal convection in a nanofluid in a layer of depth d with lower temperature T_L and upper temperature T_U they write boundary conditions involving $w = u_3, T_f, T_s$ and ϕ. The boundary conditions [320] employ are

$$w = 0, \quad \frac{\partial w}{\partial z} + \lambda_1 \frac{\partial^2 w}{\partial z^2} = 0, \quad \text{at } z = 0,$$
$$T^f = 1, \quad T^s = 1, \quad \phi = 0, \quad \text{at } z = 0, \tag{15.104}$$

and

$$w = 0, \quad \frac{\partial w}{\partial z} - \lambda_2 \frac{\partial^2 w}{\partial z^2} = 0, \quad \text{at } z = 1,$$
$$T^f = 0, \quad T^s = 0, \quad \phi = 1, \quad \text{at } z = 1. \tag{15.105}$$

These boundary conditions are very interesting and recognize the physically important fact that the fluid may slip at the boundaries, as is seen through the conditions involving λ_1 and λ_2. The aspect of slip boundary conditions is covered in more detail in chapter 9.

[320] present a z–dependent steady solution in which $v_i \equiv 0$, $T_f = T_s = 1 - z$ and $\phi = z$. They then develop a linearized instability analysis of this basic solution. They present analytical and numerical solutions for the cases of two free boundaries and also two fixed boundaries.

The work of [320] is a really useful contribution to nanofluid theory and incorporates the possibly important aspect of LTNE. They conclude that LTNE effects may be significant in some circumstances but for a dilute nanofluid they do not expect the effect to be substantial.

15.8 Generalizations

15.8.1 Extended Green–Naghdi Model

One feature we believe is important in the Green–Naghdi model of section 15.6 is the inclusion of the fourth order derivative term in the momentum equation (15.90). This term allows one to have a flattened velocity profile in steady pipe flow which is a physically observed phenomenon for a suspension, see [452]. In addition the theory of Green and Naghdi is based on a rational development from continuum thermodynamics. One very worthwhile extension of the Green and Naghdi theory would be to incorporate an equation for the conservation of nanoparticles as is done by [320] and described in section 15.7. It would be easy to adopt a heuristic approach and simply combine equations (15.90)–(15.93) together with those of [320] as given in (15.102). Such an *ad hoc* procedure would be useful as a first step. However, it really needs a careful development based on the principles of modern continuum thermodynamics. Nevertheless, I think such a development would be very worthwhile.

15.8.2 Slip Boundary Conditions

In connection with a generalization of the Green–Naghdi theory as outlined in section 15.8.1 one would need to carefully develop the correct boundary conditions. Certainly since this is for a nanofluid theory inclusion of slip boundary conditions, as is done in the work of [320], is an important consideration. One needs to also consider what boundary conditions are necessary in conjunction with the presence of the higher order velocity term $v_1 \Delta^2 v_i$ in equation (15.90). Is some sort of slip boundary condition also appropriate in this case, or is it even necessary to include the effect of higher order slip, cf. [118].

References

[1] Abid, M., Hammerschmidt, U., Köhler, J.: Temperature and moisture dependent thermophysical properties of Sander sandstone. Int. J. Thermal Sciences **86**, 88–94 (2014)

[2] Accuratus: (2009). http://www.accuratus.com/alumox.html

[3] Adler, P.M., Malevich, A.E., Mityushev, V.V.: Nonlinear correction to Darcy's law for channels with wavy walls. Acta Mechanica **224**, 1823–1848 (2013)

[4] Aegis-Ceramics: (2014), http://www.aegis-ceramics.co.uk/foamet.htm

[5] Agarwal, S., Bhadauria, B.S.: Natural convection in a nanofluid saturated rotating porous layer with thermal non-equilibrium model. Transport in Porous Media **90**, 627–654 (2011)

[6] Agrawal, N., Ugaz, V.M.: A buoyancy driven compact thermo-cycler for rapid PCR. Clinics in Laboratory Medicine **27**, 215–223 (2007)

[7] Agrawal, N., Hassan, Y.A., Ugaz, V.M.: A pocket sized convective PCR thermocycler. Angewandte Chemie International Edition **46**, 4316–4319 (2007)

[8] Aguilera, R.F., Aguilera, R.: A triple-porosity model for petrophysical analysis of naturally fractured reservoirs. Petrophysics **45**, 157–166 (2004)

[9] Al-Hadhrami, A.K., Elliott, L., Ingham, D.B., Wen, X.: Flows through horizontal channels of porous materials. Int. J. Energy Research **27**, 875–889 (2003)

[10] Alikakos, N.D., Rostamian, R.: Large time behaviour of solutions of Neumann boundary value problem for the porous medium equation. Indiana Univ. Math. J. **30**, 749–785 (1981)

[11] Allen, J., Kenward, M., Dorfman, K.: Coupled flow and reaction during natural convection PCR. Microfluidics and Nanofluidics **6**, 121–130 (2009)

[12] Altawallbeh, A.A., Bhadauria, B.S., Hashim, I.: Linear and nonlinear double-diffusive convection in a saturated anisotropic porous layer with Soret effect and internal heat source. Int. J. Heat Mass Transfer **59**, 103–111 (2013)

[13] Alvarez, F.X., Cimmelli, V.A., Jou, D., Sellitto, A.: Mesoscopic description of boundary effects in nanoscale heat transport. Nanoscale Systems, Mathematical Modelling, Theory and Applications **1**, 112–142 (2012)

© Springer International Publishing Switzerland 2015

B. Straughan, *Convection with Local Thermal Non-Equilibrium and Microfluidic Effects*,

Advances in Mechanics and Mathematics 32, DOI 10.1007/978-3-319-13530-4

[14] Ameur, D., Galliéro, G.: Slippage of binary mixtures in a nanopore. Microfluidics and Nanofluidics **15**, 183–189 (2013)

[15] Anile, A.M., Pavón, D., Romano, V.: The case for hyperbolic theories of dissipation in relativistic fluids. Tech. rep., arXiv gr-qc/9810014 (1998)

[16] Aronson, D.G., Caffarelli, L.A.: The initial trace of a solution of the porous medium equation. Trans. Amer. Math. Soc. **280**, 351–366 (1983)

[17] Aronson, D.G., Peletier, L.A.: Large time behaviour of solutions of the porous medium equation in bounded domains. J. Differential Equations **39**, 378–412 (1981)

[18] Asadzadeh, F., Nasr Esfahany, M., Etesami, N.: Natural convective heat transfer of Fe_3O_4 ethylene glycol nanofluid in electric field. Int. J. Thermal Sciences **62**, 114–119 (2012)

[19] Ashwin, T., Narasimham, G.S.V.L., Jacob, S.: CFD analysis of high frequency miniature pulse tube refrigerators for space applications with thermal non-equilibrium model. Applied Thermal Engineering **30**, 152–166 (2010)

[20] Aulisa, E., Bloshanskaya, L., Hoang, L., Ibragimov, A.: Analysis of generalized Forchheimer flows of compressible fluids in porous media. J. Math. Phys. **50**(103102), 1–43 (2009)

[21] Aulisa, E., Bloshanskaya, L., Ibragimov, A.: Long-term dynamics for well productivity index for nonlinear flows in porous media. J. Math. Phys. **52**(023506), 1–26 (2011)

[22] Avila, R., Ramos, E., Atluri, S.N.: The Chebyshev tau spectral method for the solution of the linear stability equations for Rayleigh–Bénard convection with melting. Computer Modelling in Engineering and Sciences **51**, 73–92 (2009)

[23] Avila, R., Cabello-González, Ramos, E.: A linear stability analysis of thermal convection in spherical shells with variable radial gravity based on the tau-Chebyshev method. Int. J. Heat Fluid Flow **44**, 495–508 (2013)

[24] Badur, J., Karcz, M., Lemanski, M.: On the mass and momentum transport in the Navier-Stokes slip layer. Microfluidics and Nanofluidics **11**, 439–449 (2011)

[25] Baines, P., Gill, A.: On thermohaline convection with linear gradients. J. Fluid Mech. **37**, 289–306 (1969)

[26] Bandyopadhyay, D., Reddy, P.D.S., Sharma, A., Joo, S.W., Qian, S.: Electromagnetic field induced flow and interfacial instabilities in confined stratified liquid layers. Theor. Comput. Fluid Dyn. **26**, 23–28 (2012)

[27] Banu, N., Rees, D.A.S.: Onset of Darcy–Bénard convection using a thermal non-equilibrium model. Int. J. Heat Mass Transfer **45**, 2221–2228 (2002)

[28] Barenblatt, G.I., Vazquez, J.L.: Nonlinear diffusion and image contour enhancement. Interfaces and Free Boundaries **6**, 31–54 (2004)

[29] Barenblatt, G.I., Zheltov, I.P.: Fundamental equations of filtration of homogeneous liquids in fissured rock. Soviet Physics Doklady **5**, 522–525 (1960)

[30] Bargmann, S., Greve, R., Steinmann, P.: Simulation of cryovolcanism on Saturns moon Enceladus with the Green–Naghdi theory of thermoelasticity. Bulletin of Glaciological Research **26**, 23–32 (2008)

[31] Barletta, A.: Local energy balance, specific heats and the Oberbeck–Boussinesq approximation. Int. J. Heat Mass Transfer **52**, 5266–5270 (2009)

[32] Barletta, A.: Instability of mixed convection in a vertical porous channel with uniform wall flux. Phys. Fluids **25**(084108), 1–14 (2013)

[33] Barletta, A.: Buoyancy opposed Darcy's flow in a vertical circular duct with uniform wall heat flux: a stability analysis. Transport in Porous Media **102**, 261–274 (2014)

[34] Barletta, A., Celli, M.: Local thermal non-equilibrium flow with viscous dissipation in a plane horizontal porous layer. Int. J. Thermal Sciences **55**, 53–60 (2012)

[35] Barletta, A., Rees, D.A.S.: Local thermal non-equilibrium effects in the Darcy–Bénard instability with isoflux boundary conditions. Int. J. Heat Mass Transfer **55**, 384–394 (2012)

[36] Barletta, A., Storesletten, L.: Adiabatic eigenflows in a vertical porous channel. J. Fluid Mech. **749**, 778–793 (2014)

[37] Bassom, A.P., Blyth, M.G., Papageorgiou, D.T.: Using surfactants to stabilize two-phase pipe flows of core-annular type. J. Fluid Mech. **704**, 333–359 (2012)

[38] Basu, R., Layek, G.C.: Cross-diffusive effects on the onset of double-diffusive convection in a horizontal saturated porous fluid layer heated and salted from above. Chinese Physics B **22**, 054,702–394 (2013)

[39] Batchelor, G.: An Introduction to Fluid Dynamics. Cambridge University Press, Cambridge (1967)

[40] Baytas, A.C.: Thermal non-equilibrium natural convection in a square enclosure filled with a heat-generating solid phase, non-Darcy porous medium. Int. J. Energy Research **27**, 975–988 (2003)

[41] Bear, J., Gilman, A.: Migration of salts in the unsaturated zone caused by heating. Transport in Porous Media **19**, 139–156 (1995)

[42] Beirão da Veiga, H., Crispo, F.: Concerning the $W^{k,p}$—inviscid limit for 3-D flows under a slip boundary condition. J. Math. Fluid Mech. **13**, 117–135 (2011)

[43] Beirão da Veiga, H., Crispo, F.: The 3-D inviscid limit result under slip boundary conditions. A negative answer. J. Math. Fluid Mech. **14**, 55–59 (2012)

[44] Belmiloudi, A.: Parameter identification problems and analysis of the impact of porous media in biofluid heat transfer in biological tissues during thermal therapy. Nonlinear Analysis, Real World Applications **11**, 1345–1363 (2010)

[45] Bennethum, L.S., Giorgi, T.: Generalized Forchheimer equation for two-phase flow based on hybrid mixture theory. Trans. Porous Media **26**, 261–275 (1997)

[46] Bera, P., Khalili, A.: Influence of Prandtl number on stability of mixed convective flow in a vertical channel filled with a porous medium. Physics of Fluids **18**(124103), 1–10 (2006)

[47] Berjano, E.J., Alió, J.L., Saiz, J.: Modelling for radio-frequency conductive keratoplasty: implications for the maximum temperature reached in the cornea. Physiological Measurement **26**, 157–172 (2005)

[48] Berlengiero, M., Emanuel, K.A., von Hardenberg, J., Provenzale, A., Spiegel, E.A.: Internally cooled convection: a fillip for Philip. Commun. Nonlinear Sci. Numer. Simulation **17**, 1998–2007 (2012)

[49] Berti, A., Bochicchio, I., Fabrizio, M.: Phase separation in quasi-incompressible fluids: Cahn–Hilliard model in the Cattaneo–Maxwell framework. ZAMP **66**, 135–147 (2015)

[50] Bhadauria, B.S., Agarwal, S.: Convective transport in a nanofluid saturated porous layer with thermal non-equilibrium model. Transport in Porous Media **88**, 107–131 (2011)

[51] Bhadauria, B.S., Srivastava, A.K.: Magneto-double diffusive convection in an electrically conducting-fluid-saturated porous medium with temperature modulation of the boundaries. Int. J. Heat Mass Transfer **53**, 2530–2538 (2010)

[52] Bhadauria, B.S., Hashim, I., Kumar, J., Srivastava, A.: Cross diffusion convection in a Newtonian fluid-saturated rotating porous medium. Transport in Porous Media **98**, 683–697 (2013)

[53] Bissell, J.J.: On oscillatory convection with the Cattaneo–Christov hyperbolic heat-flow model. Proc. Roy. Soc. London A **471**(20140845), 1–18 (2015)

[54] Bleustein, J.L., Green, A.E.: Dipolar fluids. Int. J. Engng. Sci. **5**, 323–340 (1967)

[55] Bogorodskii, P.V., Nagurnyi, A.P.: Under-ice meltwater puddles: a factor of fast sea ice melting in the Arctic. Doklady Earth Sciences **373**, 885–887 (2000)

[56] Bonnet, P.P., Topin, F., Tadrist, L.: Flow effects in metal foams: compressibility and pore size effects. Transport in Porous Media **73**, 223–254 (2008)

[57] Boomsma, K., Poulikakos, D., Zwick, F.: Metal foams as compact high performance heat exchangers. Mechanics of Materials **35**, 1161–1176 (2003)

[58] Böttger, P.H.M., Gusarov, A.V., Shklover, V., Patscheider, J., Sobiech, M.: Anisotropic layered media with microintrusions; thermal properties of arc-evaporation multilayer metal nitrides. Int. J. Thermal Sciences **77**, 75–83 (2014)

[59] Boussinesq, M.J.: Théorie de l'Ecoulement Tourbillonnant et Tumultueux des Liquides dans les Lits Rectilignes a Grande Section. Gauthier-Villars, Paris (1897)

[60] Brandenburg, A., Kaplya, P.J., Mohammed, A.: Non-Fickian diffusion and tau approximation from numerical turbulence. Phys. Fluids **16**, 1020–1027 (2004)

[61] Braun, D., Goddard, N.L., Lichaber, A.: Exponential DNA replication by laminar convection. Phys. Rev. Lett. **91**, 158,103–1027 (2003)

[62] Brevdo, L., Cirpka, O.: Absolute/convective instability dichotomy in a Soret driven thermosolutal convection induced in a porous layer by inclined thermal and vertical solutal gradients. Transport in Porous Media **95**, 425–446 (2012)

[63] Brinkman, H.C.: A calculation of viscous force exerted by a flowing fluid on a dense swarm of particles. Appl. Sci. Res. **1**, 27–34 (1947)

[64] Bulivcek, M., Málek, J., Rajagopal, K.R.: Navier's slip and evolutionary Navier–Stokes-like systems with pressure and shear-rate dependent viscosity. Indiana Univ. Math. J. **56**, 51–85 (2007)

[65] Buongiorno, J.: Convective transport in nanofluids. ASME J. Heat Transfer **128**, 240–250 (2006)

[66] Buonomo, B., Manca, O., Lauriat, G.: Forced convection in micro-channels filled with porous media in local thermal non-equilibrium conditions. Int. J. Thermal Sciences **77**, 206–222 (2014)

[67] Burghardt, A., Rogut, J., Gotkowska, J.: Diffusion coefficients in bidisperse porous structures. Chemical Engng. Science **43**, 2463–2476 (1988)

[68] Capone, F., De Luca, R.: Onset of convection for ternary fluid mixtures saturating horizontal porous layers with large pores. Atti Accad. Lincei **23**, 405–428 (2012)

[69] Capone, F., De Luca, R.: Ultimately boundedness and stability of triply diffusive mixtures in rotating porous layers under the action of the Brinkman law. Int. J. Non-Linear Mech. **47**, 799–805 (2012)

[70] Capone, F., Rionero, S.: Inertia effect on the onset of convection in rotating porous layers via the "auxilliary system method". Int. J. Non-Linear Mech. **57**, 192–200 (2013)

[71] Capone, F., Gentile, M., Hill, A.A.: Onset of natural convection in anisotropic porous media. In: M. Ciarletta, M. Fabrizio, A. Morro, S. Rionero (eds.) New trends in fluid and solid models, supplement, pp. 18–23. World Scientific (2010). Proceedings of the International Conference, Vietri sul Mare, 28 February–1 March, 2008

[72] Capone, F., Gentile, M., Hill, A.A.: Rionero's critical perturbations method via weighted energy for stability of convective motions in anisotropic media. In: M. Ciarletta, M. Fabrizio, A. Morro, S. Rionero (eds.) New trends in fluid and solid models, supplement, pp. 1–7. World Scientific (2010). Proceedings of the International Conference, Vietri sul Mare, 28 February–1 March, 2008

[73] Capone, F., Gentile, M., Hill, A.A.: Convection problems in anisotropic porous media with nonhomogeneous porosity and thermal diffusivity. Acta Applicandae Mathematicae **122**, 85–91 (2012)

[74] Capone, F., De Cataldis, V., De Luca, R., Torcicollo, I.: On the stability of vertical constant throughflows for binary mixtures in porous layers. Int. J. Non-Linear Mech. **59**, 1–8 (2014)

[75] Carillo, S.: Bäcklund transformations and heat conduction with memory. In: M. Ciarletta, M. Fabrizio, A. Morro, S. Rionero (eds.) New trends in fluid and solid models, supplement, pp. 8–17. World Scientific (2010). Proceedings of the International Conference, Vietri sul Mare, 28 February–1 March, 2008

[76] Carr, M.: Convection in porous media flows. Ph.D. thesis, University of Durham (2003). etheses.dur.ac.uk/4049/

[77] Carr, M.: A model for convection in the evolution of under-ice melt ponds. Continuum Mech. Thermodyn. **15**, 45–54 (2003)

[78] Carr, M., de Putter, S.: Penetrative convection in a horizontally isotropic porous layer. Continuum Mech. Thermodyn. **15**, 33–43 (2003)

[79] Cattaneo, C.: Sulla conduzione del calore. Atti Sem. Mat. Fis. Modena **3**, 83–101 (1948)

[80] Celebi, A.O., Kalantarov, V.K., Ugurlu, D.: On continuous dependence on coefficients of the Brinkman-Forchheimer equations. Appl. Math. Letters **19**, 801–807 (2006)

[81] Celli, M., Rees, D.A.S., Barletta, A.: The effect of local thermal non-equilibrium on forced convection boundary layer flow from a heated surface in porous media. Int. J. Heat Mass Transfer **53**, 3533–3539 (2010)

[82] Celli, M., Barletta, A., Storesletten, L.: Local thermal non-equilibrium effects in the Darcy-Bénard instability of a porous layer heated from below by a uniform flux. Int. J. Heat Mass Transfer **67**, 902–912 (2013)

[83] Cercignani, C.: The Boltzmann Equation and its Applications. Springer, Berlin (1988)

[84] Chand, S.: Linear stability of triple-diffusive convection in micropolar ferromagnetic fluid saturating porous medium. Appl. Math. Mech. **34**, 309–326 (2013)

[85] Chandrasekhar, S.: Hydrodynamic and Hydromagnetic Stability. Dover, New York (1981)

[86] Chen, F., Chen, C.: Onset of finger convection in a horizontal porous layer underlying a fluid layer. J. Heat Transfer **3**, 403–409 (1988)

[87] Chen, F., Wang, C.Y.: Convective instability in a porous enclosure with a horizontal conducting baffle. J. Heat Transfer **115**, 810–813 (1993)

[88] Chen, F., Wang, C.Y.: Convective instability in saturated porous enclosures with a vertical insulating baffle. Int. J. Heat Mass Transfer **36**, 1897–1904 (1993)

[89] Chen, Z., Qian, S., Abrams, W.R., Malamud, D., Bau, H.H.: Thermosiphon basec PCR reactor: experiment and modelling. Analytical Chemistry **76**, 3707–3715 (2004)

[90] Chen, X., Wang, S., Tao, J., Tan, W.: Stability analysis of thermosolutal convection in a horizontal porous layer using a thermal non-equilibrium model. Int. J. Heat Fluid Flow **32**, 78–87 (2011)

[91] Christopherson, D.: Note on the vibration of membranes. Quart. J. Math. **11**, 63–65 (1940)

[92] Christov, C.I.: On frame indifferent formulation of the Maxwell–Cattaneo model of finite-speed heat conduction. Mechanics Research Communications **36**, 481–486 (2009)

[93] Christov, C.I., Jordan, P.M.: Heat conduction paradox involving second sound propagation in moving media. Physical Review Letters **94**(154301), 1–4 (2005)

[94] Christov, I.C., Jordan, P.M.: On the propagation of second sound in nonlinear media: shock, acceleration and travelling wave results. J. Thermal Stresses **33**, 1109–1135 (2010)

[95] Christov, I.C., Jordan, P.M.: On an instability exhibited by the ballistic-diffusive heat conduction model of Xu and Hu. Proc. Royal Soc. London A **470**(20130557), 1–8 (2014)

[96] Chu, W.K.: Stability of incompressible helium II: a two fluid system. J. Phys.: Condensed Matter **12**, 8065–8069 (2000)

[97] Chu, A.K.: Instability of Navier slip flow of liquids. Comptes Rendue, Mécanique **332**, 895–900 (2004)

[98] Ciarletta, M., Straughan, B., Tibullo, V.: Anisotropic effects on poroacoustic waves. Mech. Research Communications **37**, 137–140 (2010)

[99] Conti, M., dell'Oro, F., Miranville, A.: Asymptotic behavior of a generalization of the Caginalp phase-field system. Asymptotic Analysis **81**, 297–314 (2013)

[100] Conti, M., Gatti, S., Miranville, A.: A generalization of the Caginalp phase-field system with Neumann boundary conditions. Nonlinear Analysis, Theory, Methods and Applications **87**, 11–21 (2013)

[101] Dai, W.Z., Wang, H.J., Jordan, P.M., Mickens, R.E., Bejan, A.: A mathematical model for skin burn injury induced by radiation heating. Int. J. Heat Mass Transfer **51**, 5497–5510 (2008)

[102] Damm, D.L., Fedorov, A.G.: Local thermal non-equilibrium effects in porous electrodes of the hydrogen-fueled SOFC. J. Power Sources **159**, 1153–1157 (2006)

[103] Darcy, H.: Les Fontaines Publiques de la Ville de Dijon. Dalmont, Paris (1856)

[104] Das, S., Das, T., Chakraborty, S.: Modeling of coupled momentum, heat and solute transport during DNA hybridization in a microchannel in the presence of electro-osmotic effects and axial pressure gradients. Microfluidics and Nanofluidics **2**, 37–49 (2006)

[105] Dauby, P.C., Nélis, M., Lebon, G.: Generalized Fourier equations and thermoconvective instabilities. Revista Mexicana de Fisica **48**, 57–62 (2002)

[106] de Boer, R.: Theory of Porous Media: Highlights in Historical Development and Current State. Springer, Berlin (1999)

[107] Dehghan, M., Jamal-Abad, M.T., Rashidi, S.: Analytical interpretation of the local thermal non-equilibrium condition of porous media embedded in tube heat exchangers. Energy Conversion and Management **85**, 264–271 (2014)

[108] Deléglise, M., Binétruy, C., Castaing, P., Krawczak, P.: Use of non local equilibrium theory to predict transient temperature during non-isothermal resin flow in a fibrous porous medium. Int. J. Heat Mass Transfer **50**, 2317–2324 (2007)

[109] Denison, M.F.C.: Compressibility effects on the non-linear receptivity of boundary layers to dielectric barrier discharges. Master's thesis, University of Texas Arlington (2013). http://dspace.uta.edu/handle/10106/11831

[110] Denison, M., Massa, L., Cisneros, E.: Compressibility effects on dielectric barrier discharge actuation and boundary layer receptivity. In: 52nd Aerospace Sciences Meeting, pp. 1–18 (2014)

[111] Di Benedetto, E.: Continuity of weak solutions to a general porous medium equation. Indiana Univ. Math. J. **32**, 83–118 (1983)

[112] Dincov, D.D., Parrott, K.A., Pericleous, K.A.: Heat and mass transfer in two-phase porous materials under intensive microwave heating. J. Food Engng. **65**, 403–412 (2004)

[113] Doering, C.R., Spiegel, E.A., Worthing, R.A.: Energy dissipation in a shear layer with suction. Phys. Fluids **12**, 1955–1968 (2000)

[114] Doering, C.R., Eckhardt, B., Schumacher, J.: Failure of energy stability in Oldroyd-B fluids at arbitrarily low Reynolds numbers. J. Non-Newtonian Fluid Mech. **135**, 92–96 (2006)

[115] Dongarra, J.J., Straughan, B., Walker, D.W.: Chebyshev tau–QZ algorithm methods for calculating spectra of hydrodynamic stability problems. Appl. Numer. Math. **22**, 399–435 (1996)

[116] Dragomirescu, F.I., Gheorghiu, C.I.: Analytical and numerical solutions to an electrohydrodynamic stability problem. Appl. Math. Comp. **59**, 3718–3727 (2010)

[117] Du, N., Fan, J., Wu, H., Chen, S., Liu, Y.: An improved model of heat transfer through penguin feathers and down. J. Theoretical Biology **248**, 727–735 (2007)

[118] Duan, Z.: Second-order gaseous slip flow models in long circular and noncircular microchannels and nanochannels. Microfluidics and Nanofluidics **12**, 805–820 (2012)

[119] Duan, Z., Muzychka, Y.S.: Slip flow in non-circular microchannels. Microfluidics and Nanofluidics **3**, 473–484 (2007)

[120] Dupuit, J.: Etudes Thèoriques et Pratiques sur le Mouvement des Eaux. Dunod, Paris (1863)

[121] Eckert, E.G.R., Drake, R.M.: Analysis of Heat and Mass Transfer. McGraw-Hill, New York (1972)

[122] Edwards, D.: Charge transport: through a spatially periodic porous medium: electrokinetic and convective dispersion phenomena. Phil. Trans, Roy. Soc. London A **353**, 205–242 (1995)

[123] Egorov, S.D.: Thermal convection in sections of multilayer cryogenic heat insulation. J. of Engineering Physics and Thermophysics **61**, 989–992 (1991)

[124] EngineeringToolbox: (2009). http://www.engineeringtoolbox.com/water-thermal-properties-d_162.html

[125] Eringen, A.C.: Electromagnetic theory of microstretch elasticity and bone modeling. Int. J. Engng. Sci. **42**, 231–242 (2004)

[126] Fabrizio, M., Mongiovi, M.S.: Phase transition and λ-line in liquid helium. J. Non-Equilibrium Thermodynamics **38**, 185–200 (2013)

[127] Fabrizio, M., Mongiovi, M.S.: Phase transition in liquid ^{4}He by a mean field model. J. Thermal Stresses **36**, 135–151 (2013)

[128] Falcón, N.: Compact star cooling by means of heat waves. In: RevMexAA (Serie de Conferencias) (2001). See also http://adsabs.harvard.edu/full/2001RMxAC..11..41F

[129] Falsaperla, P., Mulone, G., Straughan, B.: Rotating porous convection with prescribed heat flux. Int. J. Engng. Sci. **48**, 685–692 (2010)

[130] Falsaperla, P., Mulone, G., Straughan, B.: Inertia effects on rotating porous convection. Int. J. Heat Mass Transfer **54**, 1352–1359 (2011)

[131] Falsaperla, P., Giacobbe, A., Mulone, G.: Does symmetry of the operator of a dynamical system help stability? Acta Applicandae Mathematicae **122**, 239–253 (2012)

[132] Falsaperla, P., Giacobbe, A., Mulone, G.: Double diffusion in rotating porous media under general boundary conditions. Int. J. Heat Mass Transfer **55**, 2412–2419 (2012)

[133] Fan, J., Wang, L.: Review of heat conduction in nanofluids. J. Heat Transfer **133**(040801), 1–14 (2011)

[134] Feireisl, E., Novotny, A.: The Oberbeck–Boussinesq approximation as a singular limit of the full Navier–Stokes–Fourier system. J. Math. Fluid Mech. **11**, 274–302 (2009)

[135] Ferrari, C., Kaoui, B., L'vov V. S. Procaccia, I., Rudenko, O., ten Thije Boonkkamp, J.H.M., Toschi, F.: Analytical modelling for heat transfer in sheared flows of nanofluids. Phys. Review E **86**(016302), 1–10 (2012)

[136] Ferreira, J.A., de Oliveira, P.: Looking for the lost memory in diffusion-reaction equations. In: R. Rannacher, A. Sequeira (eds.) Advances in Mathematical Fluid Mechanics, pp. 229–251. Springer (2010)

[137] Fichot, F., Duval, F., Trégourès, N., Béchaud, C., Quintard, M.: The impact of thermal non-equilibrium and large-scale 2D/3D effects on debris bed reflooding and coolability. Nuclear Engineering and Design **236**, 2144–2163 (2006)

[138] Fife, P.C.: The Bénard problem for general fluid dynamical equations and remarks on the Boussinesq approximation. Indiana Univ. Math. J. **20**, 303–326 (1970)

[139] Fife, P.C.: A gentle introduction to the physics and mathematics of incompressible flow (2000). www.math.utah.edu/ fife/gentleb.pdf

[140] Firdaouss, M., Guermond, J.L., Le Quére, P.: Nonlinear corrections to Darcy's law at low Reynolds numbers. J. Fluid Mech. **343**, 331–350 (1997)

[141] Flavin, J.N.: The evolution to a steady state for a porous medium model. J. Math. Anal. Appl. **322**, 393–402 (2006)

[142] Flavin, J.N., Rionero, S.: Qualitative Estimates for Partial Differential Equations. CRC Press, Boca Raton (1995)

[143] Flavin, J.N., Rionero, S.: Asymptotic and other properties for a nonlinear diffusion model. J. Math. Anal. Appl. **228**, 119–140 (1998)

[144] Flavin, J.N., Rionero, S.: Nonlinear stability for a thermofluid in a vertical porous slab. Continuum Mech. Thermodyn. **11**, 173–179 (1999)

[145] Flavin, J.N., Rionero, S.: Stability properties for nonlinear diffusion in porous and other media. J. Math. Anal. Appl. **281**, 221–232 (2003)

[146] Forchheimer, P.: Wasserbewegung durch boden. Z. Vereines Deutscher Ingnieure **50**, 1781–1788 (1901)

[147] Fox, N.: Low temperature effects and generalized thermoelasticity. J. Inst. Maths. Applics. **5**, 373–386 (1969)

[148] Franchi, F., Straughan, B.: Thermal convection at low temperature. J. Non-Equilibrium Thermodynamics **19**, 368–374 (1994)

[149] Galdi, G., Straughan, B.: A nonlinear analysis of the stabilizing effect of rotation in the Bénard problem. Proc. Roy. Soc. London A **402**, 257–283 (1985)

[150] Galdi, G.P., Straughan, B.: Exchange of stabilities, symmetry and nonlinear stability. Arch. Rational Mech. Anal. **89**, 211–228 (1985)

[151] Galindo-Rosales, F., Campo-Deano, L., .Pinho, F., van Bokhorst, E., Hamersma, P., Oliveira, M., Alves, M.: Microfluidic systems for the analysis of viscoelastic fluid flow phenomena in porous media. Microfluidics and Nanofluidics **12**, 485–498 (2012)

[152] Gatti, S., Grasselli, M., Miranville, A., Pata, V.: On the hyperbolic relaxation of the one-dimensional Cahn–Hilliard equation. J. Math. Anal. Appl. **312**, 230–247 (2005)

[153] Gentile, M., Straughan, B.: Acceleration waves in nonlinear double porosity elasticity. Int. J. Engng. Sci. **73**, 10–16 (2013)

[154] Gheorghiu, C.I., Dragomirescu, F.I.: Spectral methods in linear stability. Applications to thermal convection with variable gravity field. Appl. Numer. Math. **59**, 1290–1302 (2009)

[155] Gheorghiu, C.I., Rommes, J.: Application of the Jacobi–Davidson method to accurate analysis of singular linear hydrodynamic stability problems. Int. J. Numer. Meth. Fluids **71**, 358–369 (2013)

[156] Gill, A.: A proof that convection in a porous vertical slab is stable. J. Fluid Mech. **35**, 545–547 (1969)

[157] Gilman, A., Bear, J.: The influence of free convection on soil salinization in arid regions. Trans. Porous Media **23**, 275–301 (1996)

[158] Giorgi, T.: Derivation of the Forchheimer law via matched asymptotic expansions. Trans. Porous Media **29**, 191–206 (1997)

[159] Gouin, H., Ruggeri, T.: A consistent thermodynamical model of incompressible media as a limit case of quasi-thermal-incompressible materials. Int. J. Non-Linear Mechanics **47**, 688–693 (2012)

[160] Gouin, H., Muracchini, A., Ruggeri, T.: On the Müller paradox for thermal-incompressible media. Continuum Mechanics and Thermodynamics **24**, 505–513 (2012)

[161] Govender, S.: Stability of gravity driven convection in a cylindrical porous layer subjected to vibration. Transport in Porous Media **63**, 489–502 (2006)

[162] Govender, S.: Vadasz number influence on vibration in a rotating porous layer placed far away from the axis of rotation. J. Heat Transfer **132**(112601), 1–5 (2010)

[163] Govender, S.: Coriolis effect on convection in a rotating porous layer subjected to variable gravity. Transport in Porous Media **98**, 443–450 (2013)

[164] Goyal, H., Kumar, A.A.P., Bandyopadhyay, D., Usha, R., Banerjee, T.: Instabilities of a confined two-layer flow on a porous medium: an Orr–Sommerfeld analysis. Chemical Engineering Science **97**, 109–125 (2013)

[165] Graffi, D.: Sopra alcuni fenomeni ereditari dell'elettrologia. Rend. Ist. Lomb. Sc. Lett. **19**, 151–166 (1936)

[166] Grasselli, M., Petzeltová, H., Schimperna, G.: Asymptotic behaviour of a nonisothermal viscous Cahn–Hilliard equation with inertial term. J. Differential Equations **239**, 38–60 (2007)

[167] Green, A.E., Naghdi, P.M.: A note on dipolar inertia. Q. Appl. Math. **28**, 458–460 (1970)

[168] Green, A.E., Naghdi, P.M.: An extended theory for incompressible viscous fluid flow. J. Non-Newtonian Fluid Mech. **66**, 233–255 (1996)

[169] Green, A.E., Rivlin, R.S.: Multipolar continuum mechanics. Arch. Rational Mech. Anal. **17**, 113–147 (1964)

[170] Haase, M.: The Functional Calculus for Sectorial Operators. Birkhäuser, Basel (2006)

[171] Haddad, S.A.M.: Thermal convection in a Cattaneo–Fox porous material with Guyer–Krumhansl effects. Transport in Porous Media **100**, 363–375 (2013)

[172] Haddad, S.A.M.: Stability analyses for porous convection including second sound effects. Ph.D. thesis, University of Durham (2014)

[173] Haddad, S.A.M.: Thermal instability in Brinkman porous media with Cattaneo–Christov heat flux. Int. J. Heat Mass Transfer **68**, 659–668 (2014)

[174] Haddad, S.A.M., Straughan, B.: Porous convection and thermal oscillations. Ricerche di Matematica **61**, 307–320 (2012)

[175] Hansen, U., Yuen, D.: Subcritical double-diffusive convection at infinite Prandtl number. Geophys. Astrophys. Fluid Dyn. **47**, 199–224 (1989)

[176] Harzallah, H.S., Jbara, A., Slimi, K.: Double diffusive natural convection in anisotropic porous medium bounded by finite thickness walls: validity of local thermal equilibrium assumption. Transport in Porous Media **103**, 207–231 (2014)

[177] Hennig, M., Braun, D.: Convective polymerase chain reaction around micro immersion heater. Appl. Phys. Lett. **87**(183901), 1–3 (2005)

[178] Herrera, L., Falcón, N.: Heat waves and thermohaline instability in a fluid. Physics Letters A **201**, 33–37 (1995)

[179] Hibino, K., Ishikawa, H., Ishioka, K.: Effect of a capping inversion on the stability of an Ekman boundary layer. J. Meteorological Soc. Japan **90**, 311–319 (2012)

[180] Hill, A.A., Carr, M.: The influence of a fluid-porous interface on solar pond stability. Advances in Water Resources **52**, 1–6 (2013)

[181] Hill, A.A., Carr, M.: Stabilising solar ponds by utilising porous materials. Advances in Water Resources **60**, 1–6 (2013)

[182] Hill, A.A., Malashetty, M.S.: An operative method to obtain sharp nonlinear stability for systems with spatially dependent coefficients. Proc. Roy. Soc. London A **468**, 323–336 (2012)

[183] Hill, A.A., Morad, M.R.: Convective stability of carbon sequestration in anisotropic porous media. Proc. Roy. Soc. London A **470**(20140373), 1–8 (2014)

[184] Hills, R., Roberts, P.: On the motion of a fluid that is incompressible in a generalized sense and its relationship to the Boussinesq approximation. Stab. Appl. Anal. Cont. Media **1**, 205–212 (1991)

[185] Hill, A.A., Straughan, B.: Stability of Poiseuille flow in a porous medium. In: R. Rannacher, A. Sequeira (eds.) Advances in Mathematical Fluid Mechanics, pp. 287–293. Springer, Heidelberg (2010)

[186] Hoang, L., Ibragimov, A.: Structural stability of generalized Forchheimer equations for compressible fluids in porous media. Nonlinearity **24**, 1–41 (2011)

[187] Hoang, L., Ibragimov, A.: Qualitative study of generalized Forchheimer flows with the flux boundary condition. Advances in Differential Equations **17**, 511–556 (2012)

[188] Hoang, L., Ibragimov, A., Kieu, T.T.: One-dimensional two-phase generalized Forchheimer flows of incompressible fluids. J. Math. Anal. Appl. **401**, 921–938 (2013)

[189] Huisseune, H., De Jaeger, P., De Schampheleire, S., Ameel, B., De Paepe, M.: Simulation of an aluminium foam heat exchanger using the volume averaging technique. Procedia Materials Science **4**, 334–339 (2014)

[190] Hurle, D.T.J., Jakeman, E.: Soret driven thermo-solutal convection. J. Fluid Mech. **47**, 667–687 (1971)

[191] Imamura, T., Higuchi, T., Maejima, Y., Takagi, M., Sugimoto, N., Ikeda, K., Ando, H.: Inverse insolation dependence of Venus' cloud-level convection. Icarus **228**, 181–188 (2014)

[192] Islam, A.W., Sharif, M.A.R., Carlson, E.S.: Numerical investigation of double diffusive natural convection of CO_2 in a brine saturated geothermal reservoir. Geothermics **48**, 101–111 (2013)

[193] Islam, A.W., Lashgari, H.R., Sephemoori, K.: Double diffusion natural convection of CO_2 in a brine saturated geothermal reservoir: Study of non-model growth of perturbations and heterogeneity effects. Geothermics **51**, 325–336 (2014)

[194] Jiang, J.: Convergence to equilibrium for a fully hyperbolic phase-field model with Cattaneo heat flux law. Math. Meth. Appl. Sci. **32**, 1156–1182 (2009)

[195] Jordan, P.M., Puri, P.: Thermal stresses in a spherical shell under three thermoelastic models. J. Thermal Stresses **24**, 47–70 (2001)

[196] Joseph, D.D.: Global stability of the conduction-diffusion solution. Arch. Rational Mech. Anal. **36**, 285–292 (1970)

[197] Joseph, D.D.: Stability of Fluid Motions, vol. 2. Springer (1976)

[198] Joseph, D.D., Shir, C.C.: Subcritical convective instability. Part 1. Fluid layers. J. Fluid Mech. **26**, 753–768 (1966)

[199] Ju, Y., Li, X.: New research progress on the ultrastructure of tectonically deformed coals. Progress in Naturals Science **19**, 1455–1466 (2009)

[200] Jugjai, S., Phothiya, C.: Liquid fuels-fired porous combustor-heater. Fuel **86**, 1062–1068 (2007)

[201] Kaiser, R., Mulone, G.: A note on nonlinear stability of plane parallel shear flows. J. Math. Anal. Appl. **302**, 543–556 (2005)

[202] Kaiser R. Tilgner, A., von Wahl, W.: A generalized energy functional for plane Couette flow. SIAM J. Math. Anal. **37**, 438–454 (2005)

[203] Kalantarov, V., Zelik, S.: Smooth attractors for the Brinkman–Forchheimer equations with fast growing nonlinearities. Comm. Pure Appl. Anal. **11**, 2037–2054 (2012)

[204] Kaminski, E., Chenet, A., Jaupart, C., Courtillot, V.: Rise of volcanic plumes to the stratosphere aided by penetrative convection above large lava flows. Earth Planetary Sci. Lett. **301**, 171–178 (2011)

[205] Kandem, J.D.: On the time discrete approximation of the Brinkman–Forchheimer equations. Math. Meth. Appl. Sci. **34**, 1487–1498 (2011)

[206] Kang, H.O., Cheung, F.B.: Linear instability analysis of a water sheet trailing from a wet spacer grid in a rod bundle. Nuclear Engineering and Technology **45**, 895–910 (2013)

[207] Kang, J.R., Park, J.Y.: Uniform attractors for non-autonomous Brinkman–Forchheimer equations with delays. Acta Mechanica Sinica **29**, 993–1006 (2013)

[208] Keangin, P., Rattanadecho, P.: Analysis of heat transport on local thermal non-equilibrium in porous liver during microwave ablation. Int. J. Heat Mass Transfer **67**, 46–60 (2013)

[209] Keh, H.J., Hsu, L.Y.: Diffusioosmosis of electrolyte solutions in fibrous porous media. Microfluidics and Nanofluidics **5**, 347–356 (2008)

[210] Kelliher, J.P., Temam, R., Wang, X.: Boundary layer associated with the Darcy–Brinkman–Boussinesq model for convection in porous media. Physica D **240**, 619–628 (2011)

[211] Khandelwal, M.K., Bera, P.: A thermal non-equilibrium perspective on mixed convection in a vertical channel. Int. J. Thermal Sciences **56**, 23–34 (2012)

[212] Khayat, R.E., Ostoja-Starzewski, M.: On the objective rate of heat and stress fluxes. Connection with micro/nano scale heat convection. Discrete and Continuous Dynamical Systems-B **15**, 991–998 (2011)

[213] Khoshnood, A., Jalali, M.A.: Long-lived and unstable modes of Brownian suspensions in microchannels. J. Fluid Mech. **701**, 407–418 (2012)

[214] Kim, M.C.: Onset of buoyancy-driven convection in a liquid-saturated cylindrical anisotropic porous layer supported by a gas phase. Transport in Porous Media **103**, 31–42 (2014)

[215] Kim, S.J., Kim, D., Lee, D.Y.: On local thermal equilibrium in microchannel heat sinks. Int. J. Heat Mass Transfer **43**, 1735–1748 (2000)

[216] Kirillov, S., Dmitrenko, I., Hölemann, J., Kassens, H., Bloshkina, E.: The penetrative mixing in the Laptev sea coastal polyna pynocline layer. Continental Shelf Research **63**, 34–42 (2013)

[217] Krishnamurti, R.: Convection induced by selective absorption of radiation: A laboratory model of conditional instability. Dynamics of Atmospheres and Oceans **27**, 367–382 (1997)

[218] Krishnan, M., Ugaz, V.M., Burns, M.A.: PCR in a Rayleigh–Bénard convection cell. Science **298**, 793 (2002)

[219] Krishnan, M., Agrawal, N., Burns, M.A., Ugaz, V.M.: Reactions and fluidics in miniaturized natural convection systems. Analytical Chemistry **76**, 6254–6265 (2004)

[220] Kumar, A., Bera, P., Khalili, A.: Influence of inertia and drag terms on the stability of mixed convection in a vertical porous medium channel. Int. J. Heat Mass Transfer **53**, 23–24 (2010)

[221] Kumar, A., Bera, P., Kumar, J.: Non-Darcy mixed convection in a vertical pipe filled with porous medium. Int. J. Thermal Sciences **50**, 725–735 (2011)

[222] Kumar, A.A.P., Goyal, H., Banerjee, T., Bandyopadhyay, D.: Instability modes of a two-layer Newtonian plane Couette flow past a porous medium. Phys. Rev. E **87**(063003), 1–8 (2013)

[223] Kumar, A.A.P., Usha, R., Banerjee, T., Bandyopadhyay, D.: Instabilities of a free bilayer flowing on an inclined porous medium. Phys. Rev. E **88**(063012), 1–6 (2013)

[224] Kuznetsov, A.V., Nield, D.A.: Effect of local thermal non-equilibrium on the onset of convection in a porous medium layer saturated by a nanofluid. Transport in Porous Media **83**, 425–436 (2010)

[225] Kuznetsov, A.V., Nield, D.A.: Thermal instability in a porous medium layer saturated by a nanofluid: Brinkman model. Transport in Porous Media **81**, 409–422 (2010)

[226] Kuznetsov, A.V., Nield, D.A.: The effects of combined horizontal and vertical heterogeneity on the onset of convection in a porous medium with vertical throughflow. Transport in Porous Media **90**, 465–478 (2011)

[227] Kuznetsov, A.V., Nield, D.A.: The onset of convection in a tridisperse porous medium. Int. J. Heat Mass Transfer **54**, 3120–3217 (2011)

[228] Kuznetsov, A.V., Nield, D.A.: The onset of double-diffusive convection in a vertical cylinder occupied by a heterogeneous porous medium with vertical throughflow. Transport in Porous Media **95**, 327–336 (2012)

[229] Kwok, L., Chen, C.: Stability of thermal convection in a vertical porous layer. J. Heat Transfer **109**, 889–893 (1987)

[230] Kwak, K., Kim, C.: Viscosity and thermal conductivity of copper oxide nanofluid dispersed in ethylene glycol. Korea-Australia Rheology Journal **17**, 35–40 (2005)

[231] Larson, V.: Stability properties of and scaling laws for a dry radiative-convective atmosphere. Q. J. Royal Meteorological Soc. **126**, 145–171 (2000)

[232] Larson, V.: The effects of thermal radiation on dry convective instability. Dynamics of Atmospheres and Oceans **34**, 45–71 (2001)

[233] Lauga, E., Cossu, C.: A note on the stability of slip channel flows. Phys. Fluids **17**(088106), 1–4 (2005)

[234] Lauga, E., Brenner, M.P., Stone, H.A.: Microfluidics: the no-slip boundary condition. In: C. Tropea, A. Yarin, J.F. Foss (eds.) Handbook of Experimental Fluid Dynamics, pp. 1219–1240. Springer (2007)

[235] Lebon, G.: Heat conduction at micro and nanoscales: A review through the prism of Extended Irreversible Thermodynamics. J. Non-Equilibrium Thermodynamics **39**, 35–59 (2014)

[236] Lebon, G., Cloot, G.: Bénard–Marangoni instability in a Maxwell–Cattaneo fluid. Physics Letters A **105**, 361–364 (1984)

[237] Lee, J.H., Hwang, K.S., Jang, S.P., Lee, B.H., Kim, J.H., Choi, S.U.S., Choi, C.J.: Effective viscosities and thermal conductivities of aqueoud nanofluids containing low volume concentrations of Al_2O_3 nanoparticles. Int. J. Heat Mass Transfer **51**, 2651–2656 (2008)

[238] Lee, J., Shivakumara, I., Ravisha, M.: Effect of thermal non-equilibrium on convective instability in a ferromagnetic fluid-saturated porous medium. Transport in Porous Media **86**, 103–124 (2011)

[239] Lee, J., Shivakumara, I.S., Mamatha, A.L.: Effect of non-uniform temperature gradients on thermogravitational convection in a porous layer using a thermal non-equilibrium model. J. Porous Media **14**, 659–669 (2011)

[240] Lefebvre, L.P., Banhart, J., Dunand, D.C.: Porous metals and metallic foams: current status and recent developments. Advanced Engineering Materials **10**, 775–787 (2008)

[241] Li, J.E., Wang, B.: Equivalent thermal conductivity of open cell ceramic foams at high temperatures. Int. J. Thermophys. **35**, 105–122 (2014)

[242] Li, Y., Liu, Y., Luo, S.G., Lin, C.H.: Decay estimates for the Brinkman–Forchheimer equations in a semi-infinite pipe. ZAMM **92**, 160–176 (2012)

[243] Lighthill, M.J.: I. Introduction, real and ideal fluids. In: L. Rosenhead (ed.) Laminar boundary layers, pp. 1–45. Oxford University Press (1963)

[244] Liu, Y.: Convergence and continuous dependence for the Brinkman–Forchheimer equations. Math. Computer Modelling **49**, 1401–1415 (2009)

[245] Liu, Y.: Structural stability for resonant porous penetrative convection. European J. Appl. Math. **23**, 761–775 (2012)

[246] Liu, H., Bussmann, M., Mostaghimi, J.: A comparison of hyperbolic and parabolic models of phase change of a pure metal. Int. J. Heat Mass Transfer **52**, 1177–1184 (2009)

[247] Liu, Y., Du, Y., Lin, C.: Convergence results for Forchheimer's equations for fluid flow in porous media. J. Math. Fluid Mech. **12**, 576–593 (2010)

[248] Liu, Y., Du, Y., Lin, C.H.: Convergence and continuous dependence results for the Brinkman equations. Applied Mathematics and Computation **215**, 4443–4455 (2010)

[249] Lombardo, S., Mulone, G.: Necessary and sufficient conditions for global nonlinear stability for rotating double-diffusive convection in a porous medium. Continuum Mech. Thermodyn. **14**, 527–540 (2002)

[250] Lombardo, S., Mulone, G., Trovato, M.: Nonlinear stability in reaction-diffusion systems via optimal lyapunov functions. J. Math. Anal. Appl. **342**, 461–476 (2008)

[251] López Molina, J.A., Rivera, M.J., Trujillo, M., Berjano, E.J.: Effect of the thermal wave in radiofrequency ablation modelling: an analytical study. Phys. Med. Biol. **53**, 1447–1462 (2008)

[252] Lorenzi, L., Lunardi, A., Metafune, G., Pallara, D.: Analytic semigroups and reaction-diffusion problems (2005). http://www.math.unipr.it/ lunardi/LectureNotes/I-Sem2005.pdf

[253] Lu, W.S., Shao, H.Y.: Generalized nonlinear subcritical symmetric instability. Adv. Atmos. Sci. **20**, 623–630 (2003)

[254] Machado, L., Lima, W., Pinto, O., Morales, C.: Relationship between cloud-to-ground discharge and penetrative clouds: a multi-channel satellite application. Atmospheric Research **93**, 304–309 (2009)

[255] Mahjoob, S., Vafai, K.: Analytical characterization of heat transport through biological media incorporating hypothermia treatment. Int. J. Heat Mass Transfer **52**, 1608–1618 (2009)

[256] Malashetty, M.S., Biradar, B.S.: The onset of double diffusive convection in a binary Maxwell fluid saturated porous layer with cross diffusion effects. Phys. Fluids **23**(064109), 1–13 (2011)

[257] Malashetty, M.S., Heera, R.: Linear and nonlinear double diffusive convection in a rotating porous layer using a thermal non-equilibrium model. Int. J. Nonlinear Mech. **43**, 600–621 (2008)

[258] Malashetty, M.S., Heera, R.: The onset of double diffusive convection in a sparsely packed porous layer using a thermal non-equilibrium model. Acta Mechanica **204**, 1–20 (2009)

[259] Malashetty, M.S., Swamy, M.: Effect of rotation on the onset of thermal convection in a sparsely packed porous layer using a thermal non-equilibrium model. Int. J. Heat Mass Transfer **53**, 3088–3101 (2010)

[260] Malashetty, M.S., Shivakumara, I.S., Kulkarni, S.: The onset of convection in an anisotropic porous layer using a thermal non-equilibrium model. Transport in Porous Media **60**, 199–215 (2005)

[261] Malashetty, M.S., Shivakumara, I.S., Kulkarni, S.: The onset of Lapwood–Brinkman convection using a thermal non-equilibrium model. Int. J. Heat Mass Transfer **48**, 1155–1163 (2005)

[262] Malashetty, M.S., Shivakumara, I.S., Kulkarni, S., Swamy, M.: Convective instability of Oldroyd-B fluid saturated porous layer heated from below using a thermal non-eqiulibrium model. Transport in Porous Media **64**, 123–139 (2006)

[263] Malashetty, M.S., Swamy, M., Kulkarni, S.: Thermal convection in a rotating porous layer using a thermal non-equilibrium model. Physics of Fluids **19**(054102), 1–16 (2007)

[264] Malashetty, M.S., Swamy, M., Heera, R.: Double diffusive convection in a porous layer using a thermal non-equilibrium model. Int. J. Thermal Sciences **47**, 1131–1147 (2008)

[265] Malashetty, M.S., Pop, I., Heera, R.: Linear and nonlinear double diffusive convection in a rotating sparsely packed porous layer using a thermal non-equilibrium model. Continuum Mechanics and Thermodynamics **21**, 317–339 (2009)

[266] Malashetty, M.S., Hill, A.A., Swamy, M.: Double diffusive convection in a viscoelastic fluid-saturated porous layer using a thermal non-equilibrium model. Acta Mechanica **223**, 967–983 (2012)

[267] Malik, S.V., Hooper, A.P.: Three-dimensional disturbances in channel flows. Phys. Fluids **19**(052102), 1–18 (2007)

[268] Manglik, A., Wicht, J., Christensen, U.R.: A dynamo model with double diffusive convection for Mercury's core. Earth and Planetary Science Letters **289**, 619–628 (2010)

[269] Mao, S., Love, N., Leanos, A., Rodriguez-Melo, G.: Correlation studies of hydrodynamics and heat transfer in metal foam heat exchangers. Applied Thermal Engineering **71**, 104–118 (2014)

[270] Marciak-Kozlowska, J., Kozlowski, M.: Heat waves versus diffusion in attosecond to yoctosecond laser interaction with matter. Lasers in Engineering **27**, 75–85 (2014)

[271] Maremonti, P.: Pointwise asymptotic stability of steady fluid motions. J. Math. Fluid Mech. **11**, 348–382 (2009)

[272] Maruthamanikandan, S., Nagouda, S.S.: Convective heat transfer in a Maxwell–Cattaneo dielectric fluid. Int. J. Computational Engng. Research **3**, 347–355 (2013)

[273] Masin, D., Herbstova, V., Bohac, J.: Properties of double porosity clayfills and suitable constitutive models (2012). http://web.natur.cuni.cz/uhigug/masin/download/mhb_16ICSMGE_Osaka.pdf

[274] Masoumi, M., Sohrabi, N., Behzadmehr, A.: A new model for calculating the effective viscosity of nanofluids. J. Phys. D: Appl. Phys. **42**(055501), 1–6 (2009)

[275] Massa, L., Jha, P.: Linear analysis of the Richtmeyer–Meshov instability in shock flame interactions. Phys. Fluids **24**(056101), 1–30 (2012)

[276] Maugin, G.A.: On the notions of viscous fluids, elastic solids and heat conduction in relativity (in french). C. R. Acad. Sci. Paris, Ser. A **276**, 1027–1030 (1973)

[277] Maugin, G.A.: Book review. J. Thermal Stresses **37**, 380–385 (2014)

[278] Maxwell, J.C.: On stresses in rarefied gases arising from inequalities of temperature. Phil. Trans. Roy. Soc. London **170**, 231–256 (1879)

[279] McKay, G.: Nonlinear stability analyses of problems in patterned ground formation and penetrative convection. Ph.D. thesis, Glasgow University (1992)

[280] McKay, G.: Patterned ground formation and solar radiation heating. Proc. Roy. Soc. London A **438**, 249–263 (1992)

[281] McKay, G.: Onset of double-diffusive convection in a saturated porous layer with time-periodic heating. Continuum Mech. Thermodyn. **10**, 241–251 (1998)

[282] McKay, G.: Double-diffusive convective motions for a saturated porous layer subject to modulated surface heating. Continuum Mech. Thermodyn. **12**, 69–78 (2000)

[283] McKibbin, R., O'Sullivan, M.J.: Onset of convection in a layered porous medium heated from below. J. Fluid Mech. **96**, 375–393 (1980)

[284] McKibbin, R., O'Sullivan, M.J.: Heat transfer in a layered porous medium heated from below. J. Fluid Mech. **111**, 141–173 (1981)

[285] McKibbin, R., Tyvand, P.A.: Anisotropic modelling of thermal convection in multilayered porous media. J. Fluid Mech. **118**, 315–339 (1982)

[286] McKibbin, R., Tyvand, P.A.: Thermal convection in a porous medium composed of alternating thick and thin layers. Int. J. Heat Mass Transfer **26**, 761–780 (1983)

[287] McKibbin, R., Tyvand, P.A.: Thermal convection in a porous medium with horizontal cracks. Int. J. Heat Mass Transfer **27**, 1007–1023 (1984)

[288] Mellor, G.: Introduction to Physical Oceanography. Springer, New York (1996)

[289] Mharzi, M., Daguenet, M., Daoudi, S.: Thermosolutal natural convection in a vertically layered fluid-porous medium heated from the side. Energy Conversion and Management **41**, 1065–1090 (2000)

[290] Mihaljan, J.: A rigorous exposition of the Boussinesq approximation applicable to a thin layer of fluid. Astrophys. J. **136**, 1126–1133 (1962)

[291] Minkowycz, W., Haji-Sheikh, A., Vafai, K.: On the departure from local thermal non-equilibrium in porous media due to a rapidly changing heat source: the Sparrow number. Int. J. Heat Mass Transfer **42**, 3373–3385 (1999)

[292] Miranville, A., Quintanilla, R.: A generalization of the Caginalp phase-field system based on the Cattaneo law. Nonlinear Analysis; Theory, Methods and Applications **71**, 2278–2290 (2009)

[293] Miranville, A., Quintanilla, R.: A Caginalp phase-field system with nonlinear coupling. Nonlinear Analysis; Real World Applications **11**, 2849–2861 (2010)

[294] Mongiovi, M.S., Zingales, M.: A non-local model of thermal energy transport: the fractional temperature equation. Int. J. Heat Mass Transfer **67**, 593–601 (2013)

[295] Morini, G.L., Lorenzini, M., Spiga, M.: A criterion for experimental validation of slip-flow models for incompressible rarefied gases through microchannels. Microfluidics and Nanofluidics **1**, 190–196 (2005)

[296] Morro, A., Caviglia, G.: A hyperbolic model of chemotaxis. Comm. Appl. Industrial Math. **4**, 1–15 (2013)

[297] Mulone, G.: On the nonlinear stability of a fluid layer of a mixture heated and salted from below. Continuum Mech. Thermodyn. **6**, 161–184 (1994)

[298] Mulone, G., Straughan, B.: Nonlinear stability for diffusion models in biology. SIAM J. Appl. Math. **69**, 1739–1758 (2009)

[299] Mulone, G., Straughan, B., Wang, W.: Stability of epidemic models with evolution. Stud. Appl. Math. **118**, 117–132 (2007)

[300] Murshed, S., Leong, K., Yang, C.: Investigations of thermal conductivity and viscosity of nanofluids. Int. J. Thermal Sciences **47**, 560–568 (2008)

[301] Nagouda, S.S., Maruthamanikandan, S.: Stability of porous medium convection in polarized dielectric fluids with non-classical heat conduction. Int. J. Materials Arch. **4**, 136–144 (2013)

[302] Nagouda, S.S., Pranesh, S.: Rayleigh–Bénard convection in a second order fluid with Maxwell–Cattaneo law. Bull. Soc. Mathematical Services and Standards **1**, 33–48 (2012)

[303] Narasimhan, A., Reddy, B.: Resonance of natural convection inside a bidisperse porous medium enclosure. J. Heat Transfer—ASME **133**(042601), 1–9 (2011)

[304] Navier, C.L.M.H.: Mémoire sur les lois du mouvement des fluides. Mémoires de l'Académie Royale des Sciences de l'Institut de France **6**, 389–440 (1823)

[305] Néel, M.: Convection forcée en milieu poreux: écarts à la loi de Darcy. C.R. Acad. Sci. Paris, série IIb **326**, 615–620 (1998)

[306] Nield, D.A.: Onset of thermohaline convection in a porous medium. Water Resources Research **4**, 553–560 (1968)

[307] Nield, D.A.: Effects of local thermal non-equilibrium in steady convection processes in saturated porous media: forced convection in a channel. J. Porous Media **1**, 181–186 (1998)

[308] Nield, D.A.: A note on modelling of local thermal non-equilibrium in a structured porous medium. Int. J. Heat Mass Transfer **45**, 4367–4368 (2002)

[309] Nield, D.A.: The stability of flow in a channel or duct occupied by a porous medium. Int. J. Heat Mass Transfer **46**, 4351–4354 (2003)

[310] Nield, D.A.: A note on local thermal non-equilibrium in porous media near boundaries and interfaces. Transport in Porous Media **95**, 581–584 (2012)

[311] Nield, D.A., Barletta, A.: Extended Oberbeck–Boussinesq approximation study of convective instabilities in a porous layer with horizontal flow and bottom heating. Int. J. Heat Mass Transfer **53**, 577–585 (2010)

[312] Nield, D.A., Bejan, A.: Convection in Porous Media, fourth edn. Springer, New York (2013)

[313] Nield, D.A., Kuznetsov, A.V.: Local thermal non-equilibrium effects in forced convection in a porous medium channel: a conjugate problem. Int. J. Heat Mass Transfer **42**, 3245–3252 (1999)

[314] Nield, D.A., Kuznetsov, A.V.: The interaction of thermal non-equilibrium and heterogeneous conductivity effects in forced convection in layered porous channels. Int. J. Heat Mass Transfer **44**, 4369–4373 (2001)

[315] Nield, D.A., Kuznetsov, A.V.: Forced convection in a bidisperse porous medium channel: a conjugate problem. Int. J. Heat Mass Transfer **47**, 5375–5380 (2004)

[316] Nield, D.A., Kuznetsov, A.V.: A two-velocity temperature model for a bidispersed porous medium: forced convection in a channel. Transport in Porous Media **59**, 325–339 (2005)

[317] Nield, D.A., Kuznetsov, A.V.: The onset of convection in a bidisperse porous medium. Int. J. Heat Mass Transfer **49**, 3068–3074 (2006)

[318] Nield, D.A., Kuznetsov, A.V.: The effect of combined vertical and horizontal heterogeneity on the onset of convection in a bidisperse porous medium. Int. J. Heat Mass Transfer **50**, 3329–3339 (2007)

[319] Nield, D.A., Kuznetsov, A.V.: Natural convection about a vertical plate embedded in a bidisperse porous medium. Int. J. Heat Mass Transfer **51**, 1658–1664 (2008)

[320] Nield, D.A., Kuznetsov, A.V.: The effect of local thermal non-equilibrium on the onset of convection in a nanofluid. J. Heat Transfer, ASME **132**(052405), 1–7 (2010)

[321] Nield, D.A., Kuznetsov, A.V.: The onset of convection in a horizontal nanofluid layer of finite depth. European J. Mechanics B/Fluids **29**, 217–223 (2010)

[322] Nield, D.A., Kuznetsov, A.V.: A three-velocity three-temperature model for a tridisperse porous medium: forced convection in a channel. Int. J. Heat Mass Transfer **54**, 2490–2498 (2011)

[323] Nield, D.A., Kuznetsov, A.V.: The onset of convection in a horizontal nanofluid layer of finite depth: a revised model. Int J. Heat Mass Transfer **77**, 915–918 (2014)

[324] Niknami, M., Khayat, R.E.: Energy growth of disturbances in a non-Fourier fluid. Int J. Heat Mass Transfer **67**, 613–626 (2013)

[325] Normahd, C., Azouni, A.: Penetrative convection in an internally heated layer of water near the maximum density point. Phys. Fluids A **4**, 243–253 (1992)

[326] Nouri-Borujerdi, A., Noghrehabadi, A.R., Rees, D.A.S.: Onset of convection in a horizontal porous channel with uniform heat generation using a thermal non-equilibrium model. Transport in Porous Media **69**, 343–357 (2007)

[327] Olusola, B.K., Yu, G., Aguilera, R.: The use of electromagnetic mixing rules for petrophysical evaluation of dual- and triple-porosity reservoirs. SPE Reservoir Evaluation and Engineering **16**, 378–389 (2013)

[328] Ouyang, Y., Yang, L.: A note on the existence of a global attractor for the Brinkman–Forchheimer equations. Nonlinear Analysis, Theory, Methods and Applications **70**, 2054–2059 (2009)

[329] Papanicolaou, N.C., Christov, C.I., Jordan, P.M.: Two-gradient convection in a vertical slot with Maxwell–Cattaneo heat conduction. In: M.D. Todorov, C.I. Christov (eds.) Applications of Mathematics in Technical and Natural Sciences, pp. 231–239. American Institute of Physics (2009)

[330] Papanicolaou, N.C., Christov, C.I., Jordan, P.M.: The influence of thermal relaxation on the oscillatory properties of two-gradient convection in a vertical slot. European Journal of Mechanics B/Fluids **30**, 68–75 (2011)

[331] Pasquale, V., Verdoya, M., Chiozzi, P.: Heat flow and geothermal resources in Northern Italy. Renewable and Sustainable Energy Reviews **36**, 277–285 (2014)

[332] Patil, P.M., Rees, D.A.S.: The onset of convection in a porous layer with multiple horizontal solid partitions. Int. J. Heat Mass Transfer **68**, 234–246 (2014)

[333] Payne, L.E., Straughan, B.: Convergence and continuous dependence for the Brinkman–Forchheimer equations. Stud. Appl. Math. **102**, 419–439 (1999)

[334] Payne, L.E., Straughan, B.: A naturally efficient numerical technique for porous convection stability with non-trivial boundary conditions. Int. J. Num. Anal. Meth. Geomech. **24**, 815–836 (2000)

[335] Payne, L.E., Straughan, B.: Unconditional nonlinear stability in temperature-dependent viscosity flow in a porous medium. Stud. Appl. Math. **105**, 59–81 (2000)

[336] Payne, L.E., Straughan, B.: Decay for a Keller–Segel chemotaxis model. Stud. Appl. Math. **123**, 337–360 (2009)

[337] Pearlstein, A.J., Harris, R.M., Terrones, G.: The onset of convective instability in a triply diffusive fluid layer. J. Fluid Mech. **202**, 443–465 (1989)

[338] Pedlosky, J.: Ocean Circulation Theory. Springer, Berlin, Heidelberg (1996)

[339] Peng, T., O'Neill, D.P., Payne, S.P.: A two-equation coupled system for determination of liver tissue temperature during thermal ablation. Int. J. Heat Mass Transfer **54**, 2100–2109 (2011)

[340] Petit, F., Fichot, F., Quintard, M.: Écoulement diphasique en milieu poreux: modèle à non-équilibre local. Int. J. Thermal Sciences **38**, 239–249 (1999)

[341] Phuoc, T.X., Massoudi, M.: Experimental observations of the effects of shear rates and particle concentration on the viscosity of Fe_2O_3-deionized water nanofluids. Int. J. Thermal Sciences **48**, 1294–1301 (2009)

[342] Pieters, G., van Duijn, C.: Transient growth in linearly stable gravity-driven flow in porous media. Eur. J. Mech. B/Fluids **25**, 83–94 (2006)

[343] Pilgrim, N.J., Batty, W., Kelsall, R.W., Snowden, C.M.: Nanoscale electrothermal co-simulation: compact dynamic models of hyperbolic heat transport and self-consistent device Monte Carlo. Microelectronics Journal **35**, 823–830 (2004)

[344] Platten, J.K.: The Soret effect: a review of recent experimental results. J. Applied Mechanics (ASME) **73**, 5–15 (2006)

[345] Plesa, A.C., Tosi, N., Breuer, D.: Can a fractionally crystallized magma ocean explain the thermo-chemical evolution of Mars? Earth and Planetary Science Letters **403**, 225–235 (2014)

[346] Postelnicu, A.: The onset of Darcy–Brinkman convection in a porous layer using a thermal nonequilibrium model. part II. Int. J. Thermal Sciences **47**, 1587–1594 (2008)

[347] Postelnicu, A.: The effect of a horizontal pressure gradient on the onset of a Darcy–Bénard convection in thermal non-equilibrium conditions. Int. J. Heat Mass Transfer **53**, 68–75 (2010)

[348] Postelnicu, A., Rees, D.A.S.: The onset of Darcy–Brinkman convection in a porous layer using a thermal nonequilibrium model-part I: stress-free boundaries. Int. J. Energy Research **27**, 961–973 (2003)

[349] Pranesh, S., Kiran, R.V.: Study of Rayleigh–Bénard magneto convection in a micropolar fluid with Maxwell–Cattaneo law. Applied Mathematics **1**, 470–480 (2010)

[350] Priezjev, N.V.: Molecular dynamics simulations of oscillatory Couette flows with slip boundary conditions. Microfluidics and Nanofluidics **14**, 225–233 (2013)

[351] Pritchard, D., Richardson, C.N.: The effect of temperature-dependent solubility on the onset of thermosolutal convection in a horizontal porous layer. J. Fluid Mech. **571**, 59–95 (2007)

[352] Proctor, M.R.E.: Steady subcritical thermohaline convection. J. Fluid Mech. **105**, 507–521 (1981)

[353] Prudhomme, M., Jasmin, S.: Inverse solution for a biochemical heat source in a porous medium in the presence of natural convection. Chem. Engng. Sci. **61**, 1667–1675 (2006)

[354] Puri, P., Jordan, P.M.: Stokes's first problem for a dipolar fluid with nonclassical heat conduction. J. Engng. Math. **36**, 219–240 (1999)

[355] Puri, P., Jordan, P.M.: Wave structure in Stokes' second problem for a dipolar fluid with nonclassical heat conduction. Acta Mechanica **133**, 145–160 (1999)

[356] Qin, Y., Kaloni, P.: A nonlinear stability problem of convection in a porous vertical slab. Phys. Fluids A **5**, 2067–2069 (1993)

[357] Qin, Y., Kaloni, P.N.: Spatial decay estimates for plane flow in the Brinkman–Forchheimer model. Quart. Appl. Math. **56**, 71–87 (1998)

[358] Rahman, M.M., Al-Lawatia, M.A., Eltayeb, I.A., Al-Salti, N.: Hydromagnetic slip flow of water based nanofluids past a wedge with convective surface in the presence of heat generation or absorption. Int. J. Thermal Sciences **57**, 172–182 (2012)

[359] Rajagopal, K.R., Ruzika, M., Srinivasa, A.R.: On the Oberbeck–Boussinesq approximation. Mathematical Models and Methods in Applied Sciences **6**, 1157–1167 (1996)

[360] Rajagopal, K.R., Saccomandi, G., Vergori, L.: On the Oberbeck–Boussinesq approximation for fluids with pressure dependent viscosities. Nonlinear Analysis, Real World Applications **10**, 1139–1150 (2009)

[361] Rajagopal, K.R., Saccomandi, G., Vergori, L.: A systematic approximation for the equations governing convection-diffusion in a porous medium. Nonlinear Analysis, Real World Applications **11**, 2366–2375 (2010)

[362] Rees, D.A.S.: The stability of Prandtl–Darcy convection in a vertical porous layer. Int. J. Heat Mass Transfer **31**, 1529–1534 (1988)

[363] Rees, D.A.S.: The onset of Darcy–Brinkman convection in a porous layer: an asymptotic analysis. Int. J. Heat Mass Transfer **45**, 2213–2220 (2002)

[364] Rees, D.A.S.: Microscopic modelling of the two-temperature model for conduction in heterogeneous media: three-dimensional media. In: Proc. 4th International Conference on Applications of Porous Media, Istanbul (2009)

[365] Rees, D.A.S.: Microscopic modelling of the two-temperature model for conduction in heterogeneous media. J. Porous Media **13**, 125–143 (2010)

[366] Rees, D.A.S.: The effect of local thermal non-equilibrium on the stability of convection in a vertical porous channel. Transport in Porous Media **87**, 459–464 (2011)

[367] Rees, D.A.S., Genc, G.: The onset of convection in porous layers with multiple horizontal partitions. Int. J. Heat Mass Transfer **54**, 3081–3089 (2011)

[368] Rees, D.A.S., Bassom, A.P., Siddheshwar, P.G.: Local thermal non-equilibrium effects arising from the injection of a hot fluid into a porous medium. J. Fluid Mech. **594**, 379–398 (2008)

[369] Rionero, S.: Asymptotic and other properties of some nonlinear diffusion models. In: B. Straughan, R. Greve, H. Ehrentraut, Y. Wang (eds.) Continuum Mechanics and Applications in Geophysics and the Environment, pp. 56–78. Springer (2001)

[370] Rionero, S.: Onset of convection in porous materials with vertically stratfied porosity. Acta Mechanica **222**, 261–272 (2011)

[371] Rionero, S.: Stability of ternary reaction-diffusion dynamical systems. Atti Accad. Lincei **22**, 245–268 (2011)

[372] Rionero, S.: Absence of subcritical instabilities and global nonlinear stability for porous ternary diffusive-convective fluid mixtures. Physics of Fluids **24**(104101), 1–17 (2012)

[373] Rionero, S.: Global nonlinear stability for a triply diffusive convection in a porous layer. Continuum Mech. Thermodyn. **24**, 629–641 (2012)

[374] Rionero, S.: Multicomponent diffusive-convective fluid motions in porous layers: Ultimately boundedness, absence of subcritical instabilities, and global stability for any number of salts. Physics of Fluids **25**(054104), 1–23 (2013)

[375] Rionero, S.: Soret effects on the onset of convection in rotating porous layers via the "auxilliary system method". Ricerche di Matematica **62**, 183–208 (2013)

[376] Rionero, S.: Triple diffusive convection in porous media. Acta Mechanica **224**, 447–458 (2013)

[377] Rionero, S., Straughan, B.: Convection in a porous medium with internal heat source and variable gravity effects. Int. J. Engng. Sci. **28**, 497–503 (1990)

[378] Rionero, S., Torcicollo, I.: On an ill-posed problem in nonlinear heat conduction. Transport Theory and Statistical Physics **29**, 173–186 (2000)

[379] Roberts, P.: An Introduction to Magnetohydrodynamics. Longman, London (1967)

[380] Roberts, P.H.: Convection in horizontal layers with internal heat generation. theory. J. Fluid Mech. **30**, 33–49 (1967)

[381] Rukolaine, S.A., Samsonov, A.M.: Local immobilization of particles in mass transfer described by a Jeffreys-type equation. Phys. Rev. E **88**(062116), 1–12 (2013)

[382] Sadeghi, A., Baghani, M., Saidi, M.H.: Gaseous slip flow forced convection through ordered microcylinders. Microfluidics and Nanofluidics **15**, 73–85 (2013)

[383] Sadeghi, M., Sadeghi, A., Saidi, M.H.: Buoyancy effects on gaseous slip flow in a vertical rectangular microchannel. Microfluidics and Nanofluidics **16**, 207–224 (2014)

[384] Saidane, A., Aliouat, S., Benzohra, M., Ketata, M.: A transmission line matrix (TLM) study of hyperbolic heat conduction in biological materials. J. Food Engineering **68**, 491–496 (2005)

[385] Samanta, A.: Shear wave instability for electrified falling films. Phys. Rev. E **88**(053002), 1–8 (2013)

[386] Saravanan, S.: Thermal non-equilibrium porous convection with heat generation and density maximum. Transport in Porous Media **76**, 35–43 (2009)

[387] Saravanan, S., Brindha, D.: Linear and nonlinear stability limits for centrifugal convection in an anisotropic layer. Int. J. Non-Linear Mechanics **46**, 65–72 (2011)

[388] Saravanan, S., Brindha, D.: Onset of centrifugal filtration convection: departure from thermal equilibrium. Proc. Roy. Soc. London A **469**(20120655), 1–15 (2013)

[389] Saravanan, S., Sivakumar, T.: Onset of thermovibrational filtration convection: Departure from thermal equilibrium. Phys. Rev. E **84**(026307), 1–14 (2011)

[390] Savino, R., Paterna, D.: Thermodiffusion in nanofluids under different gravity conditions. Physics of Fluids **20**(017101), 1–10 (2008)

[391] Scott, N.L.: Convection in a saturated Darcy porous medium with an exothermic chemical surface reaction and Soret effect. Int. Comm. Heat Mass Transfer **39**, 1331–1335 (2012)

[392] Scott, N.L.: Stability studies of porous media including surface reactions. Ph.D. thesis, Durham University (2013). etheses.dur.ac.uk/9421

[393] Scott, N.L., Straughan, B.: A nonlinear stability analysis of convection in a porous vertical channel including local thermal nonequilibrium. J. Math. Fluid Mech. **15**, 171–178 (2013)

[394] Sekar, R., Raju, K., Vasanthakumari, R.: A linear analytica; study of Soret-driven ferrothermohaline convection in an anisotropic porous medium. J. Magn. Magn. Mater. **331**, 122–128 (2013)

[395] Shiina, Y., Hishida, M.: Critical Rayleigh number of natural convection in high porosity anisotropic horizontal porous layers. Int. J. Heat Mass Transfer **53**, 1507–1513 (2010)

[396] Shir, C., Joseph, D.: Convective instability in a temperature and concentration field. Arch. Rational Mech. Anal. **30**, 38–80 (1968)

[397] Shivakumara, I.S., Mamatha, A.L., Ravisha, M.: Effects of variable viscosity and density maximum on the onset of Darcy–Bénard convection using a thermal non-equilibrium model. J. Porous Media **13**, 613–622 (2010)

[398] Shivakumara, I.S., Lee, J., Chavaraddi, K.B.: Onset of surface tension driven convection in a fluid overlying a layer of an anisotropic porous medium. Int. J. Heat Mass Transfer **54**, 994–1001 (2011)

[399] Shivakumara, I.S., Lee, J., Ravisha, M., Reddy, R.G.: The effects of local thermal nonequilibrium and MFD viscosity on the onset of Brinkman ferroconvection. Meccanica **47**, 1359–1378 (2012)

[400] Shivakumara, I.S., Suma, S.P., Indira, R., Gangadharaiah, Y.H.: Effect of internal heat generation on the onset of Marangoni convection in a fluid layer overlying a layer of anisotropic porous medium. Transport in Porous Media **92**, 727–743 (2012)

[401] Shivakumara, I.S., Kumar, S., Naveen, S.B.: Linear and weakly nonlinear triple diffusive convection in a couple stress fluid layer. Int. J. Heat Mass Transfer **68**, 542–553 (2014)

[402] Shojaeian, M., Shojaeian, M.: Analytical solution of mixed electromagnetic/pressure driven gaseous flows in microchannels. Microfluidics and Nanofluidics **12**, 553–564 (2012)

[403] Siddheshwar, P.G., Bhadauria, B.S., Suthar, O.P.: Synchronous and asynchronous boundary temperature modulations of Bénard–Darcy convection. Int. J. Non-Linear Mechanics **49**, 84–89 (2013)

[404] Soltani, P., Johari, M.S., Zarrebini, M.: Effect of a 3D fiber orientation on permeability of realistic fibrous porous networks. Powder Technology **254**, 44–56 (2014)

[405] Soret, C.: Sur l'état d'équilibre que prend au point vue de sa concentration une dissolution saline primitivement homogène dont deux perties sont portées a des températures différentes. Archive des Sciences Physiques et Naturelles **2**, 48–61 (1879)

[406] Spiegel, E., Veronis, G.: On the Boussinesq approximation for a compressible fluid. Astrophys. J. **131**, 442–447 (1960)

[407] Spille, A., Rauh, A., Bühring, H.: Critical curves of plane Poiseuille flow with slip boundary conditions. Nonlinear Phenomena In Complex Systems **3**, 171–173 (2000)

[408] Stebel, J.: On shape stability of incompressible fluids subject to Navier's slip condition. J. Math. Fluid Mech. **14**, 575–589 (2012)

[409] Straughan, B.: A nonlinear analysis of convection in a porous vertical slab. Geophys. Astrophys. Fluid Dyn. **42**, 269–275 (1988)

[410] Straughan, B.: Explosive Instabilities in Mechanics. Springer, Heidelberg (1998)

[411] Straughan, B.: A sharp nonlinear stability threshold in rotating porous convection. Proc. Roy. Soc. London A **457**, 87–93 (2001)

[412] Straughan, B.: Surface tension driven convection in a fluid overlying a porous layer. J. Computational Phys. **170**, 320–337 (2001)

[413] Straughan, B.: Effect of property variation and modelling on convection in a fluid overlying a porous layer. Int. J. Num. Anal. Meth. Geomech. **26**, 75–97 (2002)

[414] Straughan, B.: The Energy Method, Stability, and Nonlinear Convection, *Appl. Math. Sci.*, vol. 91, second edn. Springer, New York (2004)

[415] Straughan, B.: Resonant porous penetrative convection. Proc. Roy. Soc. London A **460**, 2913–2927 (2004)

[416] Straughan, B.: Global nonlinear stability in porous convection with a thermal non-equilibrium model. Proc. Roy. Soc. London A **462**, 409–418 (2006)

[417] Straughan, B.: Stability and Wave Motion in Porous Media, *Appl. Math. Sci.*, vol. 165. Springer, New York (2008)

[418] Straughan, B.: On the Nield–Kuznetsov theory for convection in bidispersive porous media. Transport in Porous Media **77**, 159–168 (2009)

[419] Straughan, B.: Oscillatory convection and the Cattaneo law of heat conduction. Ricerche di Matematica **58**, 157–162 (2009)

[420] Straughan, B.: Green–Naghdi fluid with non-thermal equilibrium effects. Proc. Roy. Soc. London A **466**, 2021–2032 (2010)

[421] Straughan, B.: Porous convection with Cattaneo heat flux. Int. J. Heat Mass Transfer **53**, 2808–2812 (2010)

[422] Straughan, B.: Structure of the dependence of Darcy and Forchheimer coefficients on porosity. Int. J. Engng, Sci. **48**, 1610–1621 (2010)

[423] Straughan, B.: Thermal convection with the Cattaneo–Christov model. Int. J. Heat Mass Transfer **53**, 95–98 (2010)

[424] Straughan, B.: Continuous dependence on the heat source in resonant porous penetrative convection. Studies in Applied Mathematics **127**, 302–314 (2011)

[425] Straughan, B.: Heat Waves, *Appl. Math. Sci.*, vol. 177. Springer, New York (2011)

[426] Straughan, B.: Tipping points in Cattaneo–Christov thermohaline convection. Proc. Roy. Soc. London A **467**, 7–18 (2011)

[427] Straughan, B.: Triply resonant penetrative convection. Proc. Roy. Soc. London A **468**, 3804–3823 (2012)

[428] Straughan, B.: Porous convection with local thermal non-equilibrium effects and Cattaneo effects in the solid. Proc. Roy. Soc. London A **469**(20130187), 1–12 (2013)

[429] Straughan, B.: Stability and uniqueness in double porosity elasticity. Int. J. Engng. Science **65**, 1–8 (2013)

[430] Straughan, B.: Anisotropic inertia effect in microfluidic porous thermosolutal convection. Microfluidics and Nanofluidics **16**, 361–368 (2014)

[431] Straughan, B.: Bidispersive poroacoustic waves. Int. J. Engng. Science **77**, 30–34 (2014)

[432] Straughan, B.: Nonlinear stability of convection in a porous layer with solid partitions. J. Math. Fluid Mech. **16**, 727–736 (2014)

[433] Straughan, B.: Resonant penetrative convection with an internal heat source/sink. Acta Applicandae Mathematicae **132**, 561–581 (2014)

[434] Straughan, B., Franchi, F.: Bénard convection and the Cattaneo law of heat conduction. Proc. Roy. Soc. Edinburgh A **96**, 175–178 (1984)

[435] Straughan, B., Harfash, A.J.: Instability of Poiseuille flow in a porous medium with slip boundary conditions. Microfluidics and Nanofluidics **15**, 109–115 (2013)

[436] Straughan, B., Tracey, J.: Multi-component convection-diffusion with internal heating or cooling. Acta Mechanica **133**, 219–238 (1999)

[437] Straughan, B., Walker, D.W.: Anisotropic porous penetrative convection. Proc. Roy. Soc. London A **452**, 97–115 (1996)

[438] Straughan, B., Walker, D.W.: Multi component diffusion and penetrative convection. Fluid Dyn. Res. **19**, 77–89 (1997)

[439] Sun, P.: A Dirichlet–Robin iteration-by-subdomain method for an anisotropic, non-isothermal two-phase transport model of PEM fuel cell with micro-porous layer. J. Comp. Appl. Math. **270**, 241–256 (2014)

[440] Sunil, Sharma, P., Mahajan, A.: Onset of Darcy–Brinkman double diffusive convection in a magnetized ferrofluid layer using a thermal non-equilibrium model: A nonlinear stability analysis. J. Geophys. Engng. **88**, 417–430 (2011)

[441] Sunil, Sharma, P., Mahajan, A.: Onset of Darcy–Brinkman ferroconvection in a rotating porous layer using a thermal non-equilibrium model: A nonlinear stability analysis. Transport in Porous Media **88**, 421–439 (2011)

[442] Svanadze, M.: Dynamical problems of the theory of elasticity for solids with double porosity. Proc. Appl. Math. Mech. **10**, 309–310 (2010)

[443] Svanadze, M.: Plane waves and boundary value problems in the theory of elasticity for solids with double porosity. Acta Applicandae Mathematicae **122**, 461–471 (2012)

[444] Svanadze, M.: Uniqueness theorems in the theory of thermoelasticity for solids with double porosity. Meccanica **49**, 2099–2108 (2014)

[445] Szczygiel, J.: Diffusion in a bidispersive grain of a reforming catalyst. Comput. Chem. **23**, 121–134 (1999)

[446] Szczygiel, J.: Enhancement of reforming efficiency by optimising the porous structure of reforming catalyst: theoretical considerations. Fuel **85**, 1579–1590 (2006)

[447] Szczygiel, J.: Control of transport phenomena in the interior of the reforming catalyst grain: a new approach to the optimisation of the reforming process. Fuel Processing Technology **92**, 1434–1448 (2011)

[448] Takahashi, F.: Double diffusive convection in the Earth's core and the morphology of the geomagnetic field. Phys. Earth Planetary Interiors **226**, 83–87 (2014)

[449] Teng, H., Zhao, T.S.: An extension of Darcy's law to non-Stokes flow in porous media. Chem. Engineering Science **55**, 2727–2735 (2000)

[450] Thompson, J.A., Bau, H.H.: Porous bead-based microfluidic assay: theory and confocal microscope imaging. Microfluidics and Nanofluidics **12**, 625–637 (2012)

[451] Tikhomolov, E.: Large-scale vortical flows and penetrative convection in the Sun. Nuclear Physics A **758**, 709c–712c (2005)

[452] Travis, K.P., Todd, B.D., Evans, D.J.: Poisueille flow of molecular liquids. Physica A **240**, 315–327 (1997)

[453] Tung, M.M., Trujillo, M., López Molina, J.A., Rivera, M.J., Berjano, E.J.: Modelling the heating of biological tissue based on the hyperbolic heat transfer equation. Mathematical and Computer Modelling **50**, 665–672 (2009)

[454] Tzou, D.Y.: Thermal instabilty of nanofluids in natural convection. Int. J. Heat Mass Transfer **51**, 2967–2979 (2008)

[455] Ugurlu, D.: On the existence of a global attractor for the Brinkman–Forchheimer equations. Nonlinear Analysis, Theory, Methods and Applications **68**, 1986–1992 (2008)

[456] Vadasz, P.: Coriolis effect on gravity-driven convection in a rotating porous layer heated from below. J. Fluid Mech. **376**, 351–375 (1998)

[457] Vadasz, P.: Lack of oscillations in dual-phase-lagging heat conduction for a porous slab subject to heat flux and temperature. Int. J. Heat Mass Transfer **48**, 2822–2828 (2005)

[458] Vadasz, P.: Heat conduction in nanofluid suspensions. ASME J. Heat Transfer **128**, 465–477 (2006)

[459] Vadasz, P.: On the paradox of heat conduction in porous media subject to lack of local thermal equilibrium. Int. J. Heat Mass Transfer **50**, 4131–4140 (2007)

[460] Vadasz, P.: Basic natural convection in a vertical porous layer differentially heated from its sidewalls subject to lack of local thermal equilibrium. Int. J. Heat Mass Transfer **54**, 2387–2396 (2011)

[461] Vadasz, P.: Small Nield number convection in a porous layer heated from below via a constant heat flux and subject to lack of local thermal equilibrium. J. Porous Media **15**, 249–258 (2012)

[462] Vadasz, P., Nield, D.A.: Extending the Duhamel theorem to dual phase applications. Int. J. Heat Mass Transfer **51**, 1475–1479 (2008)

[463] Vadasz, J.J., Govender, S., Vadasz, P.: Heat transfer enhancement in nanofluids suspensions: possible mechanisms and explanations. Int. J. Heat Mass Transfer **48**, 2673–2683 (2005)

[464] Valus, J., Schneider, P.: Transport characteristics of bidisperse porous α–aluminas. Applied Catalysis **16**, 329–341 (1985)

[465] van den Berg, A.P., Yuen, D.A., Beebe, G., Christiansen, M.: The dynamical impact of electronic thermal conductivity on deep mantle convection of exosolar planets. Physics of the Earth and Planetary Interiors **178**, 136–154 (2010)

[466] van Duijn, C.J., Galiano, G., Peletier, M.A.: A diffusion-convection problem with drainage arising in the ecology of mangroves. Interfaces and Free Boundaries **3**, 15–44 (2001)

[467] van Duijn, C.J., Pieters, G.J.M., Wooding, R.A., van der Ploeg, A.: Stability criteria for the vertical boundary layer formed by throughflow near the surface of a porous medium. In: Environmental Mechanics: Water, Mass and Energy Transfer in the Biosphere, *Geophysical Monographs*, vol. 192, pp. 155–169. American Geophysical Union (2002)

[468] Veronis, G.: Penetrative convection. Astrophys. J. **137**, 641–663 (1963)

[469] Veronis, G.: On finite amplitude instability in thermohaline convection. J. Marine Res. **23**, 1–17 (1965)

[470] Virto, L., Carbonell, M., Castilla, R., Gamez-Montero, P.J.: Heating of saturated porous media in practice: several causes of local thermal non-equilibrium. Int. J. Heat Mass Transfer **52**, 5412–5422 (2009)

[471] Wang, X.Q.: Molecular dynamics study of temperature behaviour in a graphene nanoribbon. Chemical Physics Letters **591**, 248–252 (2014)

[472] Wang, L., Xin, Z., Zang, A.: Vanishing viscous limits for 3D Navier-Stokes equations with a Navier slip boundary condition. J. Math. Fluid Mech. **14**, 791–825 (2012)

[473] Webber, M.: The destabilizing effect of boundary slip on Bénard convection. Math. Meth. Appl. Sci. **29**, 819–838 (2006)

[474] Webber, M.: Instability of fluid flows, including boundary slip. Ph.D. thesis, Durham University (2007). etheses.dur.ac.uk/2308

[475] Webber, M.: Instability of thread-annular flow with small characteristic length to three-dimensional disturbances. Proc. Roy. Soc. London A **464**, 673–690 (2008)

[476] Webber, M., Straughan, B.: Stability of pressure driven flow in a microchannel. Rend. Circolo Matem. Palermo **78**, 343–357 (2006)

[477] Wheeler, E.K., Benett, W., Stratton, P., Richards, J., Chen, A., Christian, A., Ness, K.D., Ortega, J., Li, L.G., Weisgraber, T.H., Goodson, K., Milanovich, F.: Convectively driven polymerase chain reaction thermal cycler. Analytical Chemistry **76**, 4011–4016 (2004)

[478] Whitaker, S.: Flow in porous media. I: A theoretical derivation of Darcy's law. Transport in Porous Media **1**, 3–25 (1986)

[479] Whitaker, S.: The Forchheimer equation: A theoretical development. Transport in Porous Media **25**, 27–62 (1996)

[480] Wong, K.F.V., Kurma, T.: Transport properties of alumina nanofluids. Nanotechnology **19**(345702), 1–8 (2008)

[481] Wörner, M.: Numerical modeling of multiphase flows in microfluidics and micro process engineering: a review of methods and applications. Microfluidics and Nanofluidics **12**, 841–886 (2012)

[482] Wu, Z., Nguyen, N.T.: Convective-diffusive transport in parallel lamination micromixers. Microfluidics and Nanofluidics **1**, 208–217 (2005)

[483] Wu, H., Grasselli, M., Zheng, S.: Convergence to equilibrium for a parabolic-hyperbolic phase-field system with dynamical boundary condition. J. Math. Anal. Appl. **329**, 948–976 (2007)

[484] Yagi, A.: Abstract Parabolic Evolution Equations and Their Applications. Springer, Berlin (2010)

[485] Yang, X.H., Bai, J.X., Kang, K.J., Lu, T.J.: Experimental investigations of natural convection in wire-woven bulk Kagome. Transport in Porous Media **105**, 1–22 (2014)

[486] Ye, C., Li, B., Sun, W.: Quasi-steady-state and steady-state models for heat and moisture transport in textile assemblies. Proc. Roy. Soc. London A **466**, 2875–2896 (2010)

[487] Yin, C., Fu, C., Tan, W.: Stability of thermal convection in a fluid-porous system saturated with an Oldroyd-B fluid heated from below. Transport in Porous Media **99**, 327–347 (2013)

[488] Yin, C., Niu, J., Fu, C., Tan, W.: Thermal convection of a viscoelastic fluid in a fluid-porous system subjected to a horizontal plane Couette flow. Int. J. Heat Fluid Flow **44**, 711–718 (2013)

[489] Yong, X., Zhang, L.T.: Slip in nanoscale shear flow: mechanisms of interfacial friction. Microfluidics and Nanofluidics **14**, 299–308 (2013)

[490] You, Y., Zhao, C., Zhou, S.: The existence of uniform attractors for 3-D Brinkman–Forchheimer equations. Discrete and Continuous Dyn. Systems **32**, 3787–3800 (2012)

[491] Yuan, P.: Numerical analysis of temperature and thermal dose response of biological tissues to thermal non-equilibrium during hypothermia therapy. Med. Eng. Phys. **30**, 135–143 (2008)

[492] Yuan, P.: Numerical analysis of an equivalent heat transfer coefficient in a porous model for simulating a biological tissue in a hypothermia therapy. Int. J. Heat Mass Transfer **52**, 1734–1740 (2009)

[493] Yuan, P., Yang, C.S., Liu, S.: Temperature analysis of a biological tissue during hyperthermia therapy in the thermal non-equilibrium porous model. Int. J. Thermal Sciences **78**, 124–131 (2014)

[494] Zhang, H., Zhang, Z., Ye, H.: Molecular dynamics-based prediction of boundary slip of fluids in nanochannels. Microfluidics and Nanofluidics **12**, 107–115 (2012)

[495] Zhang, M.K., Cao, B.Y., Guo, Y.C.: Numerical studies on dispersion of thermal waves. Int. J. Heat Mass Transfer **67**, 1072–1082 (2013)

[496] Zhang, W.M., Meng, G., Wei, X.: A review on slip models for gas microflows. Microfluidics and Nanofluidics **13**, 845–882 (2013)

[497] Zhao, C.Y.: Review on thermal transport in high porosity cellular metal foams with open cells. Int. J. Heat Mass Transfer **55**, 3618–3632 (2012)

[498] Zhao, S., Duncan, S.: On the accuracy of the calculation of transient growth in plane Poiseuille flow. J. Numerical Methods in Fluids **74**, 857–871 (2014)

[499] Zhao, W., France, D.M., Yu, W., Kim, T., Singh, D.: Phase change material with graphite foam for applications in high temperature latent heat storage systems of concentrated solar power plants. Renewable Energy **69**, 134–146 (2014)

[500] Zorrilla, S.E., Rubiolo, A.C.: Mathematical modeling for immersion chilling and freezing of foods. Part I: Model development. J. Food Engng. **66**, 329–338 (2005)

[501] Zorrilla, S.E., Rubiolo, A.C.: Mathematical modeling for immersion chilling and freezing of foods. Part II: Model solution. J. Food Engng. **66**, 339–351 (2005)

Index

© Springer International Publishing Switzerland 2015
B. Straughan, *Convection with Local Thermal Non-Equilibrium and Microfluidic Effects*,
Advances in Mechanics and Mathematics 32, DOI 10.1007/978-3-319-13530-4

Printed in the United States
By Bookmasters